Basics of

Electrical Energy for Engineers

Basics of
Electrical Energy for Engineers

V. Ramesh Babu

Associate Professor,
VNR Vignana Jyothi Institute of Engineering & Technology,
Hyderabad, Telangana.

BSP **BS Publications**

An Imprint of **BSP Books Pvt., Ltd.**
4-4-309/316, Giriraj Lane,
Sultan Bazar,
Hyderabad - 500 095

Basics of **Electrical Energy for Engineers**
by **V. Ramesh Babu**

© 2023, *by Publisher*

Disclaimer: The authors and the publishers have taken due care to provide the authentic, reliable and up to date information related to the subject. However, neither the authors nor the publisher shall be responsible for any liability for any damage caused as a result of use of this book. The respective user must check the accuracy from other sources too.

Published by:

BSP **BS Publications**
An Imprint of **BSP Books Pvt. Ltd.**
4-4-309/316, Giriraj Lane, Sultan Bazar,
Hyderabad - 500 095
Phone: 040 - 23445688
e-mail : info@bspbooks.net
www.bspbooks.net

ISBN: 978-93-91910-22-8 (Hardback)

Dedicated to

My mother

Savithri

who is the sole cause of all I am

- V. Ramesh Babu

Preface

The All India Council for Technical Education (AICTE) has suggested the model curriculum for all technical courses that offered in all professional institutes across the India in 2018. It has asserted the importance and necessity of having basic knowledge in Electrical Engineering for the branches like Information Technology, Computer Science & Engineering and its allied branches like Artificial Intelligence, Machine Learning, Data Science, Cyber Security, IOT etc. The frame work for this course is very generic and covers wide range of aspects in electrical engineering. Subsequently, it is a big question mark for the students and faculty to what extent or depth they need to prepare for. As the topics prescribed by AICTE do not get dealt in a single text book, they needs to refer to multiple learning resources which is a difficult task for the students at UG first year level for whom the subject is meant for.

This book has been authored with an intent of offering a solution for the above challenges. It lucidly deals with the concepts of Basic Electric Circuits, Circuit Theorems, Circuit Reduction Techniques, Basic Electric Machines, Electronics Devices, Op-Amps, Transducers and other Remote Monitoring and Control principles. The contents are readily presented to the extent of students' requirements. This book has been organized in 6 units. The short answer and long answer questions also have been included at the end of each unit for students' reference.

- Author

Contents

CHAPTER 4

Alternating Current Machines

CHAPTER 5

Semi-Conductor Devices

CHAPTER 6

Op-Amps, Transducers and Data Acquisition

CHAPTER 1

Introduction to Electrical Energy and DC Circuits

1.1 Introduction

Electricity is one of the most important blessings that science (so called, almighty) has given to mankind. It has become a part of modern life so much so that one cannot think of a world without it. Electrical energy is the primary source of readily usable energy in the world and it drives almost all present day activities. It has played a vital role in building up of present day civilization. The modern society is so much dependent upon the use of electrical energy that it has become a part and parcel of our life. The following are the reasons which made the Electrical Energy superior than all other forms of energies.

1. **Convenient:** Electrical energy is a very convenient form of energy. It can be easily converted into other forms of energy like light, heat, mechanical rotation etc.

2. **Easily controllable and more flexible**: The electrically operated machines provide simple control and ease of operation. For instance, an electric motor can be started or stopped by the flick of a switch

3. **Inexpensive:** Electric energy is much cheaper than other forms of energy. Thus, it is economical to use this form of energy for domestic, commercial and industrial purposes

4. **Clean:** Electric energy is not associated with smoke, fumes or poisonous gases and hence provides clean working conditions.

5. **Easily Transmittable:** The electric energy can be transmitted conveniently and efficiently from one place to other.

1.1.1 The Role of Electrical Energy in Modern Life

Electricity is an essential part of modern life and is a major contributor for a global economy. The modern society is so dependent upon the use of electrical energy that it has become a part of our life. Per capita electrical energy consumption is one of the main metrics for assessing economic growth and life quality of anybody. The greater the per capita consumption of electrical energy in a country, the higher is the standard of living of its people. People use electricity for lighting, heating, cooling, and refrigeration and for operating appliances, computers, electronics, machinery, and public transportation systems. The present-day advancement in science and technology has made it possible to convert electrical energy into any desired form. This has given electrical energy a place of pride in the modern world.

Modern means of transportation and communication have been revolutionized by it. Electric trains and battery cars are quick and non-pollutive means of travel. Electricity also provides means of amusement. Radio, television and cinema, which are the most popular forms of entertainment are the result of electricity. Modern equipment like computers and robots have electricity as one of the most important pillars of their development. Electricity plays a pivotal role in the fields of medicines and surgery too, such as X-ray and ECG. The use of electricity is increasing day by day. The residential uses of electrical energy are mainly Heating and Cooling. Computers and office equipment account for largest share of commercial sector electricity consumption. The industrial sector uses electricity to operate machine drives (motors), lights, computers and office equipment, and equipment for facility heating, cooling, and ventilation.

The other major application of electrical energy in modern life is for Electric Vehicles (EVs). These vehicles have brought a remarkable change in the present mobility system. The EVs are gaining popularity as these vehicles are environmental friendly and do not produce any pollutants. The Electric Vehicles are cheap and economical for a given drive range.

The use of Renewable Energy Systems is growing at rapid pace in recent days. The Electrical Energy which got generated from Renewable Energy Sources like solar, wind, hydel is most acceptable as the corresponding conversion mechanisms do not cause for environmental degradation. The Renewable Energy systems are forming the skeleton of future energy scenario.

1.1.2   Electrical Energy vs Computer Science Engineering

The Computer Science and Engineering is one sub field among Electronics, Communication and Instrumentation of Electrical Engineering. The engineering era has begun with Civil Engineering which mostly deals with the static systems like bridges and buildings. The rotational energy has made the Mechanical Engineering revolutionary and it has given birth to Electrical Engineering with the invention of generator.   The Electrical Engineering in association with other engineering disciplines formed the basis for the growth and promotion of Computer Science and Engineering. It is the primary responsibility of any computer science techy to appreciate the electrical energy which makes a computer system functional. The study of electrical engineering greatly helps for the inculcation of necessary logical ability in computer science engineers and it is useful in delivering optimal codes/programs.

1.2   Over View of Electrical Energy Generation, Transmission and Utilization

As we aware that energy is available in nature in different forms like heat, wind, solar, light, chemical, mechanical etc. But, according to the law of energy conservation, energy can be neither generated nor destroyed. But it can be converted from one form to other. *Generation of electrical power means converting all forms of existing energies into electrical form.* This process has to go on continuously to meet the consumer demand at all times. Hence, the input can be any form of energies, being the output always electrical energy. This conceptualized in Figure 1.1.

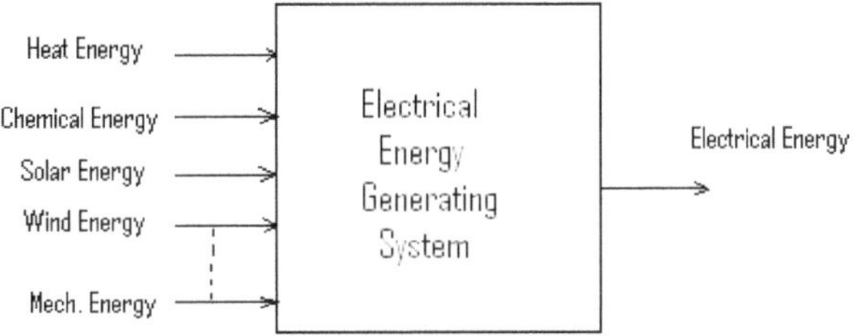

Figure 1.1 Electrical energy generation

An efficient, convenient and conventional way of generating electrical energy is by conversion of mechanical energy into electrical energy. *The device which supplies the required mechanical energy is called Prime-mover.* Since the steam turbines and hydraulic turbines are most generally used prime-movers, the location of generating units is always constrained to the availability of resources like coal, water at an elevated point etc. Unfortunately, this is not the area where the electrical power is needed. Hence, the electrical power generated must be transmitted to the load centers using a conductor system which is called *Transmission System*. Upon receiving the power at load centers, it is *distributed* to various consumers and *utilized*. The entire system which consists of Electrical Power Generation, Transmission, Distribution and finally Utilization is called an *Electrical Power System*.

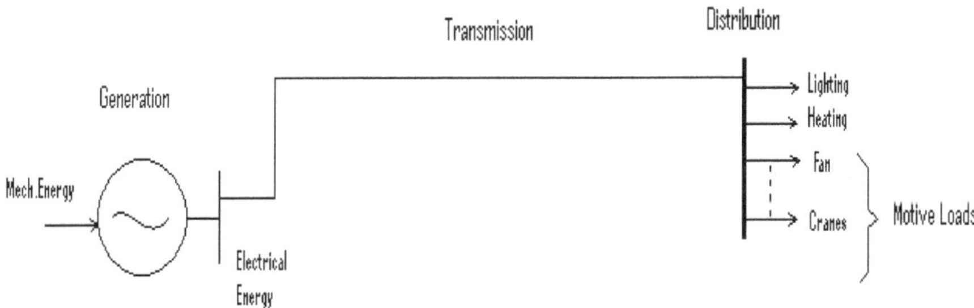

Figure 1.2 Electrical power system

As it is cleared from the figure 1.2 that the Power System starts with mechanical power and it gets converted into electrical form. The electrical power thus generated is transmitted to load center using transmission network and then it is utilized for different purposes. It is also cleared from above figure 1.2 that other than lighting and heating loads, all loads involve motion or mechanical power at their output and hence, they convert electrical power into mechanical power. These loads are called *Motive Loads* and these loads constitute 80 % of total existing loads. So, an electrical power system mostly ends with mechanical energy. Hence, the process of 'Electromechanical energy conversion' is very essential in a power system. *Electrical Machines are responsible for this conversion process in an electric power system. If a machine converts Mechanical Energy into Electrical Energy, that machine is called 'Electric Generator' and if a machine converts Electrical Energy into Mechanical Energy, that machine is called 'Electric Motor'.*

1.3 Electric Circuit

The electric circuit is a simple arrangement that consists of Source or Electromotive Force (battery or generator), Load (like lamp, fan, heater, motor etc which consumes electrical energy), controlling switch and connecting wires. The purpose of a circuit is to transfer electrical energy to the load. The schematic of a circuit is shown in figure 1.3. The electric current is circulated in the circuit by EMF (source) against the Resistance (R). The current (I) is always proportional to the voltage (V).

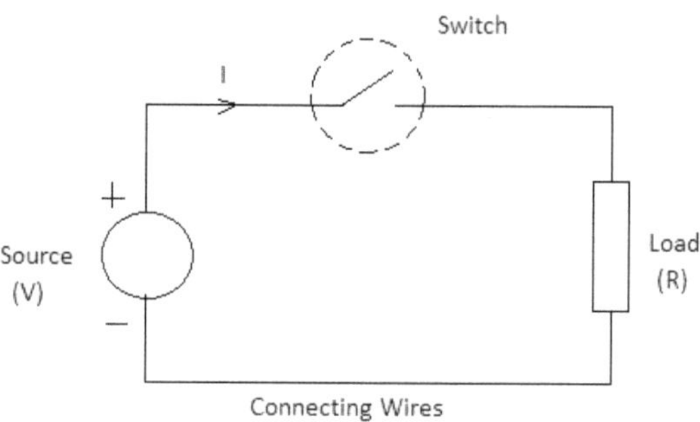

Figure 1.3 Basic circuit schematic

The interconnection of two or more circuit elements is called Electric Network. In circuits, the current should pass through all the elements whereas in networks, it need not be so. A Circuit can be a Network. But, a Network need not be a Circuit.

An electric circuit has got lot of analogy with a magnetic circuit. In magnetic circuit, an Magneto Motive Force (MMF) is used to pass the magnetic flux (Ø) through an iron core which has got finite value of Reluctance(S) which opposes the flux. Similarly, in a mechanical system, the Driving Torque (T) is responsible to keep the system at a speed (N) against the friction. Hence, the EMF, MMF and Torque are analogous. And the current, flux and speed are similar. Also, the resistance, reluctance and friction are of same kind.

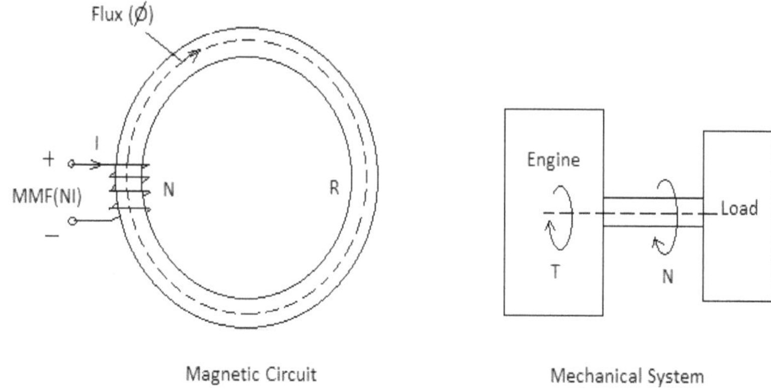

Figure 1.4 Analogy between electrical, magnetic and mechanical systems

1.3.1 Circuit Elements

A Circuit Element can be represented by any or combination of three basic parameters 1. Resistance 2. Inductance and 3. Capacitance.

(a) Resistance

Electrical Resistance is the property of a material by which it opposes the flow of electrons or current through it. When the electrons flow through a conducting material, these electrons collide with the other atoms and lose their energy and hence the potential. The voltage drop (V) in a conducting material is proportional to the flow of current (I) at given temperature.

$$\text{Voltage drop (V)} \propto I$$

or $\qquad V = I \times R$

Where the constant 'R' is called Resistance. The units for Resistance is Ohms. As the current through a Resistance increases, the voltage drop also increases linearly as shown in figure 1.5. Hence, the Resistance is a Linear Element.

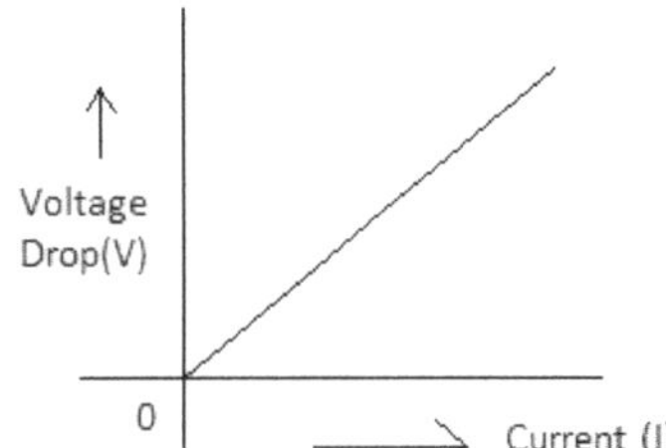

Figure 1.5 V-I Characteristics of a Resistance

The resistance of a material is proportional to its length (l) and inversely proportional to its cross-sectional area (A).

$$R \propto \frac{l}{A} \qquad \text{or} \qquad R \propto \rho \frac{l}{A}$$

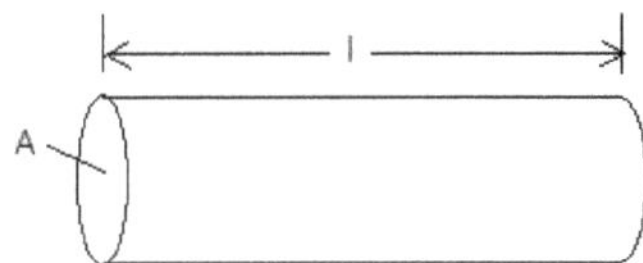

Figure 1.6 Circular Conductor

Where ρ is its Specific Resistance or Resistivity whose units are Ohm-meters. The resistivity of metals increases as the temperature increases and hence, the resistance changes with temperature.

The power absorbed by the resistance $P = V \times I$ or $P = I^2 \times R$ or $P = \dfrac{V^2}{R}$

The corresponding energy consumed in the resistance will be appeared as heat and it is expressed as

$$W = \int_0^t P.dt = P.T \text{ joules}$$

Hence, Resistance is a Passive element which consumes energy and can't store or deliver energy. And also, a resistor should be properly designed to dissipate heat produced, otherwise it melts. This is the reason the resistors used for electronic circuits of low power ratings are less in size than that used for high power electrical circuits.

(b) Inductance

Inductance is the property of a circuit element that is related to the magnetic flux. It is evident that a current carrying conductor produces magnetic flux in concentric form around it. The direction of flux is given by Right Hand Thumb Rule which is illustrated below in figure 1.7.

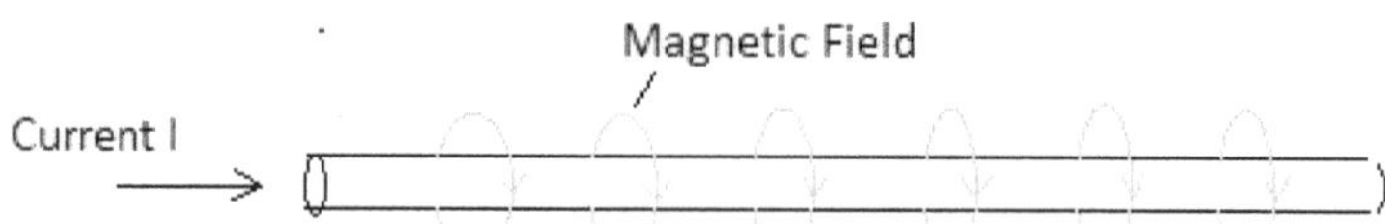

Figure 1.7 Flux lines around a current carrying conductor

When the conductor of negligible resistance is made into a coil form, all the flux lines get gathered and produces a resultant flux whose axis is called the Coil Axis. When this flux (Ø) links with the coil having 'N' number of turns, the coil get the flux linkages (Ψ). The flux linkages are given by $\varphi = N\text{Ø}$. The units are Wb-Turns.

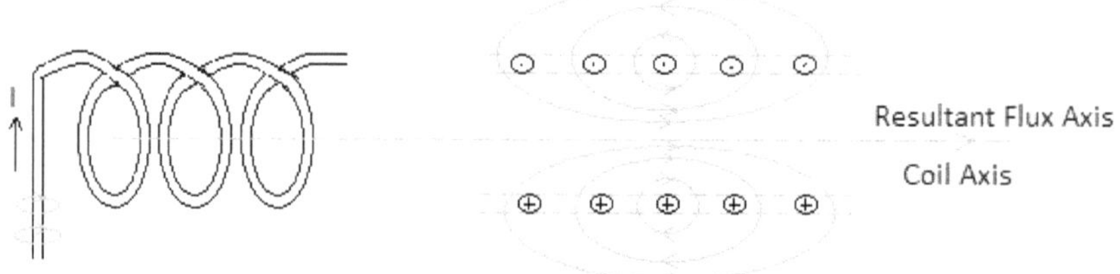

Figure 1.8 Representation of a Coil or Winding Axis

According to the Faraday's Laws of electromagnetic induction, whenever flux linkages with the coil change with respect to time, an emf is induced in it. This emf is given by

$$e \propto \frac{d\varphi}{dt}$$

Where $\varphi \propto \text{Ø}$ and $\text{Ø} \propto i$

Then, $e \propto \dfrac{d\varphi}{dt}$ or $e = L\dfrac{d\varphi}{dt}$

The constant 'L' is called Inductance whose units are 'Henry(H)'. One Henry is the inductance of a coil when it produces one volt in it if the current passing through it changes at the rate of one Ampere per Second. As per Lenz's Law, this emf opposes the cause of its generation. Here, the cause is the change in current. Hence, the changes in currents are opposed by the emf produced in the coil. This property exhibited by the coil is called its Inductance. When the current passing through the coil is steady or constant, $\frac{di}{dt} = 0$ and hence no emf is generated.

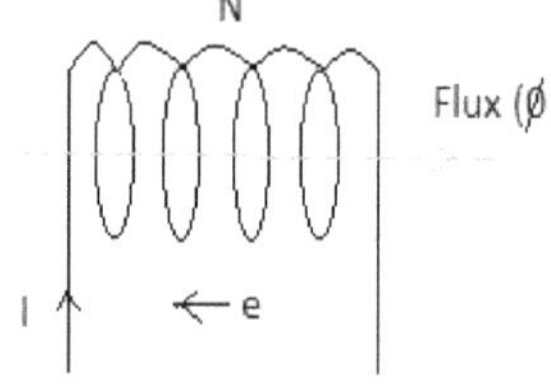

Figure 1.9 Emf induced in a coil opposes changes in current

Hence, Inductance is the property of a coil by which the changes in currents are opposed.

Like a flywheel having huge inertia opposes the changes in speeds, an Inductor opposes the changes in currents. Hence, Inductance is called *'Electrical Inertia'*.

The electrical energy absorbed by an inductor is given by

$$W = \int_0^t P.dt = \int_0^t e.i.dt = \int_0^t L \times \frac{di}{dt} \times dt = \int_0^t L \times di = \frac{1}{2} \times LI^2$$

Since, the coil is made with a conductor having negligible resistance and assumed to be ideal. It does not consume any energy. It stores the electrical energy absorbed by it in the form of magnetic field. Hence, inductor is an Active Element.

Therefore, the energy stored by an inductor $= W = \frac{1}{2} \times LI^2$ Joules

Hence, the important points related to an inductor are summarized as

1. The induced emf across an inductor is zero if $\frac{di}{dt} = 0$. Hence, a pure inductor behaves as a short circuit to direct current (DC)

2. The current in an inductor cannot change instantaneously as the emf induced become infinite which does not allow the current to change

3. An inductor stores the energy only when the current passing through it changes

4. A pure inductor does not consume any power or energy. It stores energy.

(c) Capacitance

When a conductor is formed into a coil, it exhibits the property of inductance. Similarly, when a conductor is made into two plates and these plates are separated by a thin insulating material called Dielectric, a Capacitor is formed and it is shown in figure 1.10.

When a capacitor is connected across a source (battery), it gets charged ie., the charge is of opposite polarities get transferred to the plates. Thus, an electrostatic field is established between the plates. The charge accumulated the plates is proportional to the voltage 'V'.

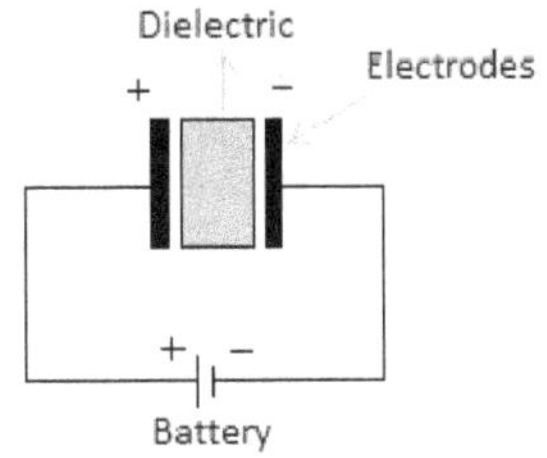

Figure 1.10 Capacitor

$$Q \propto V$$

or $\qquad Q = CV$

Where the proportionality constant 'C' is called Capacitance. Capacitance is the capability of an element to store electric charge on it. For a given voltage, as the capacitance increases, the charge on capacitor plates also increases. The units for capacitance are 'Farads'. The capacitance is said to be one farad provided one coulomb of charge can be stored with one volt across the capacitor.

The charging current 'I' is given by $i = \dfrac{dq}{dt} = \dfrac{d(CV)}{dt} = C \times \dfrac{dV}{dt}$

The energy absorbed by the capacitor

$$W = \int_0^t P.dt = \int_0^t V.i.dt = \int_0^t V \times C \times \frac{dV}{dt} \times dt = \frac{1}{2} CV^2$$

Neglecting the resistance of connecting wires and electrodes and assuming no leakage currents through dielectric, the energy input to the capacitor will be stored in capacitor in the form of electrostatic field.

Energy stored in the capacitor $= \dfrac{1}{2} CV^2$ joules

As the capacitor stores energy and does not consume energy, it is called an Active Element.

The important points related to a capacitor are summarized as follows

1. The current in a capacitor is zero if the voltage across it is a constant $\left(\dfrac{dV}{dt} = 0 \right)$. Hence, it acts as an open circuit for Direct Currents (DC)

2. The voltage across a capacitor can't change instantaneously as infinite current is produced

3. Capacitor stores energy only when the voltage across it changes

4. A pure capacitor does not consume energy. It stores energy only.

1.3.2   Electrical Energy Sources

The purpose of an Electrical Energy source is to supply electrical energy or power to a load. These sources are broadly classified into Voltage Sources and Current Sources. The Voltage Sources supply different values of currents at given voltage whereas the Current Sources supply different values of voltages at a constant current.

The other classification of electrical energy sources is 1. Independent Sources and 2. Dependent Sources.

1. Independent Sources

An Independent Source can be an Independent Voltage Source or an Independent Current Sources.

(a) *Independent Voltage Source*

An Independent Voltage source is characterized by a terminal voltage which is completely independent of current supplied by it. The magnitude of voltage does not change by any variation in connected load network.

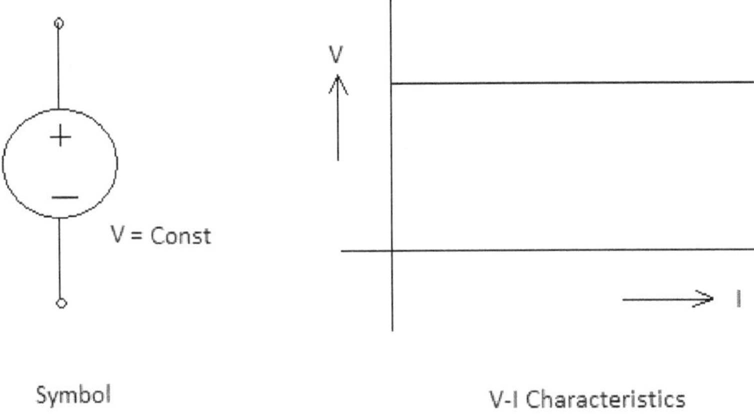

Figure 1.11 Independent Voltage source-symbol and V-I Characteristics

(b) Independent Current Source

In this source, the current supplied is completely independent of voltage across it. The strength of current does not change by the variation in the load network.

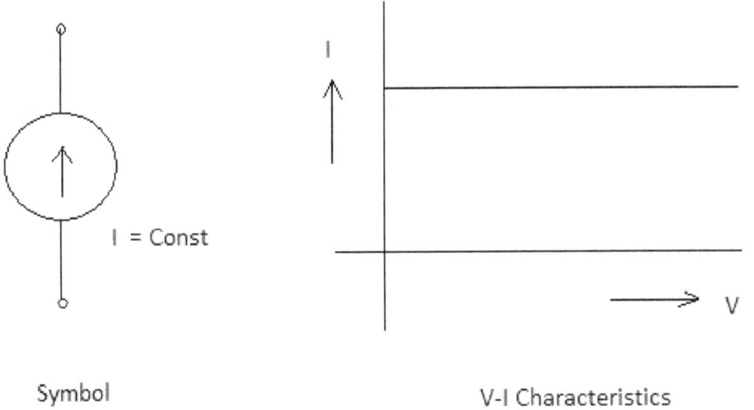

Figure 1.12 Independent Current Source-Symbol and V-I Characteristics

2. Dependent Sources

In Dependent Sources, the voltage of a voltage source and the current of a current source are controlled by a voltage or a current that exist at some other location/point of the circuit. These sources are also called as Controlled Sources. These are represented as

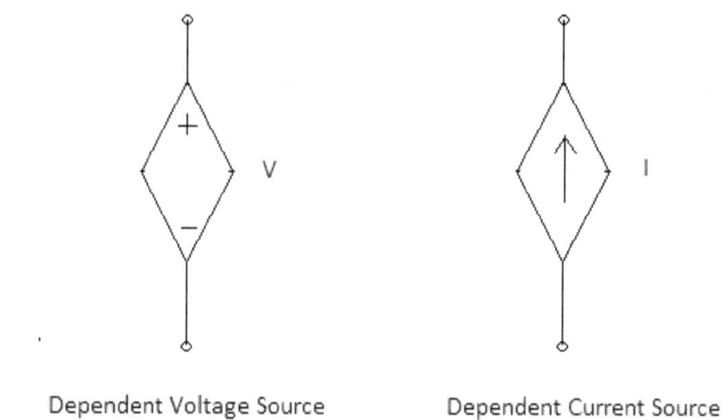

Figure 1.13 Symbols of Dependent Voltage and Current Sources

Sine, the control of Dependent Sources is achieved by a voltage or current, there are four possible dependent sources.

(a)   Voltage Controlled Voltage Source (VCVS)

(b)   Current Controlled Voltage Source (CCVS)

(c)   Voltage Controlled Current Source (VCCS)

(d)   Current Controlled Current Source (CCCS)

(a)   Voltage Controlled Voltage Source

In this case, the source voltage V_2 is set by a control voltage V_1.

$$V_2 = \mu V_1$$

Where μ is a constant having no units. For such controlled source, the voltage- current characteristics become a set of curves as shown in figure 1.14.

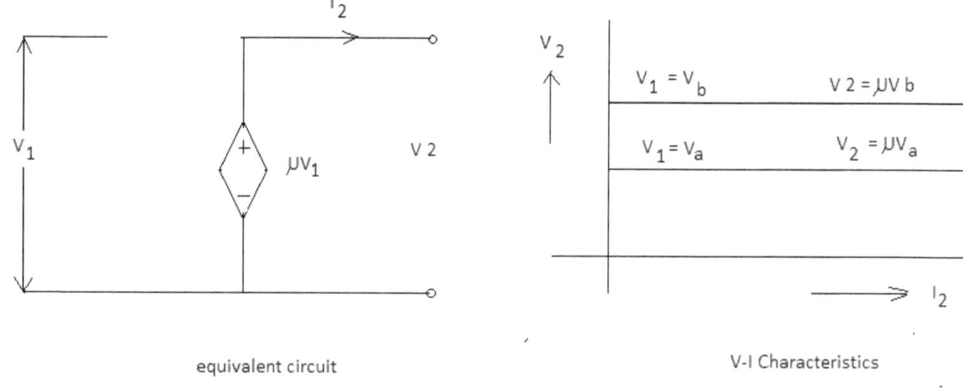

Figure 1.14 Equivalent Circuit and V-I Characteristics of Voltage controlled Voltage source

(b) Current Controlled Voltage Source

Here, the control signal which sets voltage is current i_1. The source voltage is given by $V_2 = R_m i_1$. Where R_m is a constant and its units are Ohms or Volt per Amp. The volt-current characteristics of this source are shown in figure 1.15.

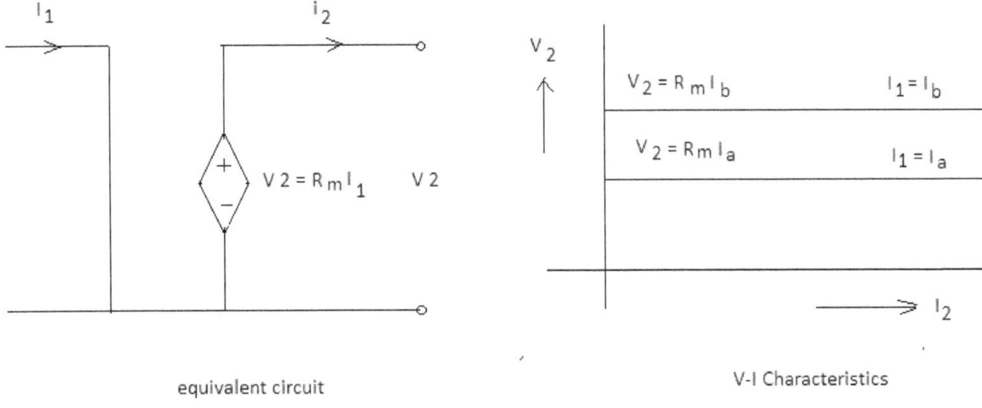

Figure 1.15 Equivalent Circuit and V-I Characteristics of Current controlled Voltage source

(c) Voltage Controlled Current Source

Here, the control signal which controls source current is a voltage V_1. The source current is given by $i_2 = G_m \times V_1$. where G_m is a constant whose units are mhos or Amp per Volt. The following figure 1.16 shows the V-I characteristics of this source.

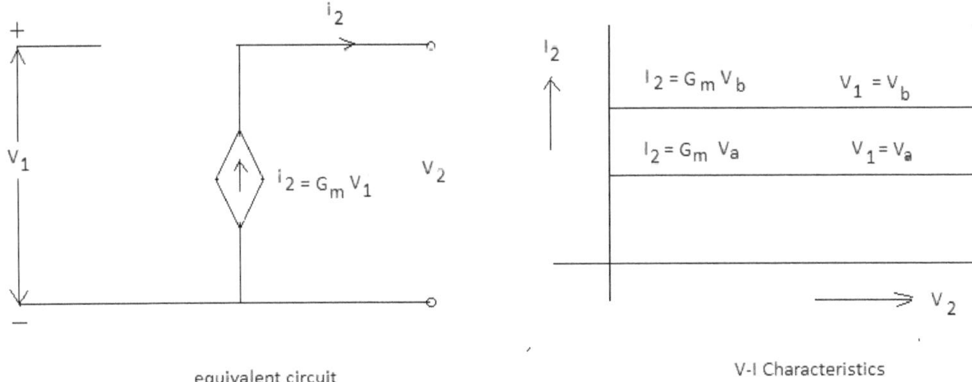

Figure 1.16 Equivalent Circuit and V-I Characteristics of Voltage controlled Current source

(d) Current Controlled Current Source

The current signal i_1 controls the current of the current source. The source current i_2 is given by $i_2 = \alpha \times i_1$. Where α is a constant having no units. The volt-current characteristics of this source are shown below in figure 1.17. Control Sources are used in many models of devices like transistors.

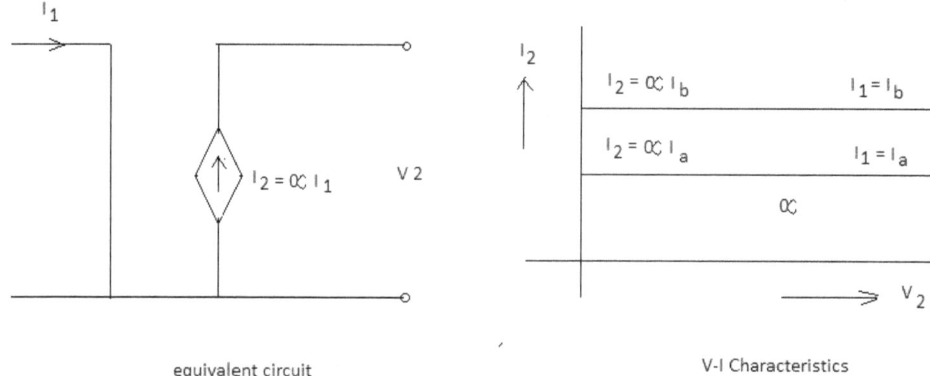

Figure 1.17 Equivalent Circuit and V-I Characteristics of Current controlled Current source

Practical Voltage Sources

In practice, no voltage source is ideal. As the current supplied by the source increases, its terminal voltage decreases. This is because its internal resistance R_s.

The terminal Voltage V_L is given by $V_L = V_S - IR_S$

As I becomes zero under no load conditions, V_L becomes equal to V_S . As the load current increases, IR_S drop increases and the terminal voltage V_L decreases as shown in figure 1.18. Hence, a practical voltage source is represented by an ideal voltage source connected in series with an internal resistance R_S .

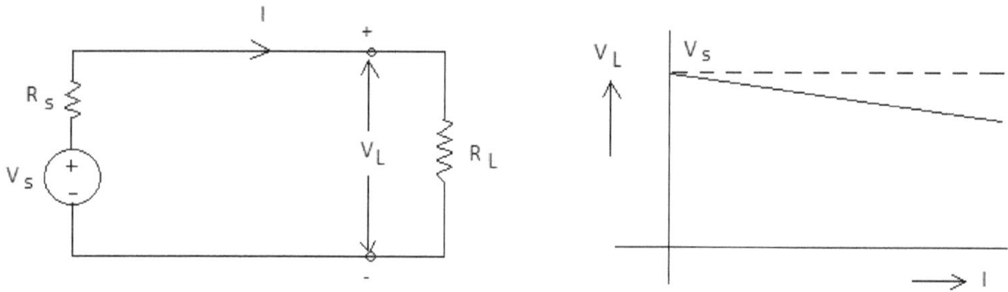

Figure 1.18 V-I Characteristics of a Practical Voltage Source

Practical Current Sources

Ideal current sources are not found in practice. As the demanded voltage by load increases by increase of load resistance, the source current starts reducing. Hence, a practical current source is represented by an ideal current source in parallel with an internal resistance R_P.

As the load resistance increases, the current through R_P increases and the load current slightly decreases as shown in figure 1.19.

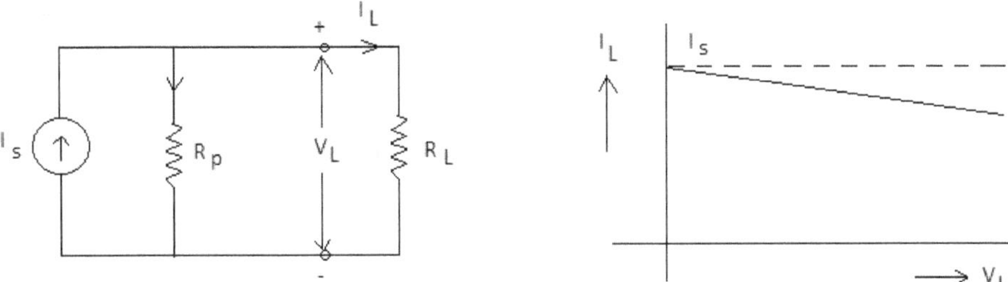

Figure 1.19 V-I Characteristics of a Practical Current Source

1.4 Source Transformation

A Practical Voltage Source can be converted into an equivalent practical current source and a practical current source can be converted into its equivalent practical voltage source so that identical values of load voltage and load currents are maintained.

Consider a practical voltage source as shown in figure 1.20 feeding a load current I_L at voltage V_L. The load current I_L is calculated as follows.

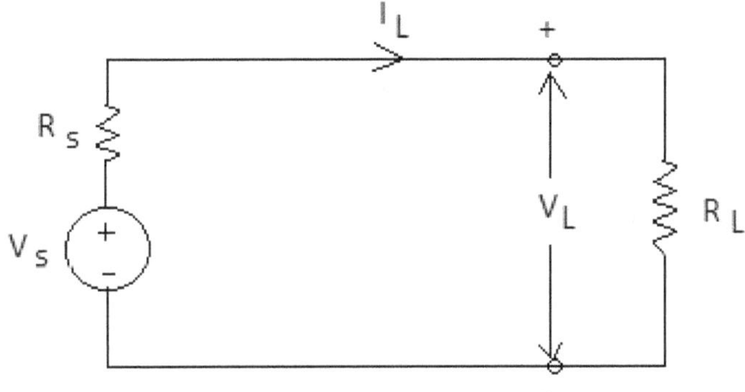

Figure 1.20 Practical Voltage Source

$$I_L = \frac{V_S}{R_S + R_L}$$

$$V_L = I_L R_L = \frac{V_S}{R_S + R_L} \times R_L$$

And also, the open circuit voltage across the load terminals that would drive the load current is

$$V_0 = V_S - R_S \times 0 = V_S$$

Similarly consider a practical current source as shown in figure 1.21 that drives the same load current I_L at same load voltage V_L.

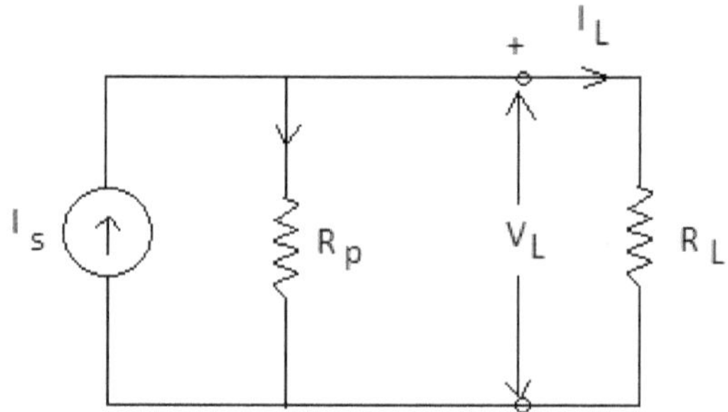

Figure 1.21 Practical Current Source

The load current I_L is given by

$$I_L = I_S \times \frac{R_P}{R_P + R_L}$$

The load voltage $V_L = I_L R_L = I_S \times \frac{R_P}{R_P + R_L} \times R_L$

The open circuit voltage across the load terminals $V_0 = I_S R_P$

By equating the open circuit voltages in both cases, $V_0 = I_S R_S = V_S$

Similarly, by equating the load voltages in both cases,

$$V_L = \frac{V_S}{R_S + R_L} \times R_L = I_S \times \frac{R_P}{R_P + R_L} \times R_L$$

$$\frac{V_S}{R_S + R_L} = I_S \times \frac{R_P}{R_P + R_L}$$

Also writing $I_S R_S = V_S$ in above equation, we get

$$R_S = R_P$$

and $\qquad V_S = I_S R_P$

Hence, a practical Voltage source of voltage V_S and internal resistance R_S can be equated to

a Practical Current Source of current $I_S \left(= \dfrac{V_S}{R_S} \right)$ with an internal resistance $R_P \left(= R_S \right)$

connected in parallel and it is shown in figure 1.22

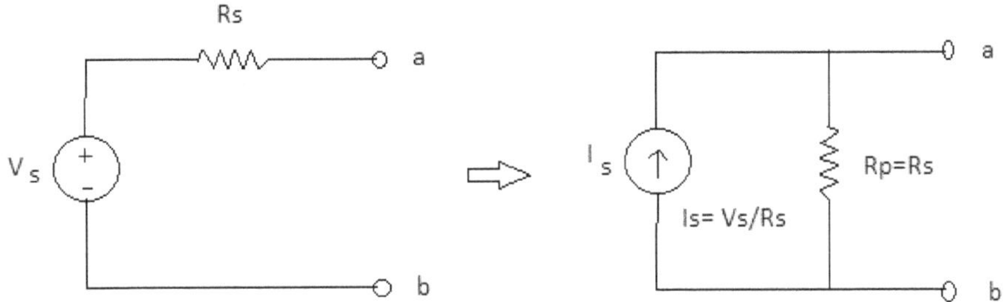

Figure 1.22 Conversion of a Voltage Source into a Current Source

Similarly, a current source of current I_S and internal resistance R_P can be replaced by a practical voltage source of voltage $V_S (= I_S R_P)$ and series connected internal resistance $R_S (= R_P)$. It is shown in figure 1.23.

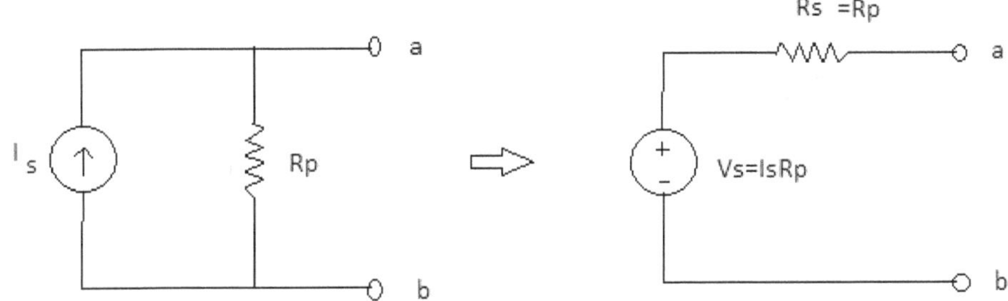

Figure 1.23 Conversion of a Current Source into a Voltage Source

Exercise 1.4.1: Determine the equivalent Voltage Source for a given Current Source.

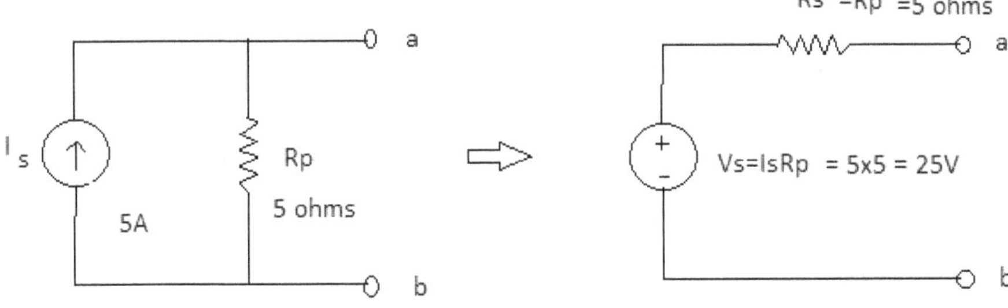

Figure 1.24 Conversion of a Current Source into a Voltage Source

Exercise 1.4.2: Find the equivalent Current Source for the given Voltage Source.

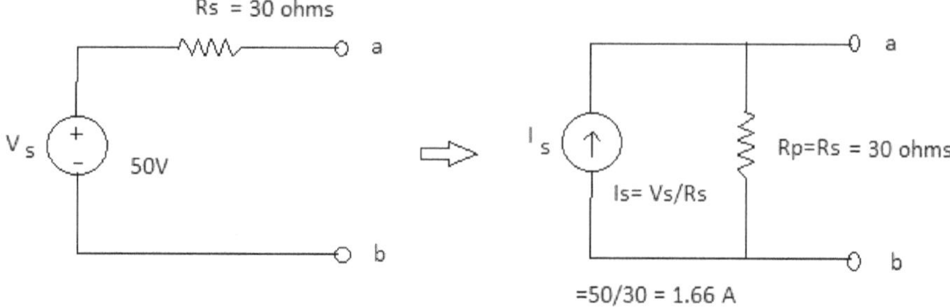

Figure 1.25 Conversion of a Voltage Source into a Current Source

1.5   Kirchhoff's Laws

The Kirchhoff's laws in association with the Ohm's Law become a powerful tool to solve and to analyze any complex electric network. These laws are

(a)   Kirchhoff's Current Law (KCL)

(b)   Kirchhoff's Voltage Law (KVL)

(a) Kirchhoff's Current Law (KCL)

The Kirchhoff's Current Law (KCL) states that the algebraic sum of the currents at a given node is zero.

Mathematically, KCL is represented as $\sum_{n=1}^{n=N} i_n = 0$

Where 'N' is the total number of branches connected to the node and i_n is current passing through the node 'n'. (Node is the point where two or many branches are connected together). The currents entering the node are generally regarded as positive currents and the currents leaving the node are treated as negative currents. Consider the node 'a' in the figure 1.26. It is associated with 5 branches carrying currents i_1, i_2, i_3, i_4 and i_5.

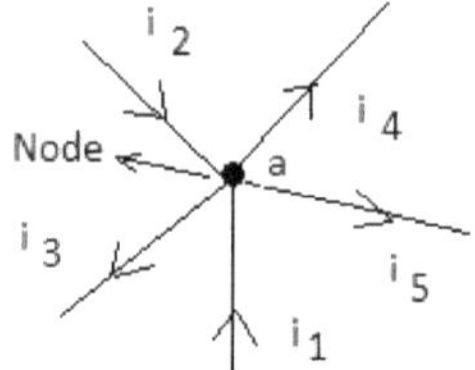

Figure 1.26 Representation of KCL

It is evident that the currents i_1 and i_2 are the currents entering and hence these are considered as positive currents and the currents i_3, i_4 and i_5 are leaving and hence these are negative currents.

The algebraic sum of the currents at node 'a' $= i_1 + i_2 - i_3 - i_4 - i_5 = 0$

Or $i_1 + i_2 = i_3 + i_4 + i_5$

The sum of the currents entering the node is equal to the sum of the currents leaving the node.

As the charge transferred towards a node in a given time is equal to the charge carried away from the node, no charge gets accumulated a node or a node does not store any net charge.

(b) Kirchhoff's Voltage Law

The KVL states that the algebraic sum of all voltages around a closed path (loop) is zero.

Mathematically. $\displaystyle\sum_{m=1}^{m=N} V_m = 0$

Where N is the number of branches in a loop and V_m is the voltage across m^{th}branch.

Consider the circuit shown in figure 1.27 to illustrate the KVL.

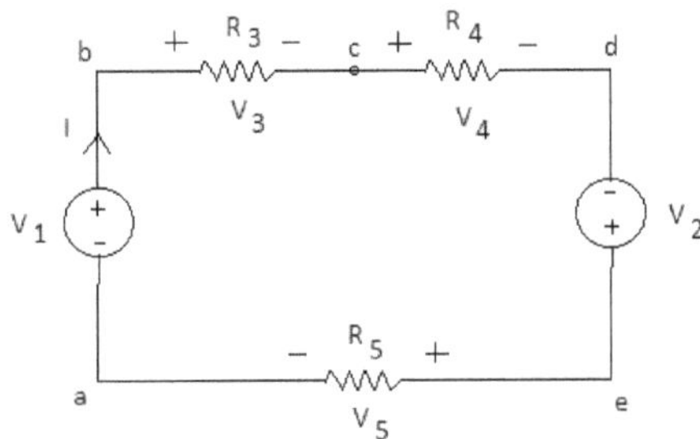

Figure 1.27 Representation of KVL

As we move from the node 'a' towards node 'b', the potential rises by V_1 (-ve of battery to +ve of battery). Hence, the voltage V_1 should be treated as +ve in the direction of a to b. From point 'b' to 'c', the potential reduces by V_3 as the potential decreases in the direction of current. Hence, V_3 is negative. Likewise, the polarities of the voltages V_2, V_4 and V_5 are found. Then, the algebraic sum of the voltages is $V_1 - V_3 - V_4 + V_2 - V_5 = 0$

Or $V_1 + V_2 = V_3 + V_4 + V_5$

Sum of the voltage rises or sources = Sum of the voltage drops

Exercise 1.5.1 :

Find the values of V_1 and V_2 in the network given in figure 1.28.

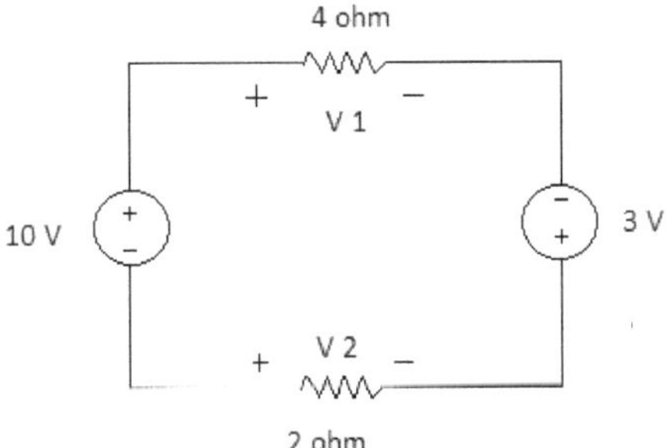

Figure 1.28

Solution:

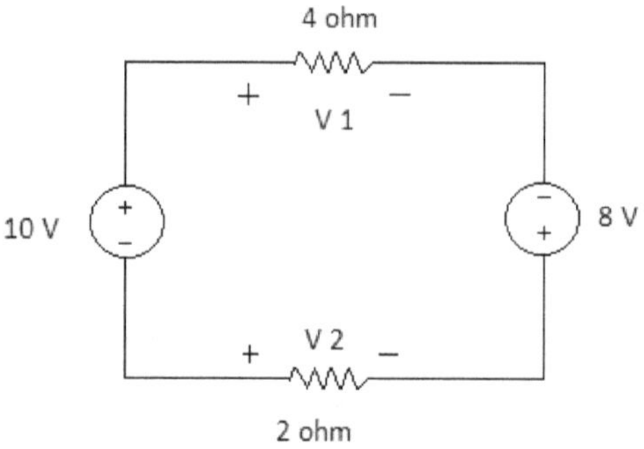

Figure 1.28 (a)

Assume the current 'I' in a particular direction. Let V_1 and V_2 be the voltage drops in the 4 ohm and 2 ohm resistances respectively for a given current 'I' direction.

Applying KVL,

$$10 - 4i + 8 - 2i = 0$$

$$18 - 6i = 0$$

$$6i = 18$$

$$i = \frac{18}{6} = 3A$$

Or

$$V_1 = 4 \times i = 4 \times 3 = 12V$$

$$V_3 = 2 \times i = 2 \times 3 = 6V$$

$$V_2 = -V_3 = -6V$$

Exercise 1.5.2:

Find the values of V_x and V_0 in the circuit shown in figure 1.29.

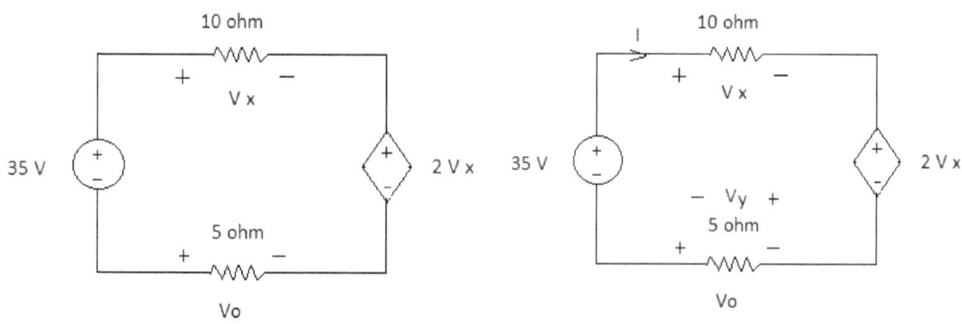

Figure 1.29   Figure 1.29 (a)

Solution:

Assume the current direction in the circuit as shown.

Let V_x and V_y be the voltage drops in the resistances 10 ohms and 5ohms respectively for a given current direction.

Applying KVL,

$$35 - 10i - 2V_x - V_y = 0$$

($2V_x$ is a dependent voltage source whose voltage is a function of voltage drop across 10 ohms resistance)

$$35 - 10i - 2(10 \times i) - V_y = 0$$
$$35 - 10i - 20i - V_y = 0$$
$$35 - 30i - 5i = 0$$
$$35 - 35i = 0$$
$$i = \frac{35}{35} = 1$$

Then, $\qquad V_x = 10i = 10 \times 1 = 10 \text{ V}$

$$V_0 = -V_y = -(5 \times 1) = -5 \text{ V}$$

Exercise 1.5.3:

Find V_0 and I_0 in the given circuit (figure1.30).

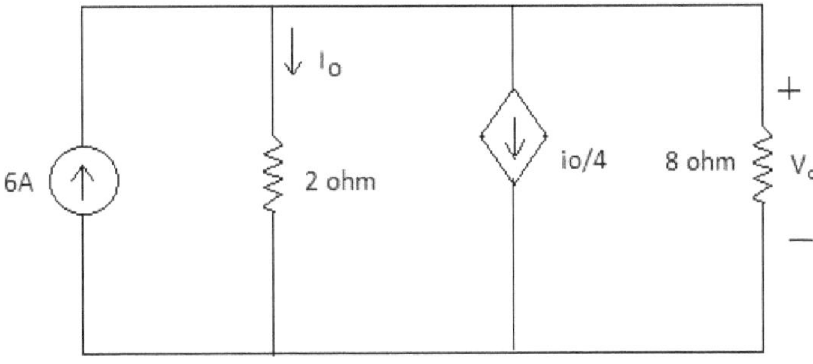

Figure 1.30

Solution:

Assuming the current 'I' through 8 ohm resistance in a particular direction to satisfy the polarity of given voltage V_0.

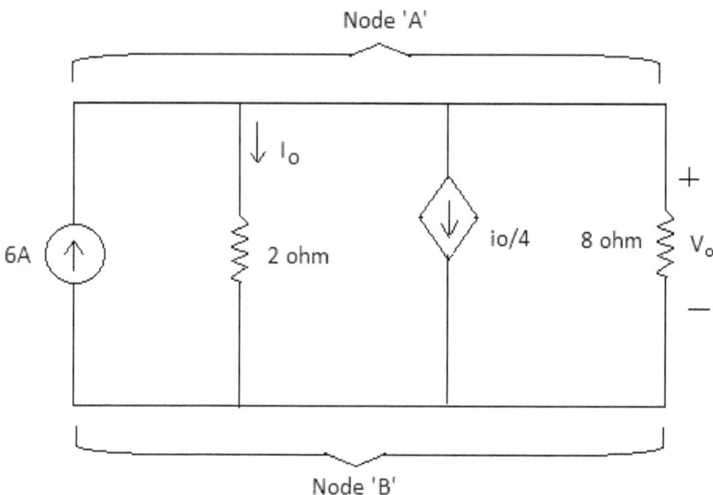

Figure 1.30 (a)

Since, 8 ohms and 2 ohms are connected in parallel,

$$2i_0 = V_0 \quad \text{or} \quad i_0 = \frac{V_0}{2} \qquad \qquad \dots(1.1)$$

Applying KCL at node 'a'

$$6 - i_0 - \frac{i_0}{4} - i = 0$$

$$6 - i_0\left[1 + \frac{1}{4}\right] - \frac{V_0}{8} = 0$$

$$6 - 1.25 i_0 - \frac{V_0}{8} = 0$$

Writing $i_0 = \frac{V_0}{2}$ in above equation,

$$6 - 1.25\left[\frac{V_0}{2}\right] - \frac{V_0}{8} = 0$$

$$6 - 1.375 \times \frac{V_0}{2} = 0$$

or $$V_0 = \frac{6 \times 2}{1.375} = 8\,V$$

And $$i_0 = \frac{V_0}{2} = \frac{8}{2} = 4\,A$$

Exercise 1.5.4:

Obtain the currents i_1, i_2 and i_3 using KCL for the network shown in figure 1.31

Solution:

Applying KCL at node 'a', $\quad 8 - i_1 - 12 = 0$

Or $\qquad\qquad i_1 = 8 - 12 = -4 \text{ mA}$

Applying KCl at node 'c', $\quad 12 + i_3 - 9 = 0$

Or $\qquad\qquad i_3 = 9 - 12 = -3 \text{ mA}$

Applying KCL at node 'b',

$$i_1 + i_2 = i_3$$

$$-4 + i_2 = -3$$

Or $\qquad\qquad i_2 = -3 + 4 = 1 \text{ mA}$

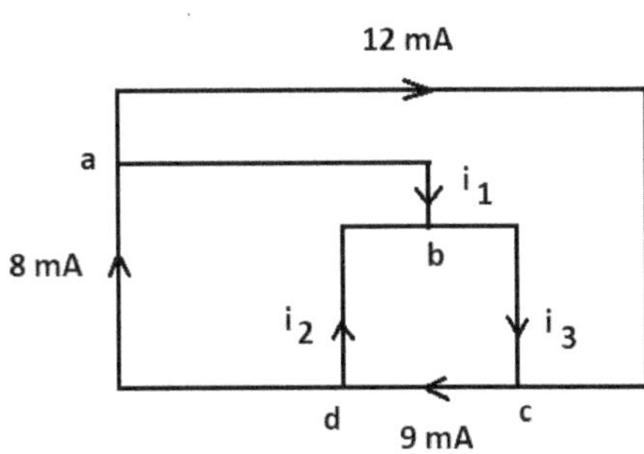

Figure 1.31

Exercise 1.5.5:

Find the values of V_1 and V_2 in the network given in figure 1.32.

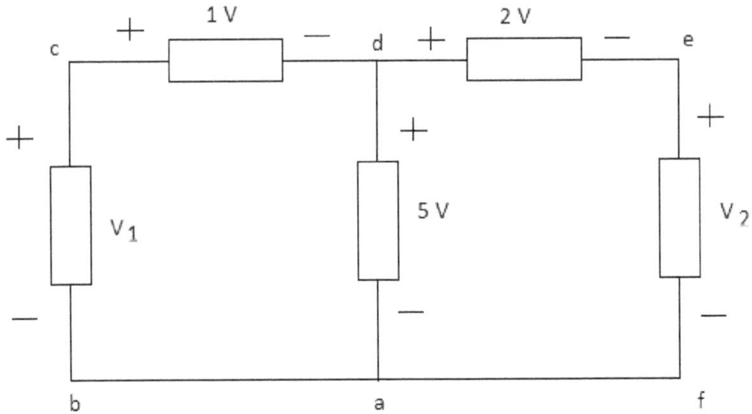

Figure 1.32

Solution:

In the loop abcda, $\quad V_1 - 1 - 5 = 0 \text{ or } V_1 = 6 \text{ V}$

In the loop adefa, $\ 5 - 2 - V_2 = 0 \text{ or } V_2 = 3 \text{ V}$

Exercise 1.5.6:

Calculate the values of V and i_x in the circuit shown in figure 1.33.

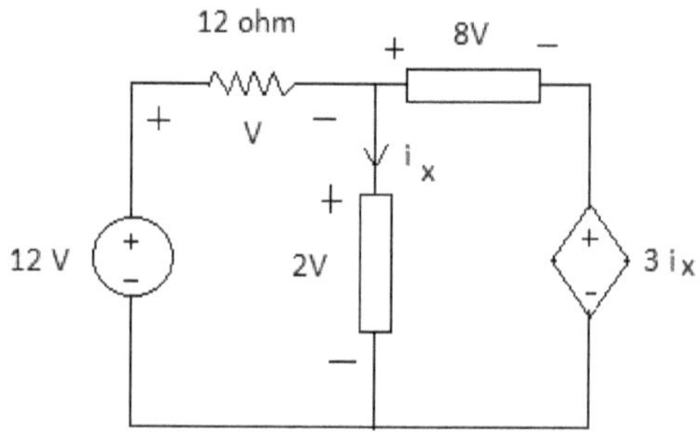

Figure 1.33

Solution:

Assuming the currents i_1 and i_2. Also identify the different nodes a, b, c and d.

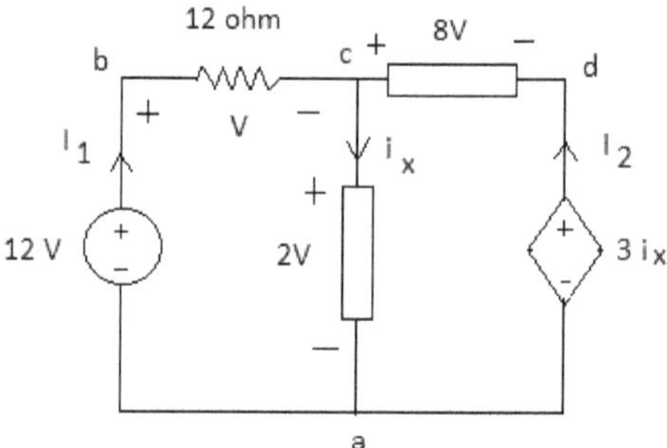

Figure 1.33 (a)

In loop abca, $12 - 12i_1 - 2 = 0$

$$10 - 12i_1 = 0$$

$$i_1 = \frac{10}{12} = \frac{5}{6} \text{ A}$$

and $\qquad V = i_1 \times 12 = \frac{5}{6} \times 12 = 10 \text{ V}$

In the loop acda,

$$2 - 8 - 3i_x = 0$$

$$3i_x = -6$$

$$i_x = \frac{-6}{3} = -2 \text{ A}$$

Exercise 1.5.7:

Find the voltages V_1, V_2 and V_3 using KVL for the circuit shown in figure 1.34

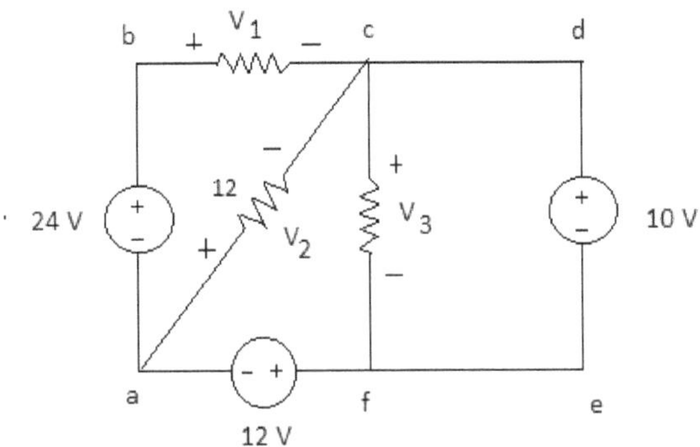

Figure 1.34

Solution:

Identify the nodes a, b, c, d, e and f.

In the loop abcdefa, $24 - V_1 - 10 - 12 = 0$

$24 - V_1 - 22 = 0$ or $V_1 = 2\,\text{V}$

In the loop abca, $24 - 2 + V_2 = 0$ or $V_2 = -22$ V

In the loop fcdef, $V_3 - 10 = 0$ or $V_3 = 10$

Exercise 1.5.8:

Find the values of V_{ab} and I in the given circuit shown in figure 1.35.

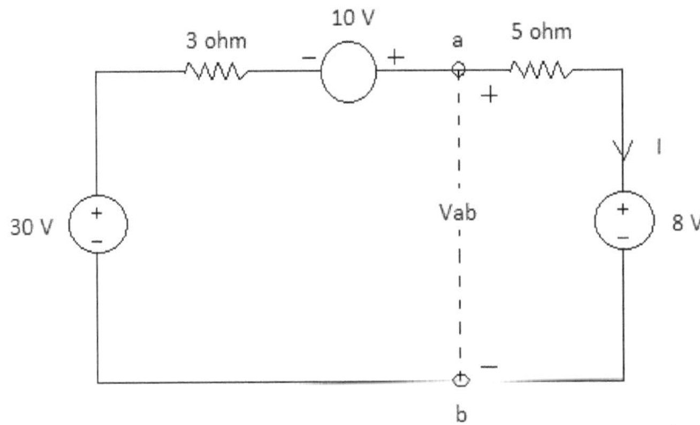

Figure 1.35

Solution:

Applying KVL,

$$30 - 3i + 10 - 5i - 8 = 0$$

$$32 - 8i = 0 \text{ or } \quad i = \frac{32}{8} = 4 \text{ A}$$

Voltage drop in 5 ohm $= 4 \times 5 = 20 \text{ V}$

V_{ab} = Potential of point 'a' with respect to point 'b'.

Then, V_{ab} (starting from a) $= +8 + 20 = 28 \text{ V}$

Exercise 1.5.9:

Determine the voltage drop across 10 ohm resistance for the circuit shown in figure 1.36.

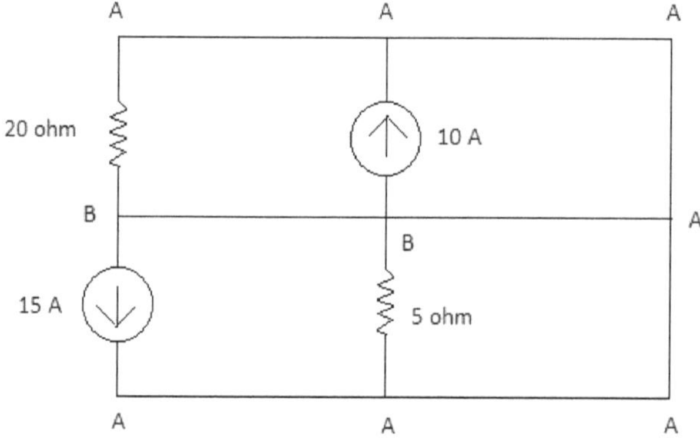

Figure 1.36

Solution:

Let us identify the points 'A' and 'B' across 10A current source initially. These points get extended to points of same potential as shown. The above circuit gets modified as shown in figure 1.36(a).

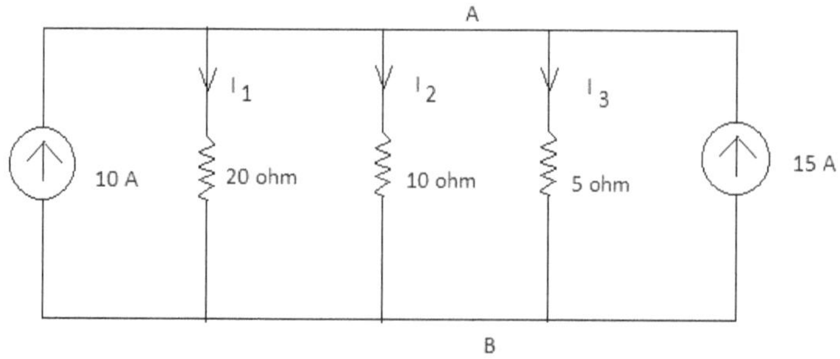

Figure 1.36 (a)

Let 'V' be the voltage across 10 ohm resistance.

$$V = i_2 \times 10$$

Applying KCL at node 'A',

$$10 - \frac{V}{20} - \frac{V}{10} - \frac{V}{5} + 15 = 0$$

$$25 - V\left[\frac{1}{20} + \frac{1}{10} + \frac{1}{5}\right] = 0$$

$$25 - V\left[\frac{1+2+4}{20}\right] = 0$$

Or $V = 71.4 \text{ V}$

Exercise : 1.5.10

Find the current flowing through different resistances using Kirchhoff's Laws for the circuit shown in figure 1.37.

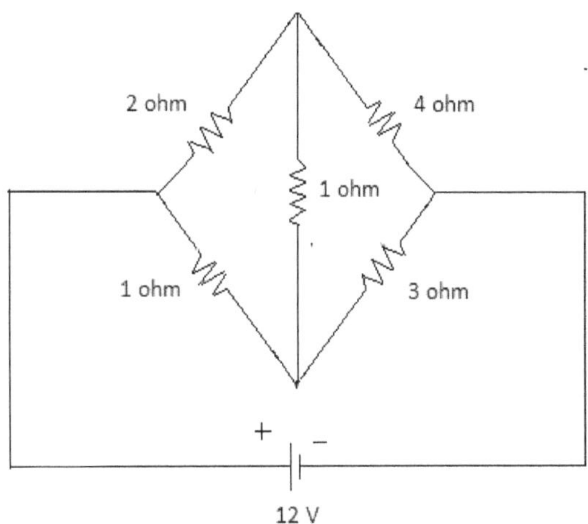

Figure 1.37

Solution:

Identify the different nodes a,b,c and d and assume the current directions through all branches.

Applying KVL for the loop abca,

$$-2i_2 - 1 \times i_3 + 1\left[i_1 - i_2\right] = 0$$

$$-2i_2 - i_3 + i_1 - i_2 = 0$$

$$i_1 - 3i_2 - i_3 = 0 \qquad\qquad ....(1)$$

Applying KVL for the loop cbdc,

$$1 \times i_3 - 4(i_2 - i_3) + 3(i_1 - i_2 + i_3) = 0$$

$$i_3 - 4i_2 + 4i_3 + 3i_1 - 3i_2 + 3i_3 = 0$$

$$3i_1 - 7i_2 + 8i_3 = 0 \qquad\qquad ....(2)$$

Applying KVL for the loop acda through 12V source,

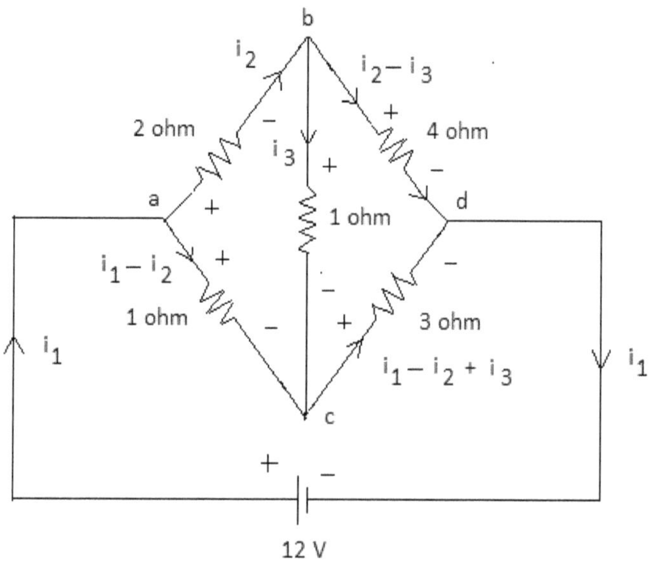

Figure 1.37(a)

$$-(i_1 - i_2) \times 1 - (i_1 - i_2 + i_3) \times 3 + 12 = 0$$
$$-i_1 + i_2 - 3i_1 + 3i_2 - 3i_3 + 12 = 0$$
$$-4i_1 + 4i_2 - 3i_3 = -12$$
$$4i_1 - 4i_2 + 3i_3 = 12 \qquad\qquad ....(3)$$

Solving the equations (1) and (2),

$$(1) \qquad \times 3 \quad 3i_1 - 9i_2 - 3i_3 = 0$$

$$-(2), \qquad -3i_1 + 7i_2 - 8i_3 = 0$$

$$-2i_2 + 11i_3 = 0 \qquad\qquad ....(4)$$

Solving equations (2) and (3),

$$(2)\times4, \quad 12i_1 - 28i_2 + 32i_3 = 0$$

$$-(3)\times3, \quad -12i_1 + 12i_2 - 9i_3 = -36$$

$$-16i_2 + 23i_3 = -36 \qquad\qquad ....(5)$$

Solving equations (4) and (5),

$$-(4)\times8, \quad 16i_2 - 88i_3 = 0$$

$$(5), \quad -16i_2 + 23i_3 = -36$$

$$0 - 65i_3 = -36 \quad i_3 = \frac{36}{65} = 0.553\,\text{A}$$

From equation (4), $-2\times i_2 + 11\times\dfrac{36}{65} = 0$ or $i_2 = \dfrac{396}{130}\,\text{A}$

From equation (1), $i_1 - 3\times\dfrac{396}{130} - \dfrac{36}{65} = 0$ or $i_1 = 9.69\,\text{A}$

The different branch currents are i_1=9.69 A, $i_2 = \dfrac{396}{130}$ A and i_3= 0.553 A

Exercise: 1.5.11

Find the voltage V_x in the circuit shown in figure 1.38.

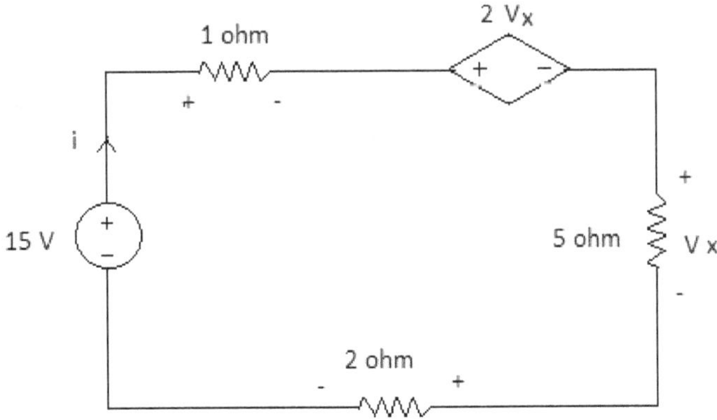

Figure 1.38

Solution:

Let us assume the current 'I' in the circuit and applying KVL,

$$15 - 1 \times i - 2V_x - 5i - 2i = 0$$

$$15 - 8i - 2V_x = 0$$

$$15 - 8 \times \frac{V_x}{5} - 2V_x = 0$$

$$15 = 1.6V_x + 2V_x$$

$$V_x = \frac{15}{3.6} = 4.16V$$

Exercise : 1.5.12

Find the voltage across all the resistances and branch currents using Kirchhoff's Laws for the circuit shown in figure 1.39.

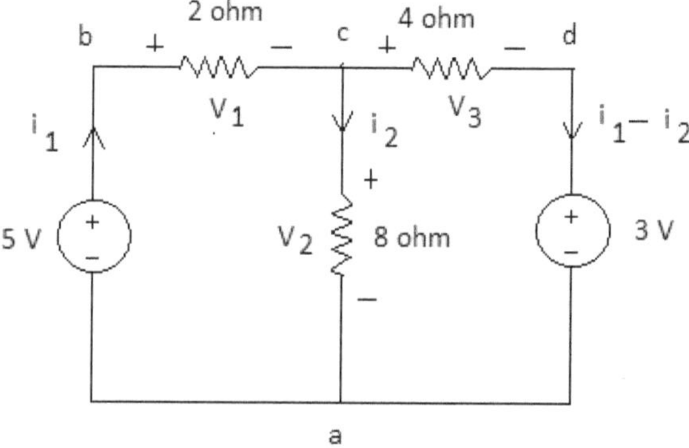

Figure 1.39

Solution:

Identify different nodes a,b,c and d. Assume the currents in different branches as shown.

Applying KCL at node c,

$$i_1 = i_2 + i_3 \text{ or } i_3 = i_1 - i_2$$

Applying KVL for the loop abca

$$5 - 2i_1 - 8i_2 = 0$$

Or $\qquad 2i_1 + 8i_2 = 5 \qquad\qquad\qquad\qquad\qquad\qquad\qquad\qquad\qquad .....(1)$

Applying KVL for the loop acda,

$$8i_2 - 4\left(i_1 - i_2\right) + 3 = 0$$

Or $\qquad -4i_1 + 12i_2 = -3 \qquad\qquad\qquad\qquad\qquad\qquad\qquad\qquad .....(2)$

Solving the equations (1) and (2),

$2\times(1),\qquad 4i_1 + 16i_2 = 10$

$(2),\quad -4i_1 + 12i_2 = -3$

$$28i_2 = 7\,\text{or}\quad i_2 = \frac{7}{28} = 0.25\,\text{A}$$

From equation (1), $\quad 2i_1 + 8\times0.25 = 5\,\text{or}\quad i_1 = \frac{3}{2} = 1.5\,\text{A}$

And also, $i_3 = i_1 - i_2 = 1.5 - 0.25 = 1.25\,\text{A}$

Then, the voltage drops across different resistances are

$$V_1 = 2\times i_1 = 2\times1.5 = 3V$$
$$V_2 = 8\times i_2 = 8\times0.25 = 2V$$
$$V_3 = 4\times i_3 = 4\times1.25 = 5V$$

Exercise 1.5.13:

What is the voltage between the points A and B in the circuit shown in figure 1.40?

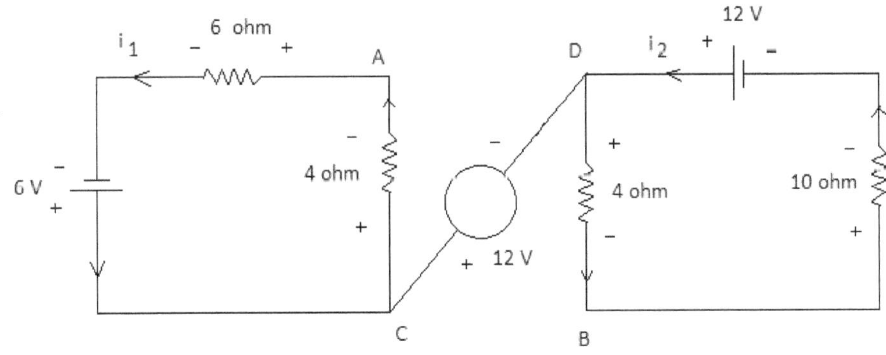

Figure 1.40

Solution:

Assume the current directions i₁ and i₂. No current flows through 12V source as there is no closed path or return path for the current which enters this branch.

Identify the polarity of voltage drops across different resistances for a given directions of the currents.

$$i_1 = \frac{6}{6+4} = 0.6 \, \text{A} \quad \text{and} \quad i_2 = \frac{12}{4+10} = \frac{6}{7} \, \text{A}$$

To find the voltage between the points A and B, start considering the potential from point B.

From B to D, the voltage rises by $4 \times i_2 = 4 \times \dfrac{6}{7} = 3.4 \, \text{V}$

From D to C, the voltage rises by 12 V

From C to A, voltage drops by $4i_1 = 4 \times 0.6 = 2.4 \, \text{V}$

Therefore, the voltage between points A and B $= V_{AB} = 3.4 + 12 - 2.4 = 13 \, \text{V}$

Or the potential of point B with respect to point A $= V_{BA} = -13 \, \text{V}$

Or the potential difference between A and B $= \mp 13 \, \text{V}$

Exercise 1.5.14:

Find the power absorbed by each element in circuit shown in figure 1.41.

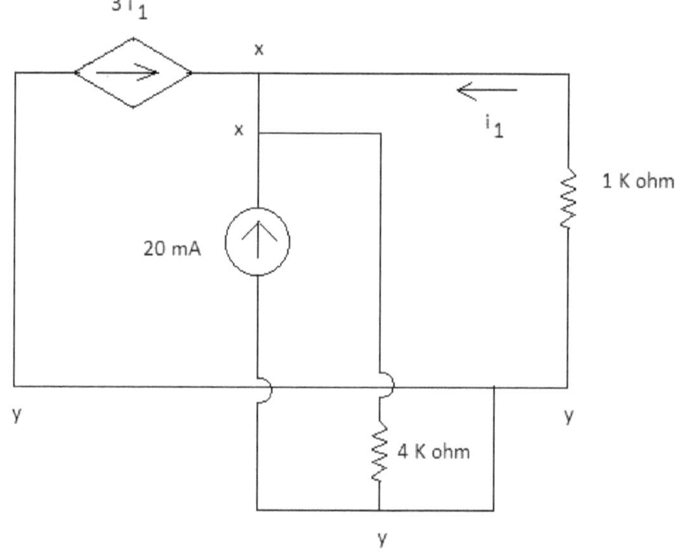

Figure 1.41

Solution:

Identify the nodes 'x' and 'y'. By keen observation, all the elements got connected in parallel between x and y. The resultant circuit becomes,

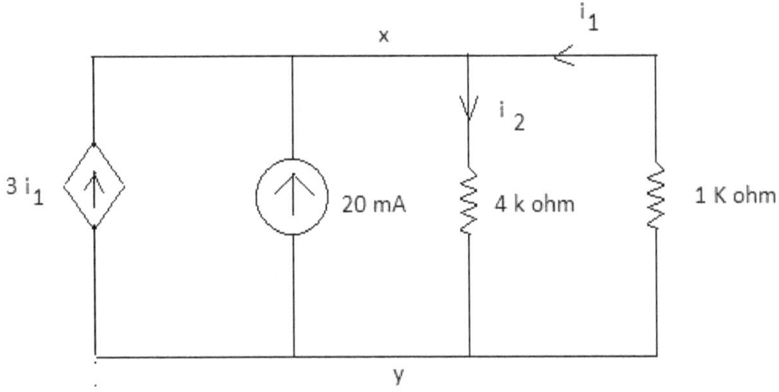

Figure 1.41 (a)

$$1000 \times i_1 = 4000 \times i_2$$

Or $\qquad i_2 = \dfrac{-1000}{4000} \times i_1 = -0.25 i_1$

Applying KCL at the node 'x',

$$3i_1 + 20 \times 10^{-3} - i_2 + i_1 = 0$$

$$3i_1 + 20 \times 10^{-3} - \left(-0.25 i_1\right) + i_1 = 0$$

$$4.25 i_1 + 20 \times 10^{-3} = 0$$

$$4.25 i_1 = -20 \times 10^{-3} \quad \text{or} \quad i_1 = -4.7 \ \text{mA}$$

$$i_2 = -\left(0.25\right) \times \left(-4.7 \times 10^{-3}\right) = 1.175 \ \text{A}$$

Power absorbed by 4 K resistance $= i_2^2 \times 4000 = \left(1.175 \times 10^{-3}\right)^2 \times 4000 = 5.5 \ \text{mW}$

Power absorbed by 1 K resistance $= i_1^2 \times 1000 = \left(4.7 \times 10^{-3}\right)^2 \times 1000 = 22 \ \text{mW}$

1.6 Problems on Series-Parallel Connections

Exercise 1.6.1:

Find the current delivered by the source for the circuit shown in figure 1.42.

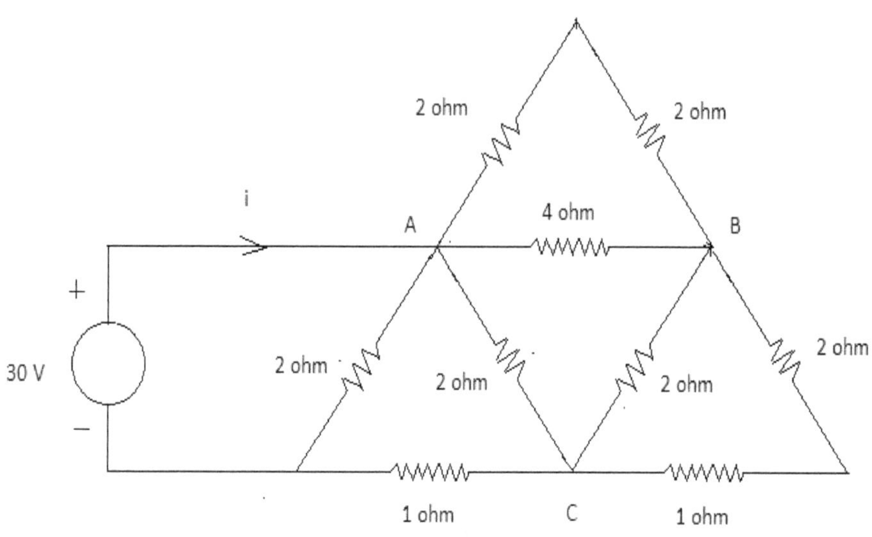

Figure 1.42

Solution:

Identify the nodes A,B and C. Between A and B, 2 ohms and 2 ohms are in series.

Between the points B and C, 2 ohms and 1 ohm are connected in series. Then, the circuit gets simplified to:

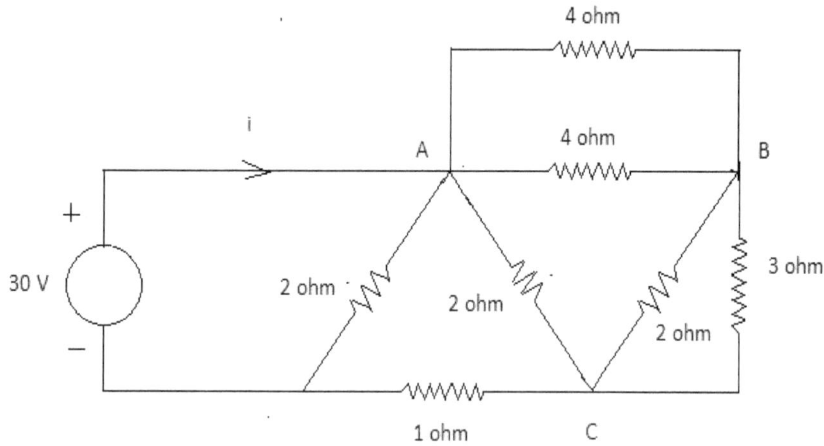

Figure 1.42 (a)

Between A and B, 4 ohms and 4 ohms are connected in parallel.

Between B and C, 2 ohms and 3 ohms are in parallel.

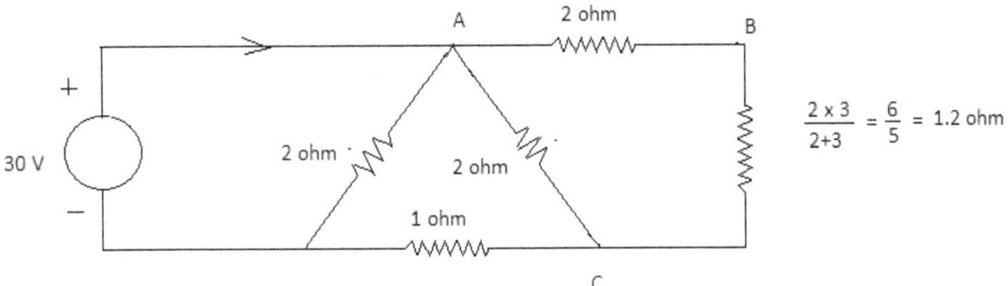

Figure 1.42 (b)

Between A and C, 2 ohms and 1.2 ohms are in series.

$$\text{Then, } R_{AC} = \frac{(2+1.2)\times 2}{2+(2+1.2)} = 1.23 \text{ ohms}$$

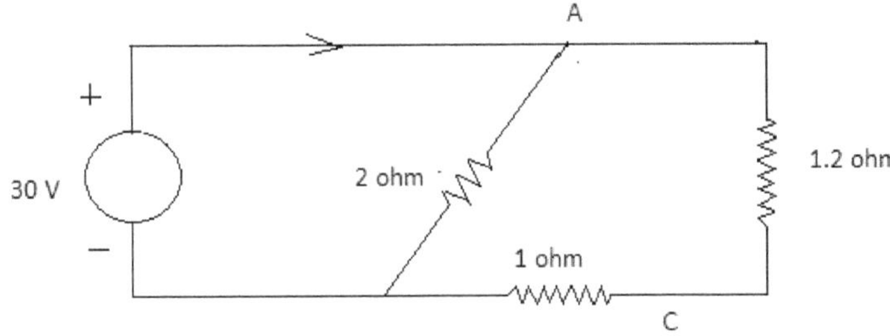

Figure 1.42 (c)

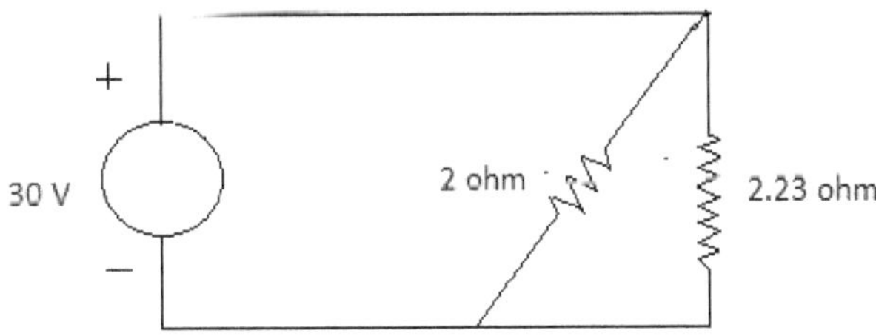

Figure 1.42 (d)

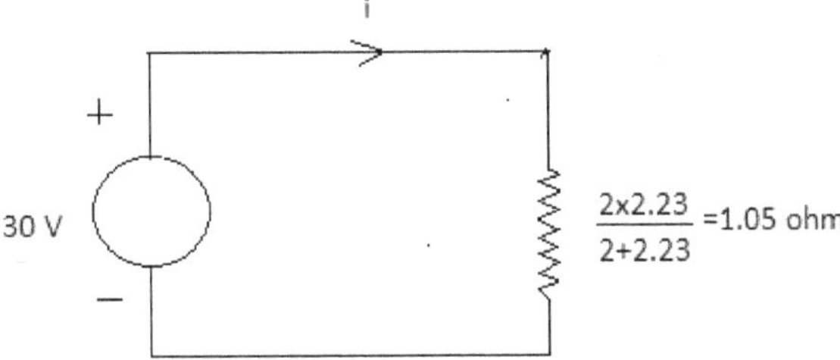

Figure 1.42 (e)

The current supplied by the source $i = \dfrac{30}{1.05} = 28.57\,\text{A}$

Exercise : 1.6.2

Find the resistance between the points 'x' and 'y'in the circuit shown in figure 1.43. Also find the total current supplied by 6 V sources when it is connected between 'x' and 'y'.

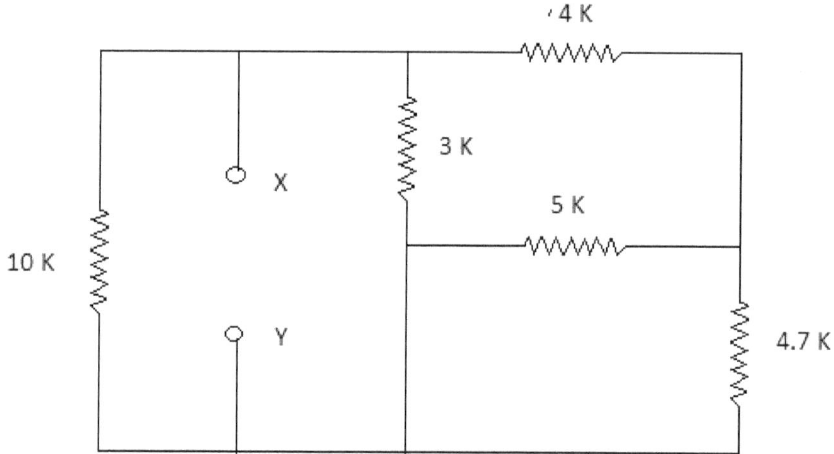

Figure 1.43

Solution:

Identify the nodes A, B, C, D and E as shown.

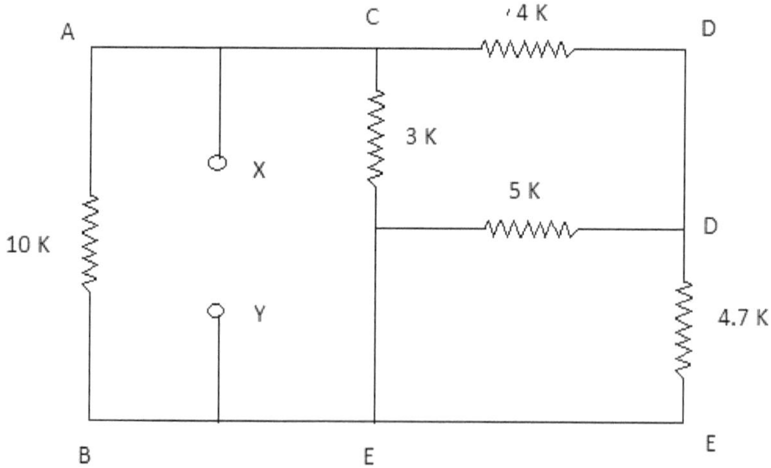

Figure 1.43 (a)

Between E and D, 5K and 4.7K resistances are connected in parallel.

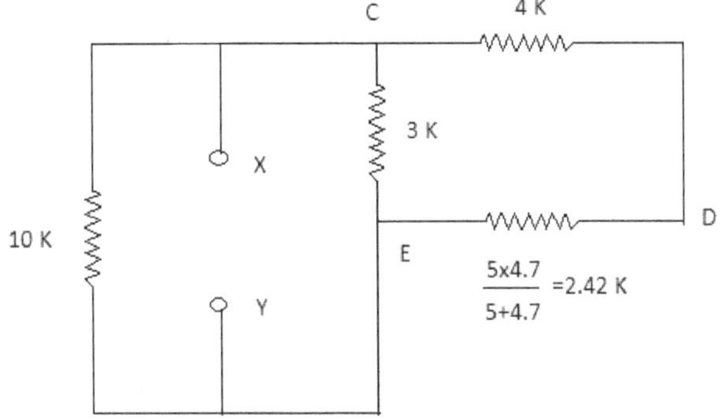

Figure 1.43 (b)

Between C and E, 4 K and 2.42 K resistances are connected in series and it is in parallel with 3 K resistance.

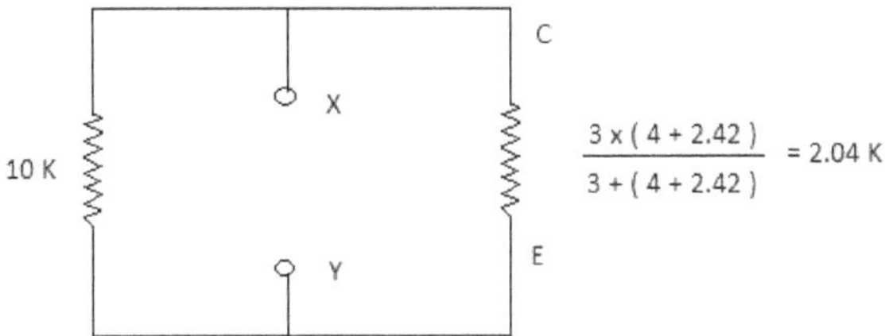

Figure 1.43 (c)

Resistance between 'x' and 'y' is obtained as follows.

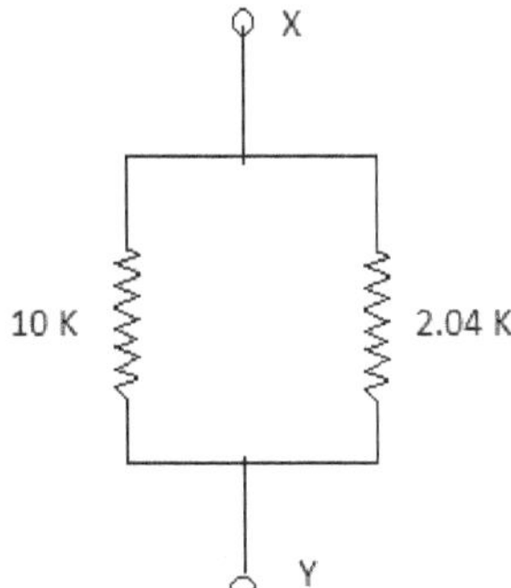

Figure 1.43 (d)

$$R_{xy} = \frac{10 \times 2.04}{10 + 2.04} = 1.698 \, \text{K Ohms}$$

When 6 V battery is connected between x and y,

$$\text{The current driven } = \frac{6}{1.698 \times 10^3} = 3.53 \text{ mA}$$

Exercise : 1.6.3

Find the resistance between 'a' and 'b' for the circuit shown in figure 1.44

Solution:

The nodes A and B are of same potential. Hence, these two nodes are one and same. Then, the above circuit gets reduced to:

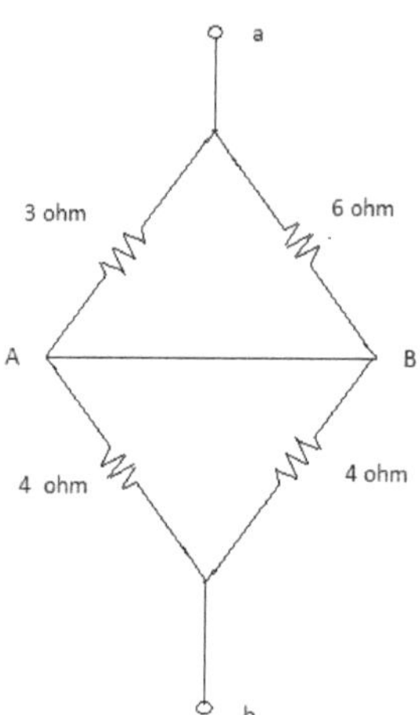

Figure 1.44

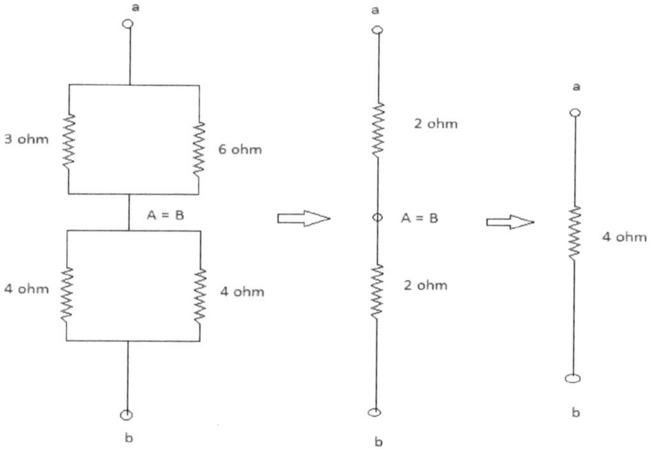

Figure 1.44 (a)

1.7   Star/Delta Transformations

Star- Delta transformations are useful to solve the networks which cannot be easily solved by regular series parallel circuit reduction technics.

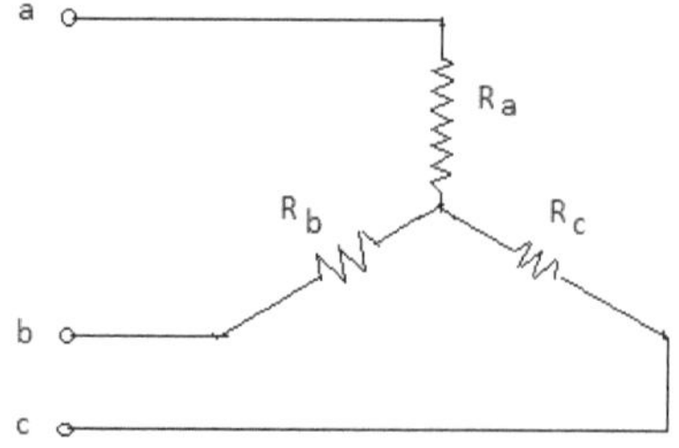

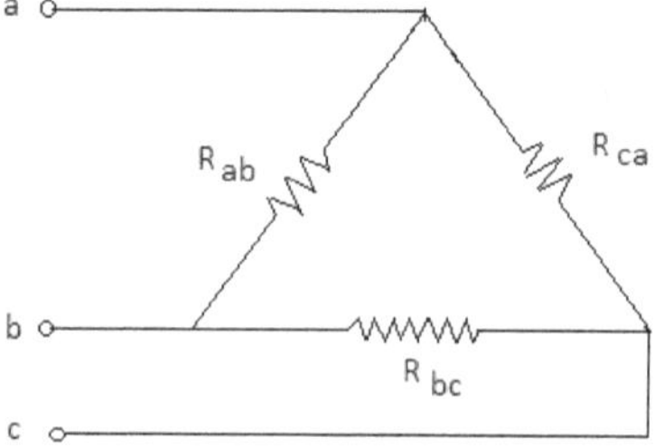

Figure 1.45 (a) Star Network (b) Delta Network

For Star Connection:

The resistance between the points 'a' and 'b' $= R_{a-b} = R_a + R_b$

The resistance between points 'b' and 'c' $= R_{b-c} = R_b + R_c$

Similarly, the resistance between 'c' and 'a' $= R_{c-a} = R_a + R_c$

For Delta Connection:

The resistance between the points 'a' and 'b' $R_{a-b} = R_{ab} \| (R_{bc} + R_{ca})$

$$= \frac{R_{ab}(R_{bc} + R_{ca})}{R_{ab} + R_{bc} + R_{ca}}$$

$$= \frac{R_{ab}R_{bc} + R_{ab}R_{ca}}{R_{ab} + R_{bc} + R_{ca}}$$

The resistance between the points 'b' and 'c' $R_{b-c} = R_{bc} \| (R_{ab} + R_{ca})$

$$= \frac{R_{bc}(R_{ab} + R_{ca})}{R_{ab} + R_{bc} + R_{ca}}$$

$$= \frac{R_{bc}R_{ab} + R_{bc}R_{ca}}{R_{ab} + R_{bc} + R_{ca}}$$

The resistance between the points 'c' and 'a' $R_{c-a} = R_{ca} \| (R_{ab} + R_{bc})$

$$= \frac{R_{ca}(R_{ab} + R_{bc})}{R_{ab} + R_{bc} + R_{ca}}$$

$$= \frac{R_{ca}R_{ab} + R_{ca}R_{bc}}{R_{ab} + R_{bc} + R_{ca}}$$

To convert the given Delta Network into equivalent Star Network:

By equating the values of R_{a-b}, R_{b-c} and R_{c-a} of both star and delta connections, we get

$$R_a + R_b = \frac{R_{ab}R_{bc} + R_{ab}R_{ca}}{R_{ab} + R_{bc} + R_{ca}} \quad \quad \dots\dots(1)$$

$$R_b + R_c = \frac{R_{bc}R_{ab} + R_{bc}R_{ca}}{R_{ab} + R_{bc} + R_{ca}} \quad \quad \dots\dots(2)$$

$$R_c + R_a = \frac{R_{ca}R_{ab} + R_{ca}R_{bc}}{R_{ab} + R_{bc} + R_{ca}} \quad \quad \dots\dots(3)$$

Equations (1) –(2),

$$R_a + R_b - R_b - R_c = R_a - R_c$$

$$\frac{R_{ab}R_{bc} + R_{ab}R_{ca}}{R_{ab} + R_{bc} + R_{ca}} - \frac{R_{bc}R_{ab} + R_{bc}R_{ca}}{R_{ab} + R_{bc} + R_{ca}} = \frac{R_{ab}R_{bc} + R_{ab}R_{ca} - R_{bc}R_{ab} - R_{bc}R_{ca}}{R_{ab} + R_{bc} + R_{ca}}$$

$$R_a - R_C = \frac{R_{ab}R_{ca} - R_{bc}R_{ca}}{R_{ab} + R_{bc} + R_{ca}} \qquad \dots\dots(4)$$

Equation (3) + (4),

$$R_a + R_c + R_a - R_c = 2R_a$$

$$= \frac{R_{ca}R_{ab} + R_{ca}R_{bc}}{R_{ab} + R_{bc} + R_{ca}} + \frac{R_{ab}R_{ca} - R_{bc}R_{ca}}{R_{ab} + R_{bc} + R_{ca}}$$

$$= \frac{2R_{ab}R_{ca}}{R_{ab} + R_{bc} + R_{ca}}$$

Or
$$R_a = \frac{R_{ab}R_{ca}}{R_{ab} + R_{bc} + R_{ca}} \qquad \dots\dots(5)$$

Similarly
$$R_b = \frac{R_{ab}R_{bc}}{R_{ab} + R_{bc} + R_{ca}} \qquad \dots\dots(6)$$

Similarly
$$R_c = \frac{R_{bc}R_{ca}}{R_{ab} + R_{bc} + R_{ca}} \qquad \dots\dots(7)$$

To convert the given Star Network into equivalent Delta Network:

Multiplying equations (5) and (6)

$$R_a \times R_b = \frac{R_{ab}R_{ca}}{R_{ab} + R_{bc} + R_{ca}} \times \frac{R_{ab}R_{bc}}{R_{ab} + R_{bc} + R_{ca}}$$

$$= \frac{R_{ab}^2 R_{bc} R_{ca}}{\left(R_{ab} + R_{bc} + R_{ca}\right)^2}$$

Multiplying equations (6) and (7)

$$R_b \times R_c = \frac{R_{ab}R_{bc}}{R_{ab} + R_{bc} + R_{ca}} \times \frac{R_{bc}R_{ca}}{R_{ab} + R_{bc} + R_{ca}}$$

$$= \frac{R_{ab} R_{bc}^2 R_{ca}}{\left(R_{ab} + R_{bc} + R_{ca}\right)^2}$$

Multiplying equations (7) and (5)

$$R_c \times R_a = \frac{R_{bc}R_{ca}}{R_{ab} + R_{bc} + R_{ca}} \times \frac{R_{ab}R_{ca}}{R_{ab} + R_{bc} + R_{ca}}$$

$$= \frac{R_{ab}R_{bc}R_{ca}^2}{\left(R_{ab} + R_{bc} + R_{ca}\right)^2}$$

$$R_aR_b + R_bR_c + R_cR_a = \frac{R_{ab}^2 R_{bc}R_{ca} + R_{ab}R_{bc}^2 R_{ca} + R_{ab}R_{bc}R_{ca}^2}{\left(R_{ab} + R_{bc} + R_{ca}\right)^2}$$

$$= \frac{R_{ab}R_{bc}R_{ca}\left(R_{ab} + R_{bc} + R_{ca}\right)}{\left(R_{ab} + R_{bc} + R_{ca}\right)^2}$$

$$= \frac{R_{ab}R_{bc}R_{ca}}{R_{ab} + R_{bc} + R_{ca}} \qquad \qquad .....(8)$$

By equation (8)/(5),

$$\frac{R_aR_b + R_bR_c + R_cR_a}{R_a} = \frac{R_{ab}R_{bc}R_{ca}}{R_{ab} + R_{bc} + R_{ca}} \times \frac{R_{ab} + R_{bc} + R_{ca}}{R_{ab}R_{ca}}$$

$$R_{bc} = \frac{R_aR_b + R_bR_c + R_cR_a}{R_a}$$

Similarly, by performing the operations, (8)/(6) and (8)/(7), we get

$$R_{ca} = \frac{R_aR_b + R_bR_c + R_cR_a}{R_b}$$

$$R_{ab} = \frac{R_uR_b + R_bR_c + R_cR_a}{R_c}$$

If all the branches have equal values of resistances (R_Δ) in a Delta network, the equivalent resistance value in each branch of Star network is

$$R_a = R_b = R_c = R_{star} = \frac{R_\Delta}{3}$$

Similarly, If all the branches have equal values of resistances (R_{Star}) in a Star network, the equivalent resistance value in each branch of Delta network is

$$R_{ab} = R_{bc} = R_{ca} = R_\Delta = 3 \times R_{star}$$

As the connection is transformed from Delta to Star, the resistance in each leg or branch decreases. Similarly, as the transformation is from Star to Delta, the resistance in each branch increases.

Exercise 1.7.1:

Convert the given Delta Network shown in figure 1.46 into equivalent Star Network.

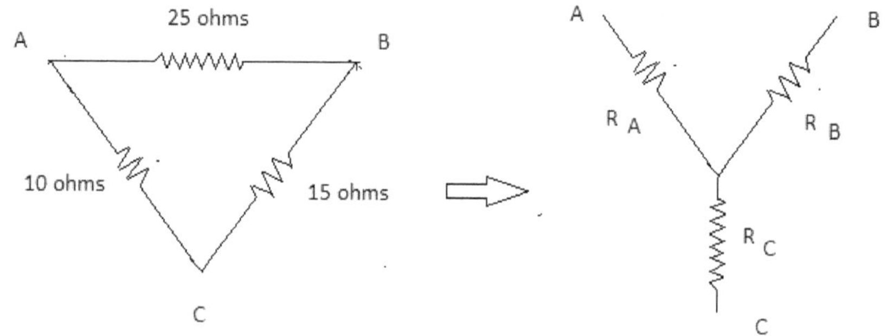

Figure 1.46

Solution:

$$R_a = \frac{25 \times 10}{25+15+10} = \frac{250}{50} = 5 \text{ ohms}$$

$$R_b = \frac{25 \times 15}{25+15+10} = \frac{375}{50} = 7.5 \text{ ohms}$$

$$R_c = \frac{10 \times 15}{25+15+10} = \frac{150}{50} = 3 \text{ ohms}$$

Exercise 1.7.2:

Find the current supplied by 120V Battery shown in figure 1.47.

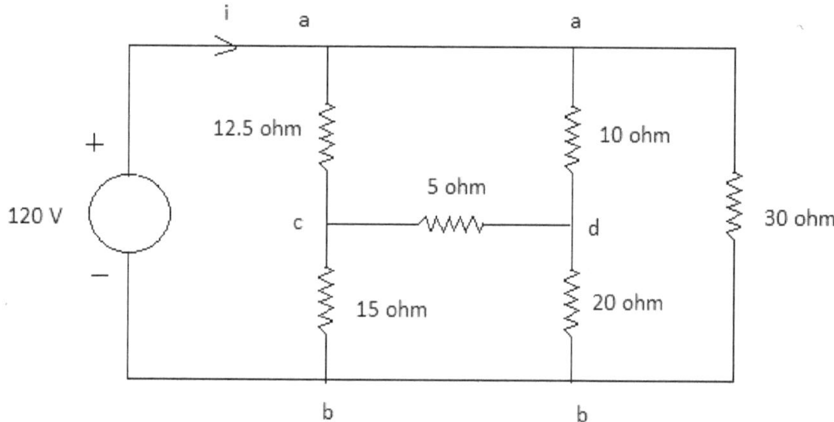

Figure 1.47

Solution:

The Star network between the points a, b, c is converted into Delta network as follows.

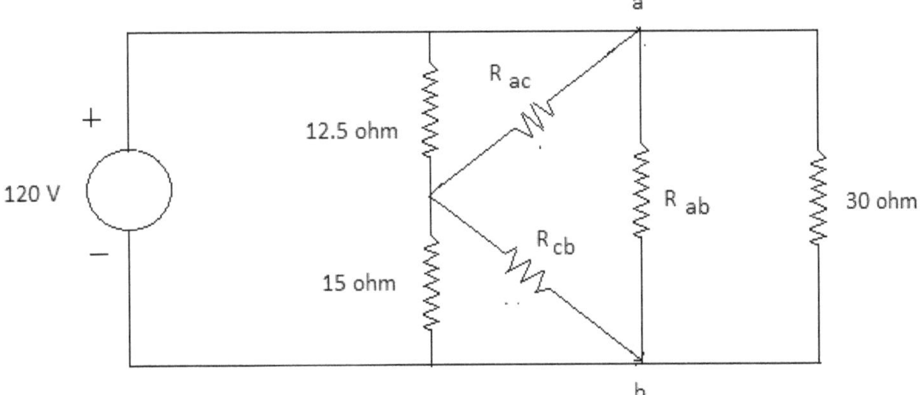

Figure 1.47 (a)

$$R_{ac} = \frac{5 \times 10 + 10 \times 20 + 20 \times 5}{20} = \frac{50 + 200 + 100}{20} = \frac{350}{20} = 17.5 \text{ ohms}$$

$$R_{ab} = \frac{5 \times 10 + 10 \times 20 + 20 \times 5}{5} = \frac{50 + 200 + 100}{5} = \frac{350}{5} = 70 \text{ ohms}$$

$$R_{bc} = \frac{5 \times 10 + 10 \times 20 + 20 \times 5}{10} = \frac{50 + 200 + 100}{10} = \frac{350}{10} = 35 \text{ ohms}$$

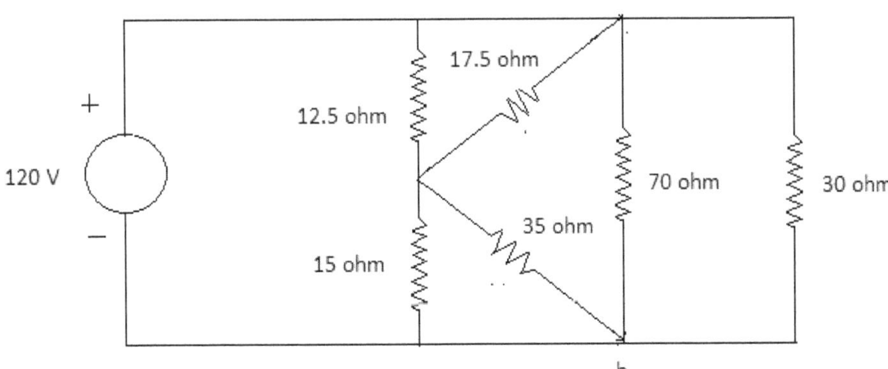

Figure 1.47 (b)

The above circuit gets reduced to

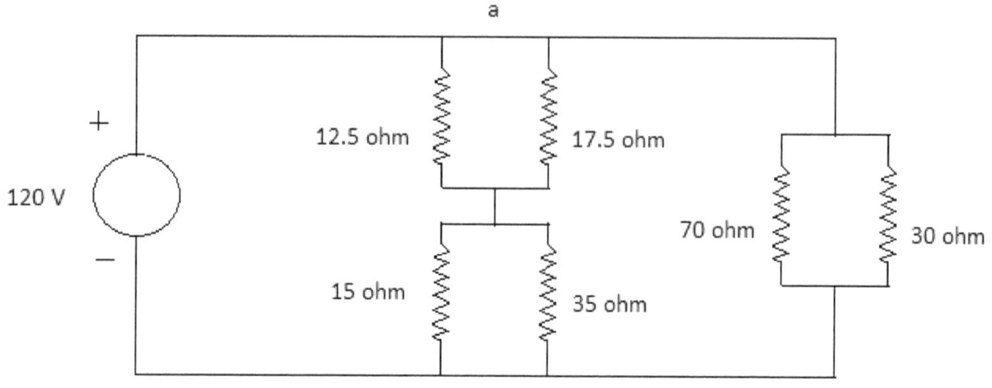

Figure 1.47 (c)

It is further reduced to

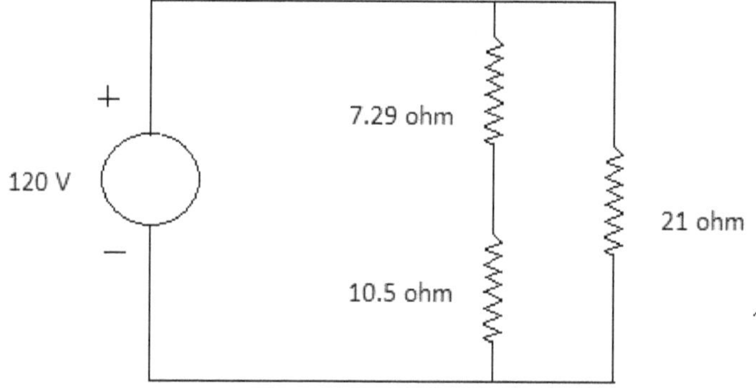

Figure 1.47 (d)

and

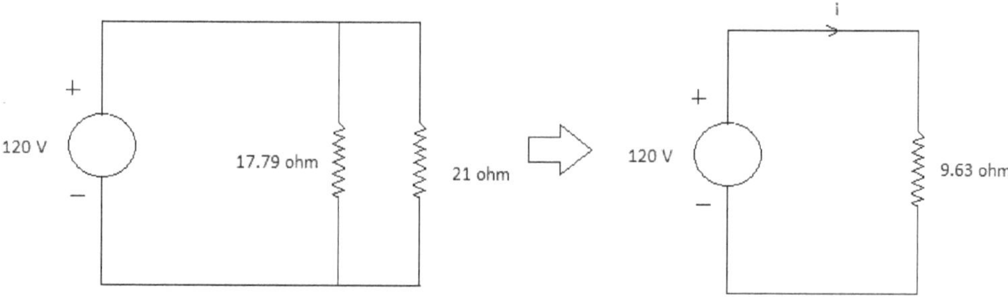

Figure 1.47 (e)

The current supplied $i = \dfrac{120}{9.63} = 12.46\,\text{A}$

Exercise: 1.7.3:

Find the current supplied by the source for the circuit shown in figure 1.48

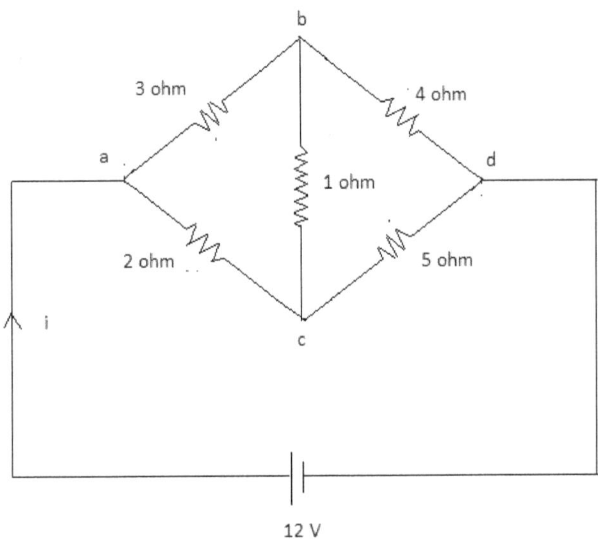

Figure 1.48

Solution:

The Delta Network between the points a,b,c is converted into equivalent Star Network.

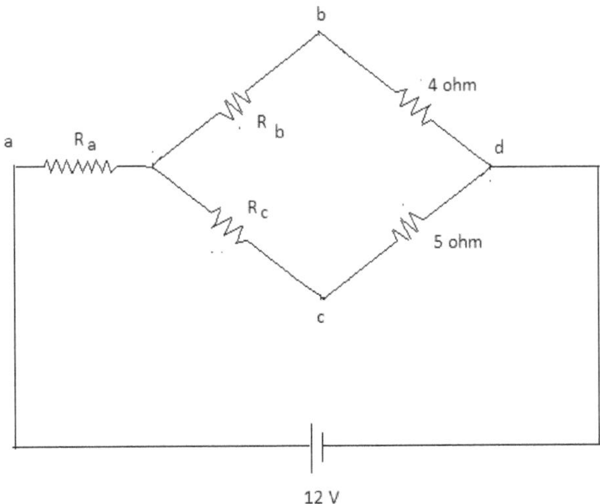

Figure 1.48 (a)

$$R_a = \frac{3 \times 2}{3+2+1} = \frac{6}{6} = 1 \text{ ohms}$$

$$R_b = \frac{3 \times 1}{3+2+1} = \frac{3}{6} = 0.5 \text{ ohms}$$

$$R_c = \frac{1 \times 2}{3+2+1} = \frac{2}{6} = 0.33 \text{ ohms}$$

By substituting these values in above network, we get

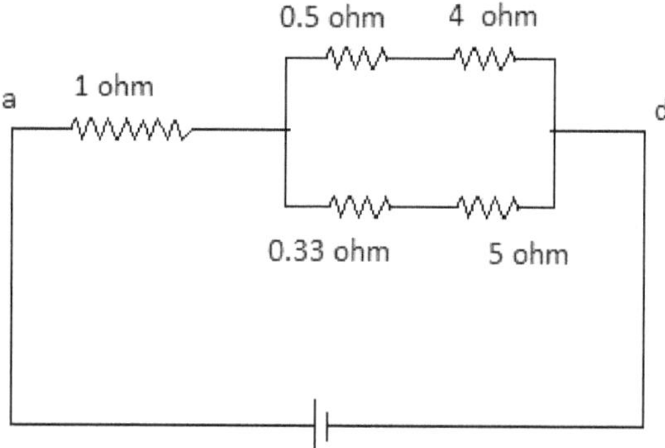

Figure 1.48 (b)

It is simplified to

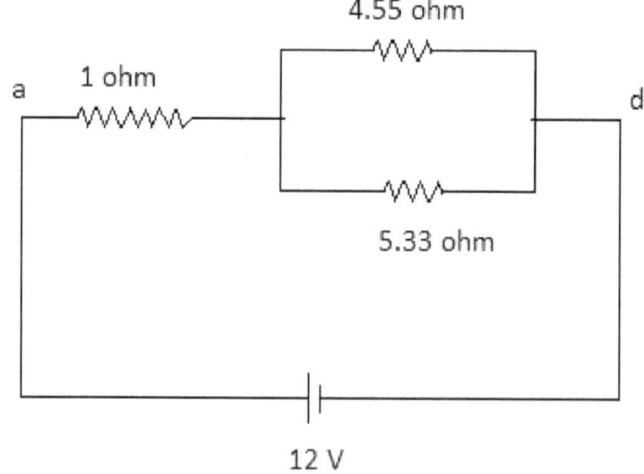

Figure 1.48 (c)

It is again further reduced to

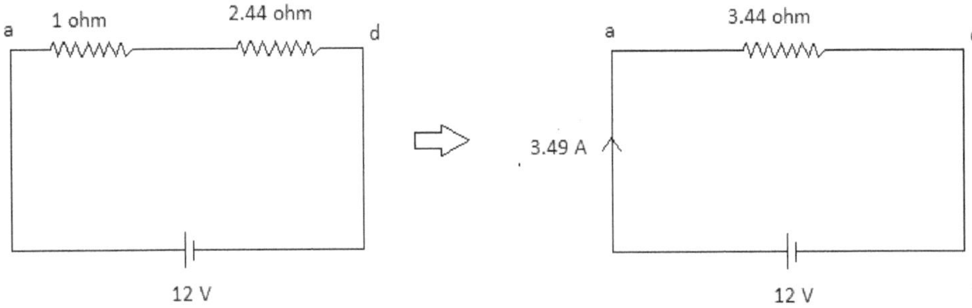

Figure 1.48 (d)

The current delivered by the source $i = \dfrac{12}{3.44} = 3.49\,\text{A}$

Exercise 1.7.4 : Find the equivalent resistance and the current 'I' for the circuit shown in figure 1.49.

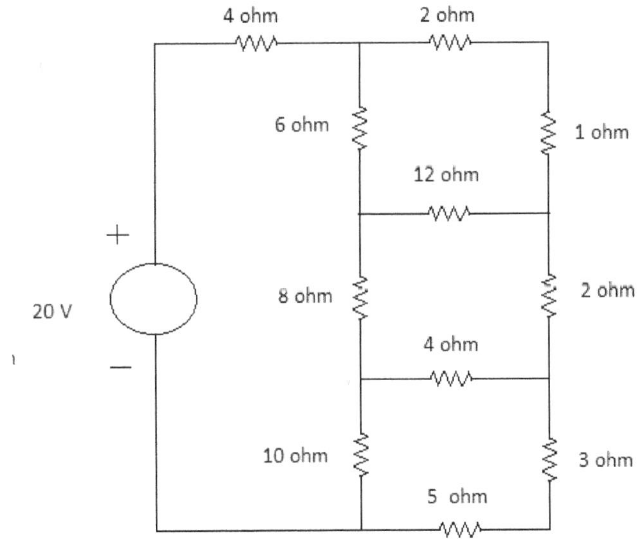

Figure 1.49

Solution:

The resistances 2 ohms and 1 ohm are connected in series. Also, 3 ohms and 5 ohms are in series. Then, the circuit gets changed to

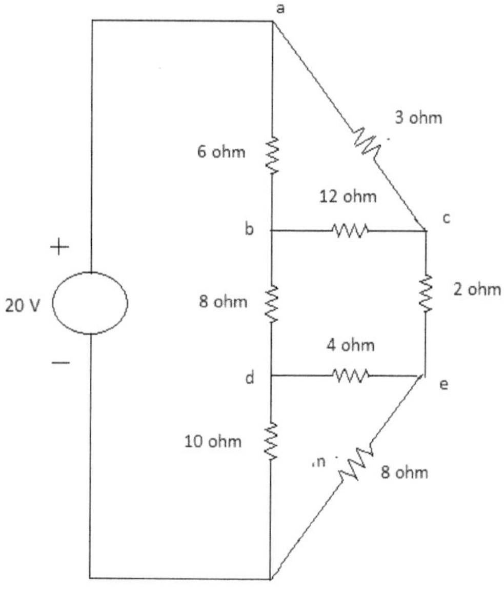

Figure 1.49 (a)

The delta connections between the points a,b,c and d,e,f are changed to equivalent star networks as follows.

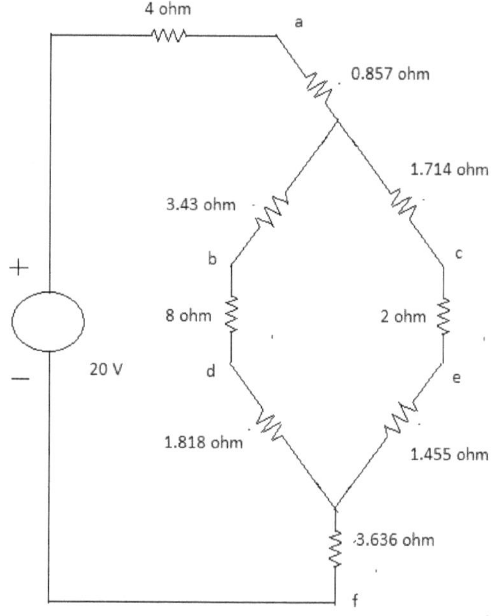

Figure 1.49 (b)

It is further reduced to

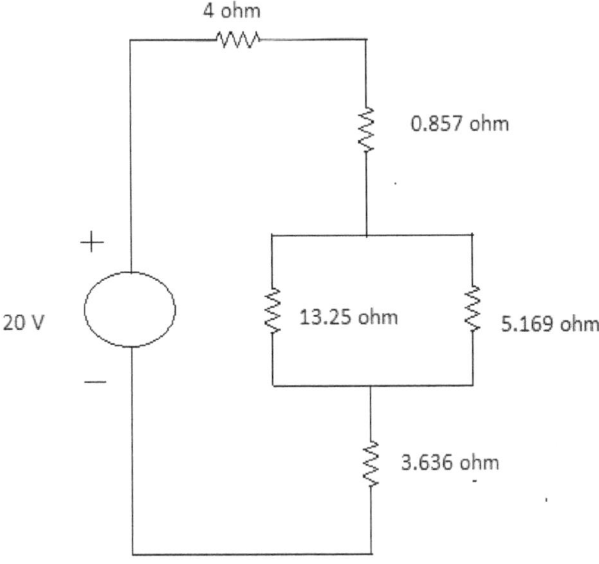

Figure 1.49 (c)

Then, it is simplified to

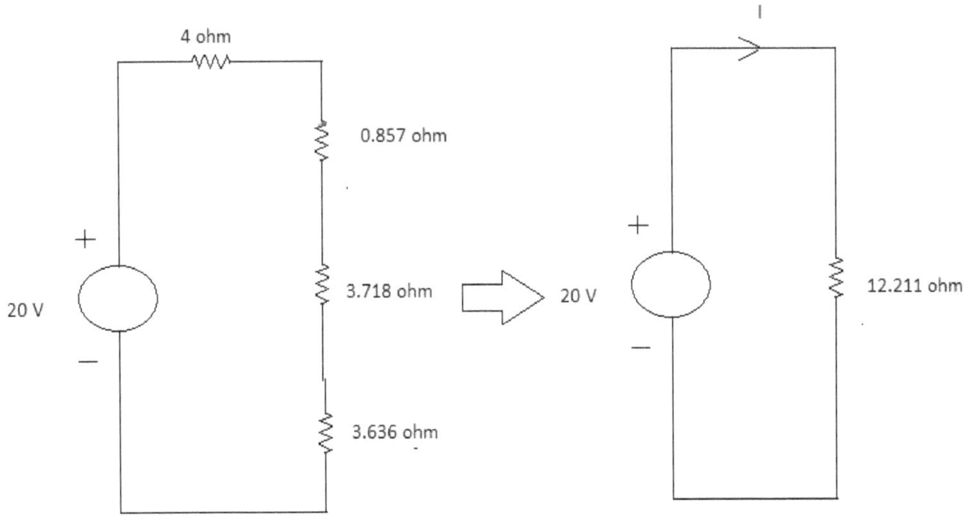

Figure 1.49 (d)

The equivalent resistance is given by $R_{eq} = 12.211\ ohm$

The current 'I' is given by $i = \dfrac{20}{12.211} = 1.638$ A

1.8 Superposition Theorem

The Superposition Theorem states that in any linear, bilateral network containing two or more sources, the response in any element is equal to the algebraic sum of responses caused by individual responses acting alone, while the other sources are non-operative. When considering the responses of individual sources, other ideal voltage sources and ideal current sources are replaced by short circuit and open circuits respectively. This theorem is valid only for linear systems.

A linear network is a network whose parameters do not change with voltage and current. That means, V-I characteristic is linear. Bilateral Network is network whose characteristics or behavior is same irrespective of the direction of current through various elements. Simple resistor is an example of a Bilateral network and Diode is an Unilateral element.

The Superposition Theorem is well illustrated through the following example.

Exercise 1.8.1: Find the voltage across the 2 ohms of the circuit shown in figure 1.50 using Superposition Theorem.

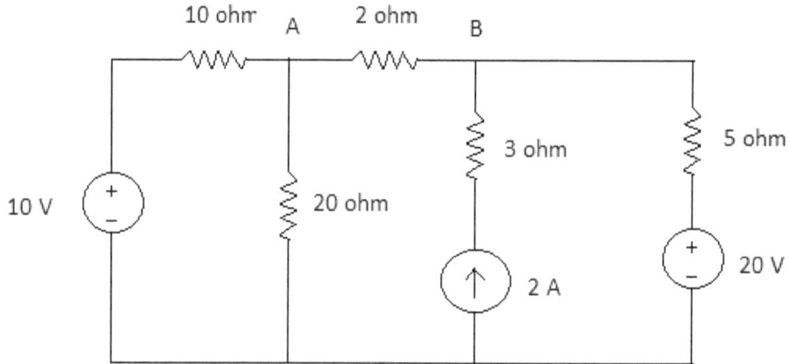

Figure 1.50

Solution:

To find the voltage across 2 ohm resistor, the current passing through this resistance should be known. The individual responses or currents in 2 ohm resistance are determined as follows.

To find the response caused by 10 V Source alone:

For this, 20V source is replaced by an open circuit and 2A current source is replaced by an open circuit as shown below.

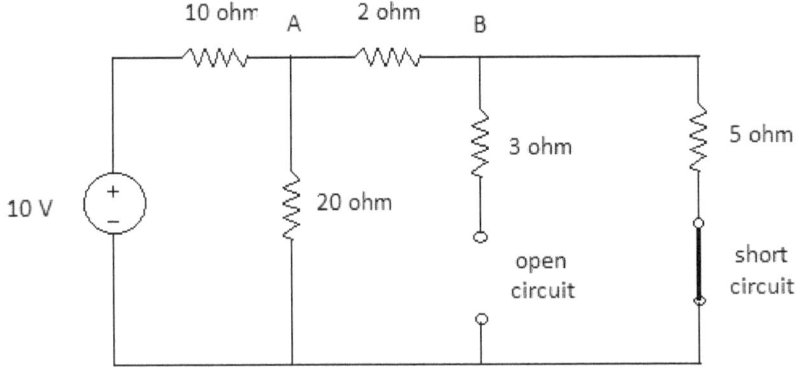

Figure 1.50 (a)

The 2 ohms and 5 ohms are connected in series and this combination is in parallel with 20 ohms resistance. Then, the resultant of this combination is given by

$$R_1 = \frac{20 \times (2+5)}{20+2+5} = 5.18\,\text{ohm}$$

The total resistance $R = 10 + 5.18 = 15.18\,\text{ohm}$

The total current supplied by 10V source is $I = \dfrac{10}{15.18} = 0.658\,\text{A}$

The current passing through the 2 ohms from A to B $i_1 = 0.658 \times \dfrac{20}{20+7} = 0.4879\,\text{A}$

To find the response caused by 2 A Source alone:

For this, the 10V and 20V voltage resources are replaced by short circuits as shown.

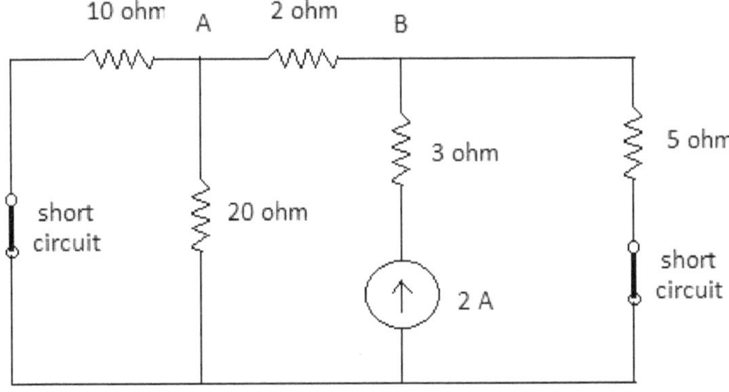

Figure 1.50 (b)

The 10 ohm resistance gets connected in shunt with 20 ohm resistance and this combination is in series with the 2 ohm resistance. Then, the resultant resistance of this branch is given by

$$R_{eq} = 2 + \frac{20 \times 10}{20 + 10} = 8.66\,\text{ohm}$$

Current passing through the 2 ohms resistance from B to A is given by

$$i_2 = 2 \times \frac{5}{5 + 8.66} = 0.732\,\text{A}$$

Since, the current flows from B to A, $i_2 = -0.732\,\text{A}$

To find the response caused by 20V Source alone:

For this, the 10V voltage source and 2Acurrent resources are replaced by open circuit and short circuits respectively as shown.

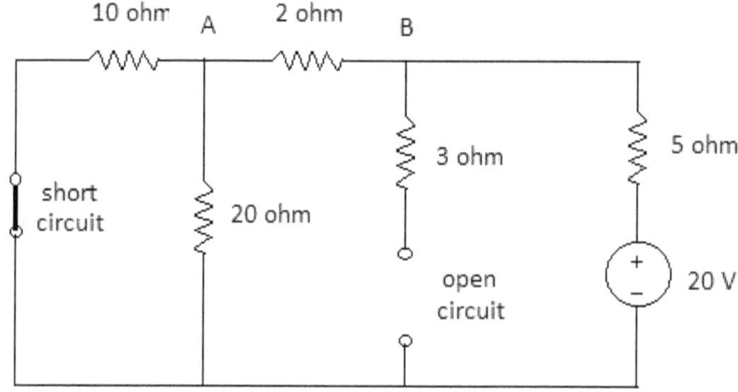

Figure 1.50 (c)

The total resistance of the circuit is given by

$$R_{eq} = \frac{10 \times 20}{10 + 20} + 2 + 5 = 13.66\,\text{ohm}$$

The current supplied by 20V battery is $I = \dfrac{20}{13.66} = 1.464\,\text{A}$

The same current passes through 2 ohms resistance also from B to A. Hence, $i_3 = -1.434\,\text{A}$

The sum of the individual responses $= 0.4879 - 0.732 - 1.434 = -1.708\,\text{A}$

Then, the voltage across 2 ohm resistance $= 1.708 \times 2 = 3.416\,\text{V}$

Or $V_{AB} = -3.416\,\text{V}$ or $V_{BA} = 3.416\,\text{V}$ (Since, the potential of B is higher than A. the net current flows from point B to A)

1.9 Thevenin's Theorem

It is quite common to find the current in a particular resistance (Generally called as Load Resistance R_L) which changes with remaining circuit leftovers same. To avoid the labor in simplifying the circuit for every value of R_L, Thevenin's Theorem is useful. This theorem replaces the leftover circuit with an equivalent circuit called Thevenin's Equivalent circuit.

Thevenin's Theorem states that a linear, bilateral two terminal circuit can be replaced by an equivalent circuit consisting of a voltage source V_{th} in series with a resistance R_{th}. Where V_{th} is the open circuit voltage between the load terminals and R_{th} is the resistance between the load terminals.

The figure 1.51 depicts how a given circuit is represented by its equivalent Thevenin's equivalent circuit.

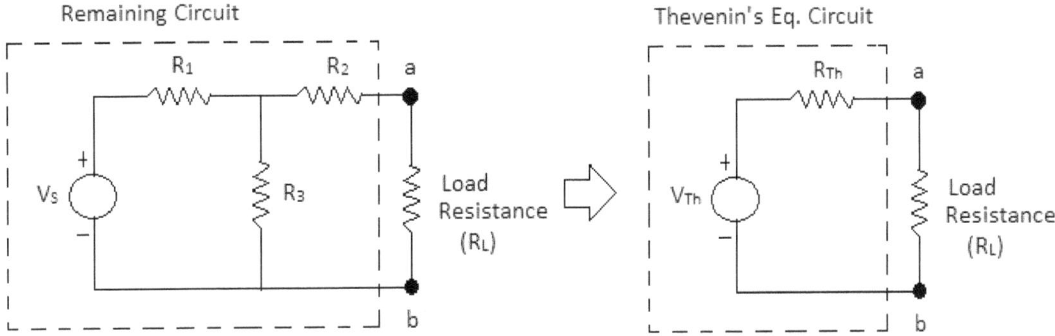

Figure 1.51 Thevenin's equivalent Circuit

The steps involved for applying Thevenin's Theorem to solve a given circuit are:

1. Finding Thevenin's Resistance R_{Th}: Remove the load resistance between the terminals 'a' and 'b'. Replace the voltage sources with the short circuits and current sources with open circuits. Find the resultant resistance between the terminals 'a and 'b. this resistance is called 'Thevenin's Resistance R_{Th}'. The procedure for the above circuit is illustrated in figure 1.52.

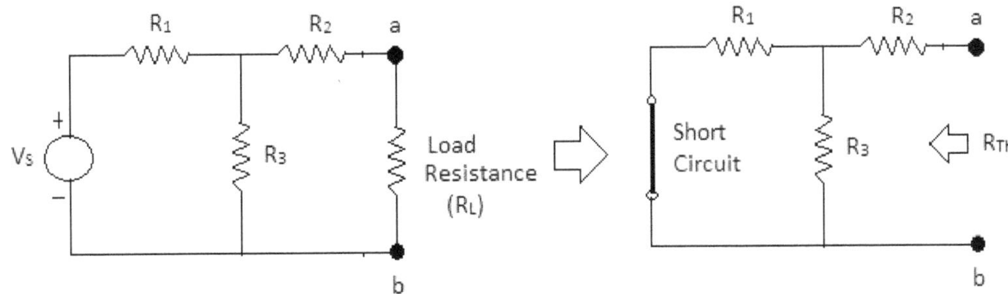

Figure 1.52 Thevenin's equivalent resistance

The Thevenin's Equivalent Resistance $R_{Th} = \dfrac{R_1 \times R_2}{R_1 + R_2} + R_2$

2. Finding Thevenin's Voltage V_{Th}: Remove the load resistance between the terminals 'a' and 'b'. Obtain the open circuit voltage between the terminals 'a' and 'b'. This open circuit voltage V_{ab} is known as 'Thevenin's Voltage V_{Th}'. For the above circuit, the Thevenin's Voltage is given by

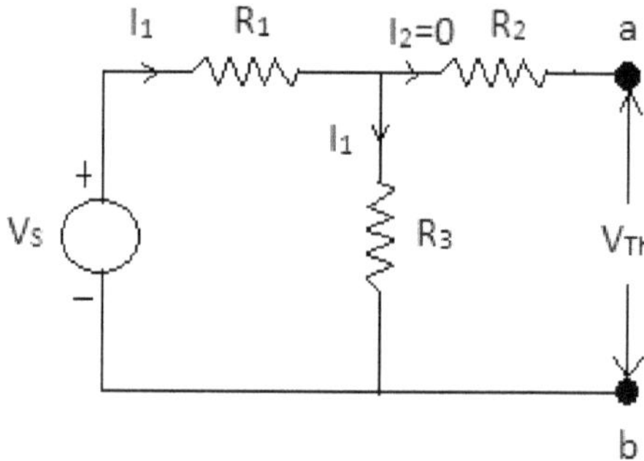

Figure 1.53 Finding Thevenin's voltage

$$I_1 = \frac{V_S}{R_1 + R_2}$$

$$V_0 = V_{ab} = I_1 \times R_3$$

$$= \frac{V_S}{R_1 + R_2} \times R_3$$

$$= V_{Th}$$

3. Obtain the Thevenin's equivalent circuit as shown in figure 1.54. This circuit remains same irrespective of changes in load resistance.

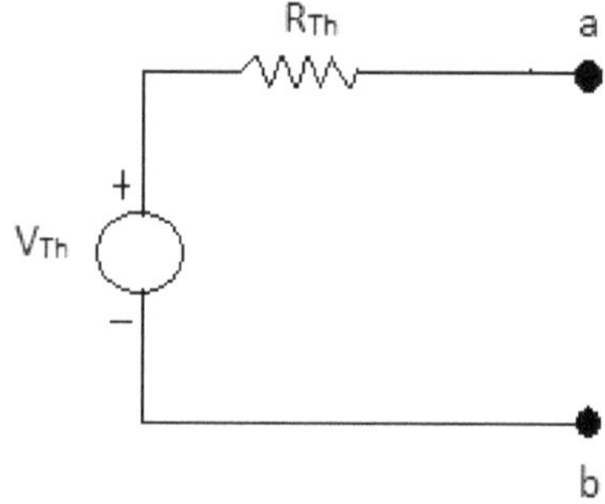

Figure 1.54 Thevenin's Equivalent Circuit

4. Reconnect the load resistance R_L and find the load current I_L.

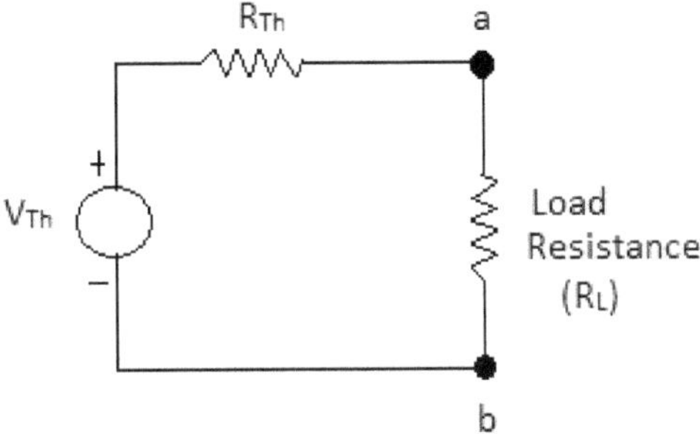

Figure 1.55 Thevenin's Equivalent Circuit including load

$$I_L = \frac{V_{Th}}{R_{Th} + R_L}$$

Example 1.9.1: Find the current passing through 4 ohms resistance using Thevenin's Theorem.

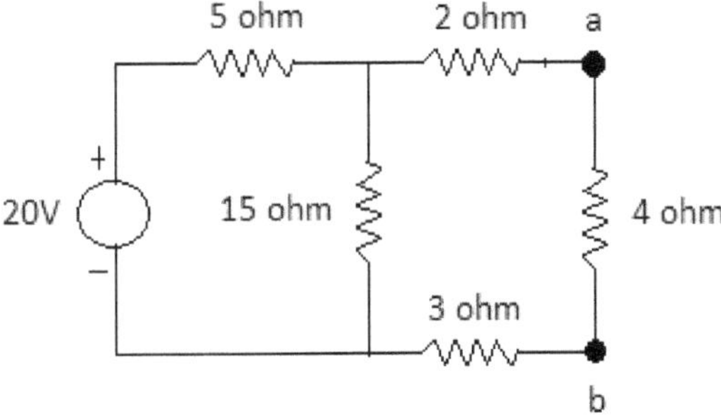

Figure 1.56

Solution:

To find the Thevenin's resistance, the resistance 4 ohm is removed from the circuit and the 20V voltage source is short circuited. The resistance between the load terminals 'a' and 'b' is

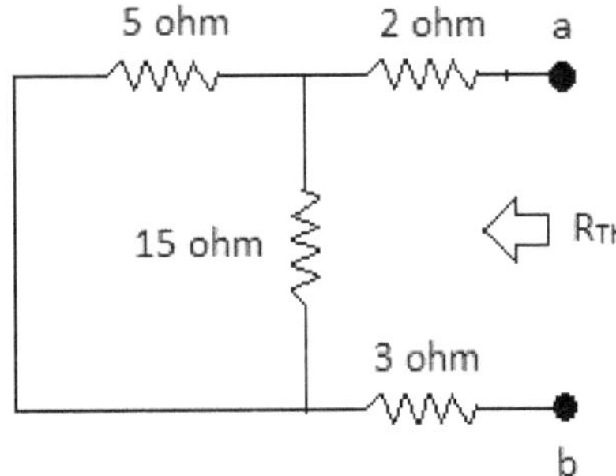

Figure 1.57(a)

$$R_{Th} = R_{ab} = \frac{5 \times 15}{5+15} + 2 + 3 = 8.75 ohms$$

To find the Thevenin's resistance, the resistance 4 ohm is removed from the circuit and the open circuit voltage is found between the load terminals 'a' and 'b'.

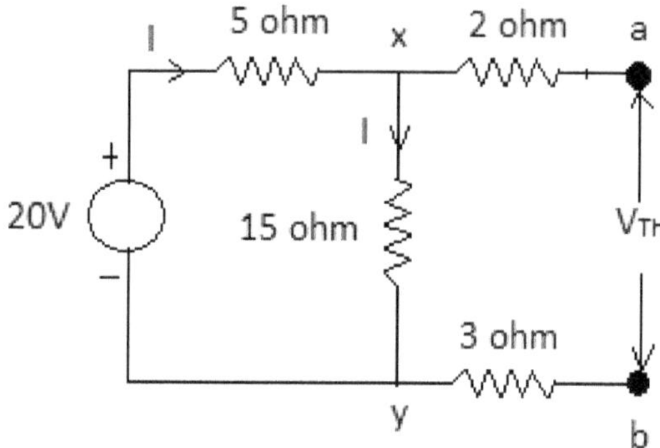

Figure 1.57(b)

$V_{Th} = V_{ab} = V_{xy}$ since no current flows through 2 and 3 ohms resistances.

$$I = \frac{20}{5+15} = 1 \text{ A}$$

The voltage across the 15 ohm resistance $V_{xy} = V_{ab} = V_{Th} = I \times 15 = 1 \times 15 = 15$ V

The Thevenin's equivalent circuit is given by

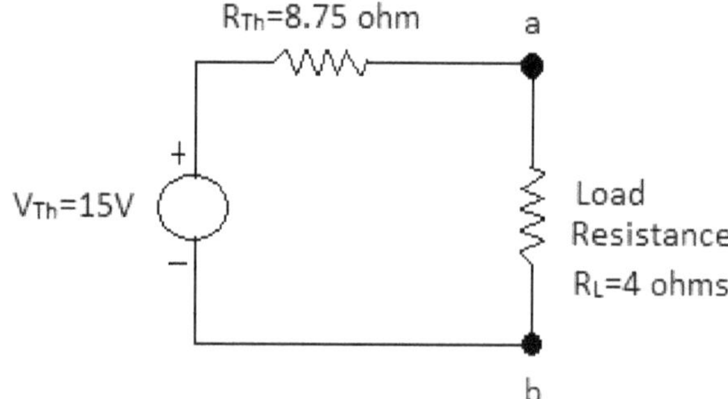

Figure 1.57(c)

The current passing through 4 ohms resistance is $I_L = \dfrac{V_{Th}}{R_{Th} + R_L} = \dfrac{15}{8.75 + 4} = 1.17$ A

Example 1.9.2: Find the Thevenin's equivalent circuit for the circuit shown in figure . Find the value of 'I' for i) R=3 ohms and ii) R=8 ohms.

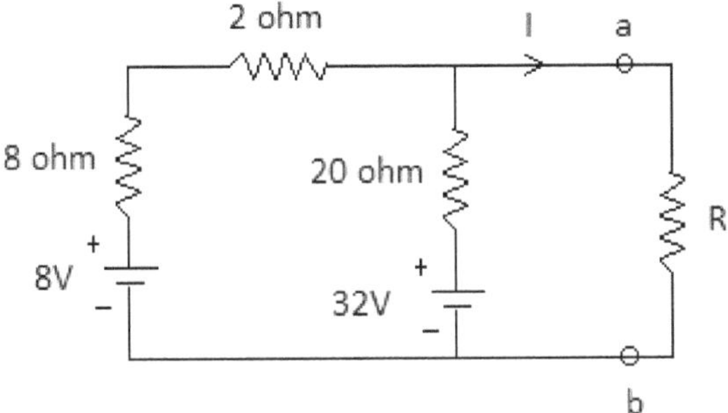

Figure 1.58

Solution:

To find the Thevenin's Resistance R_{Th}, the voltage sources are short circuited and the load resistance R is removed. The resistance between the terminals 'a' and 'b is called Thevenin's Resistance R_{Th}.

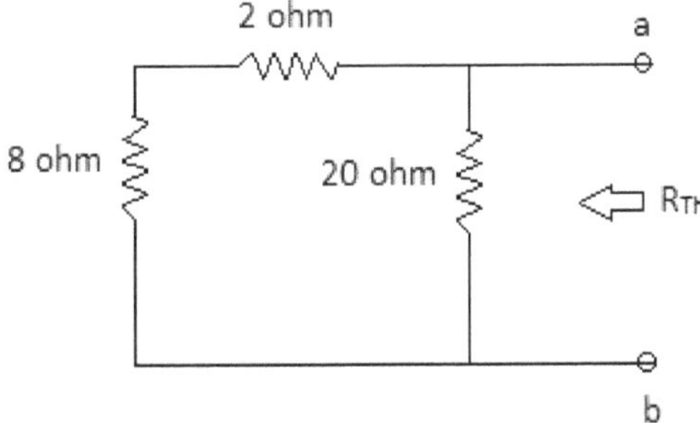

Figure 1.59(a)

$$R_{Th} = R_{ab} = \frac{(8+2) \times 20}{(8+2)+20} = \frac{200}{30} = 6.66 \, \text{ohm}$$

To find the Thevenin's Voltage V_{Th}, the load resistance R is removed and the open circuit voltage between the terminals 'a' and 'b' is found.

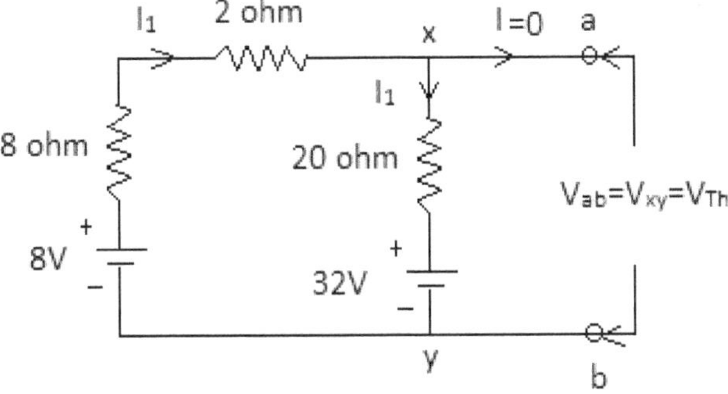

Figure 1.59(b)

$$V_{Th} = V_{ab} = V_{xy}$$

$$I_1 = \frac{8-32}{(8+2+20)} = -0.8 \text{ A}$$

$$V_{Th} = V_{ab} = V_{xy} = +32 + (-0.8 \times 20) = 16 \text{ V} \quad \text{(The potential of point 'a' with respect to point 'b')}$$

(i)   When R = 3 ohms

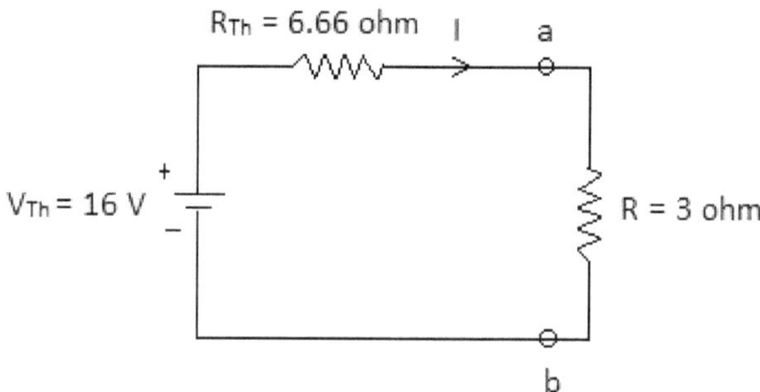

Figure 1.59(c)

$$I_L = \frac{V_{Th}}{R_{Th} + R} = \frac{16}{6.66 + 3} = 1.65 \text{ A}$$

(ii)   When R = 8 ohms

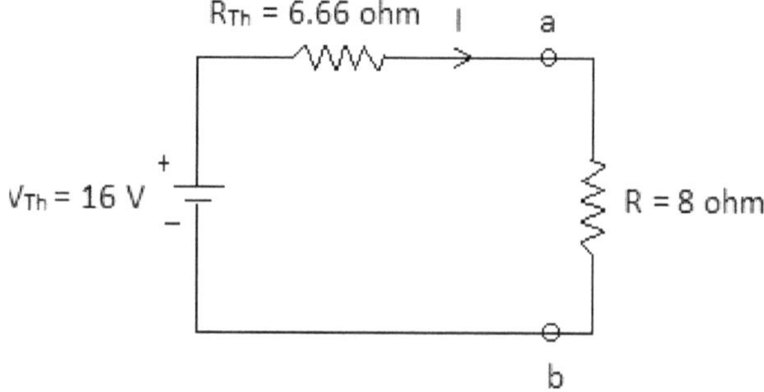

Figure 1.59(d)

$$I_L = \frac{V_{Th}}{R_{Th} + R} = \frac{16}{6.66 + 8} = 1.09 \text{ A}$$

Example 1.9.3: Find the current flowing through 2 ohms resistor connected between the points 'a' and 'b' using Thevenin's Theorem.

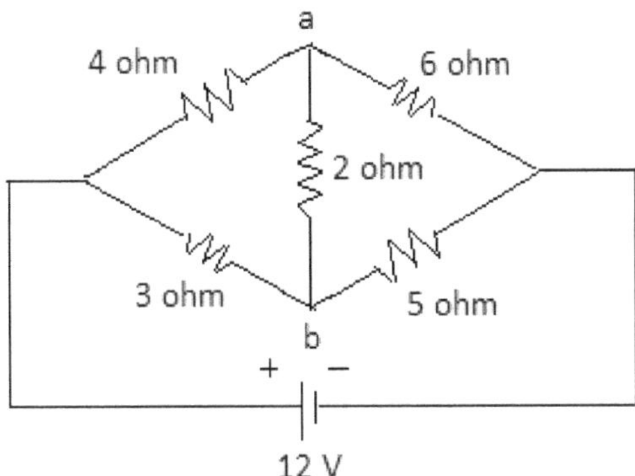

Figure 1.60

Solution:

To find the Thevenin's Resistance R_{Th}, the voltage sources are short circuited and the load resistance 2 ohms is removed. The resistance between the terminals 'a' and 'b is called Thevenin's Resistance R_{Th}.

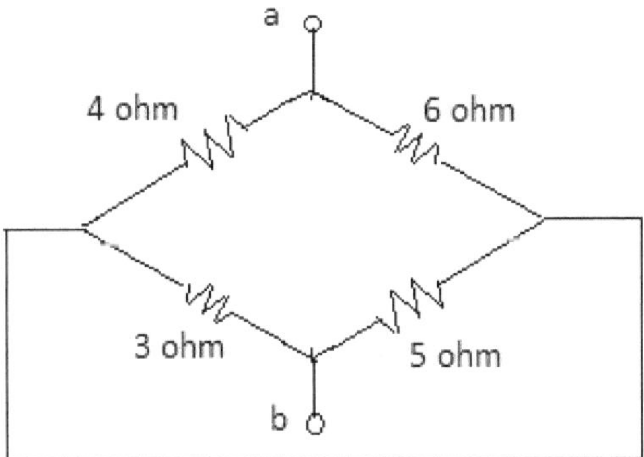

Figure 1.61(a)

The above circuit gets reduced to

Figure 1.61(b)

It further reduces to

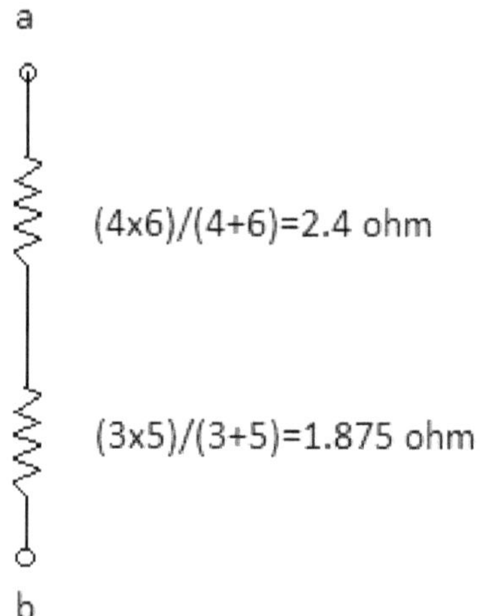

Figure 1.61(c)

The Thevenin's Resistance R_{Th} is given by $R_{Th} = R_{ab} = 2.4 + 1.875 = 4.275\,ohms$

To find the Thevenin's Voltage V_{Th}, the load resistance R is removed and the open circuit voltage between the terminals 'a' and 'b' is found.

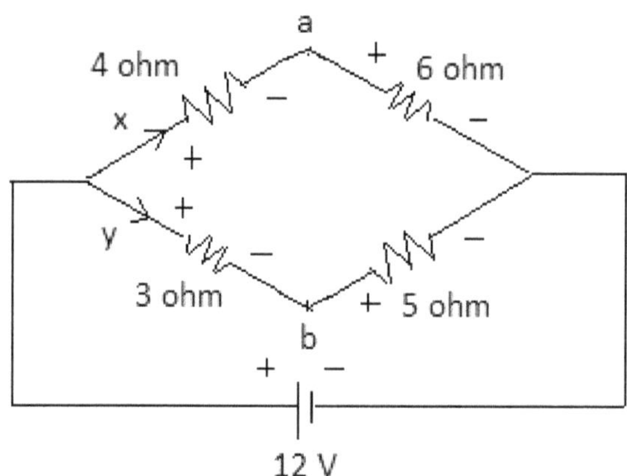

Figure 1.61(d)

$$x = \frac{12}{(4+6)} = 1.2\,A$$

$$y = \frac{12}{(3+5)} = 1.5\,A$$

The voltage drop across 4 ohms resistance $= 1.2 \times 4 = 4.8\,V$

The voltage drop across 4 ohms resistance $= 1.5 \times 3 = 4.5\,\text{V}$

The potential of point 'a' with respect to point 'b' is $+4.5V - 4.8V = -0.3V$ or the potential difference between the points 'a' and 'b' is $V_{Th} = V_{ab} = 0.3\,\text{V}$.

The Thevenin's equivalent circuit is shown in figure.

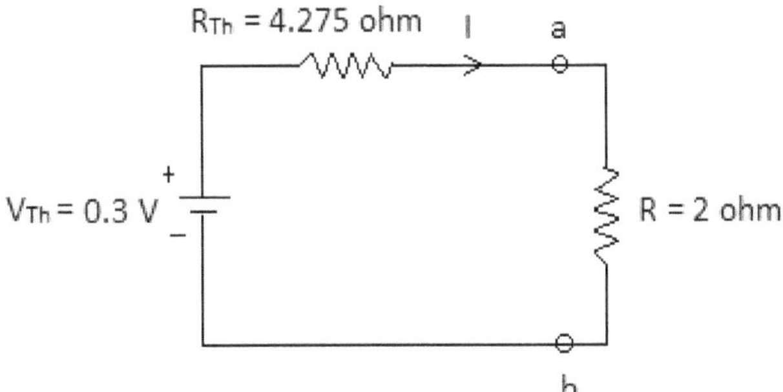

Figure 1.61(e)

The current flowing through 2 ohms resistance

$$I = \frac{V_{Th}}{R_{Th} + R} = \frac{0.3}{4.275 + 2} = \frac{0.3}{6.275} = 0.0478\,\text{A}$$

Example 1.9.4: Find the current passing through the load resistance R_L using the Thevenin's Theorem for R_L values 6 ohms and 16 ohms.

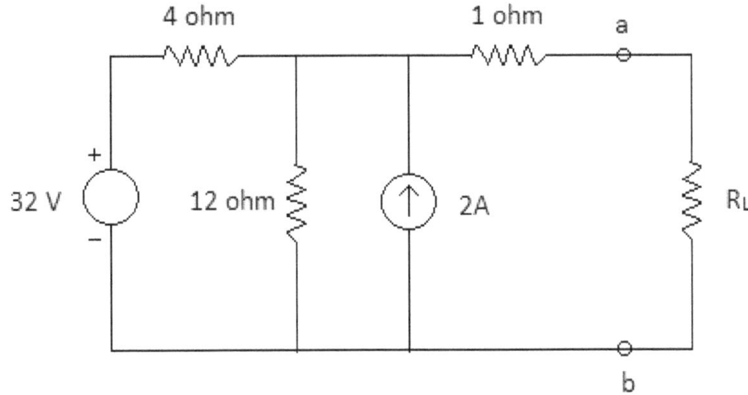

Figure 1.62

Solution:

In above circuit, the given current source I converted into its equivalent voltage source to simplify the circuit as follows.

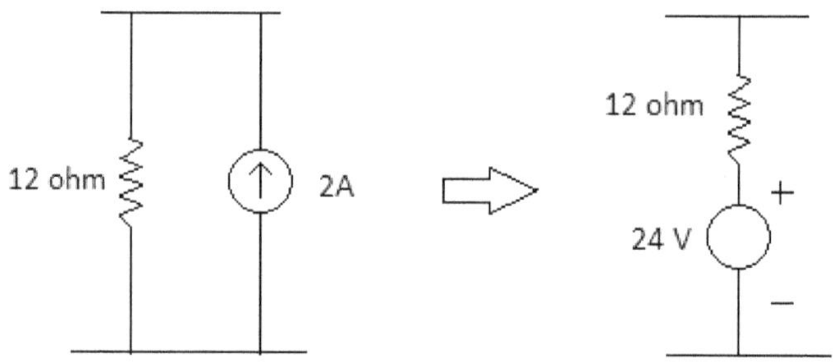

Figure 1.63(a)

By including this voltage source in above circuit in its appropriate place,

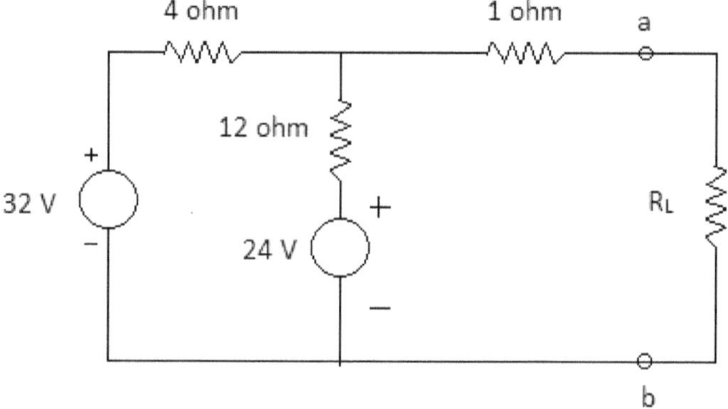

Figure 1.63(b)

To find the Thevenin's Voltage V_{Th}, the load resistance R_L is removed and the open circuit voltage between the terminals 'a' and 'b' is found.

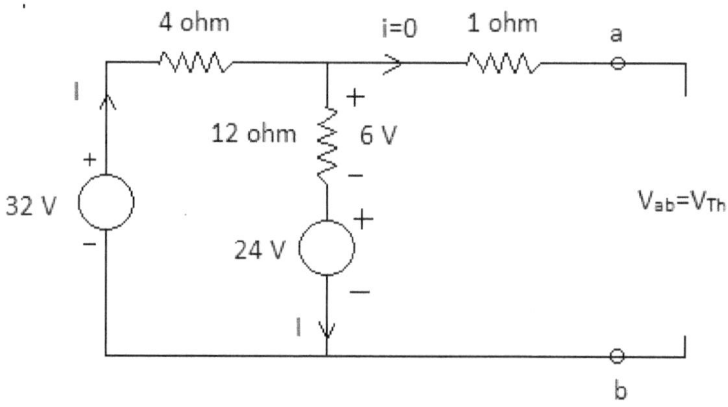

Figure 1.63(c)

$$i = \frac{(32-24)}{(4+12)} = \frac{8}{16} = 0.5\,\text{A}$$

Voltage drop across 12 ohms resistance $= 12 \times 0.5 = 6\,\text{V}$

$$V_{Th} = V_{ab} = 24 + 6 = 30\,\text{V}$$

To find the Thevenin's Resistance R_{Th}, the voltage sources are short circuited and the load resistance R_L is removed. The resistance between the terminals 'a' and 'b is called Thevenin's Resistance R_{Th}.

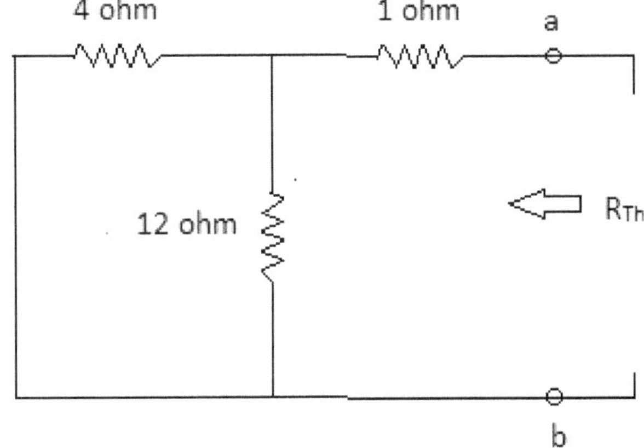

Figure 1.63(d)

Thevenin's Resistance $R_{Th} = 1 + \dfrac{(4 \times 12)}{(4+12)} = 1 + 3 = 4\,\text{ohms}$

The Thevenin's equivalent circuit is shown in figure 1.63(e).

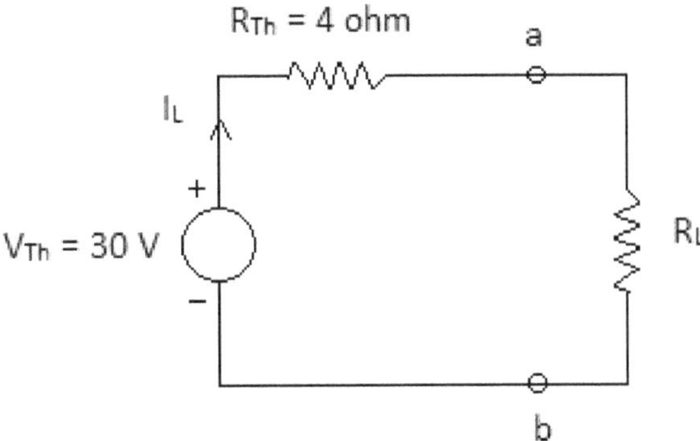

Figure 1.63(e)

(i)   When $R_L = 6$ ohms,

$$I_L = \frac{V_{Th}}{R_{Th} + R_L} = \frac{30}{4+6} = 3\,A$$

(ii)   When $R_L = 16$ ohms,

$$I_L = \frac{V_{Th}}{R_{Th} + R_L} = \frac{30}{4+16} = 1.5\,A$$

1.10   Time Response of RL and RC Circuits

1.10.1   Introduction to Transients

A network subjected to constant energy sources is said to be in steady state if the branch currents and voltages do not change with time. The networks with the sinusoidal voltage and currents (which are periodic functions of time) are also assumed to be in steady state condition. That means that the amplitude and frequency of sinusoidal quantity do not change in steady state.

Before a circuit reaches its steady state, which is different from its previous state, it passes through a transition period. This period is required for the currents and voltages to adjust themselves to their steady state modes of variation is called 'Transient State' or 'Transient Period'. The behavioral analysis of voltages and currents in this state is called 'Transient Analysis'. In general, any switching operation within the circuit itself causes transient conditions to exist in the circuit. Although transient periods are of very short spans, they result for some serious operating problems.

The application of Kirchhoff's Laws is extended to analyze the circuits under transient conditions too. Such application results for differential equations whose solution consists of two parts; Complementary Function and Particular Solution. The Complementary Function always becomes zero in relatively short time which is referred to as Transient Response or Source Free response. The particular Solution is the Steady State response or Forced Response. The study of transient analysis is started with a network having passive elements and sources having known initial voltage and currents values. At the reference time, t=0, the system state is altered by closing/opening the switch.

1.10.2 DC Response of an R-L Circuit

Consider a circuit consisting of a series R-L parameters connected to a DC source (battery) through a controlling switch 'S'. as shown in figure 1.64.

When the switch 'S' is closed at t = 0, the circuit gets subjected to a transient state. By applying KVL to the circuit in this state,

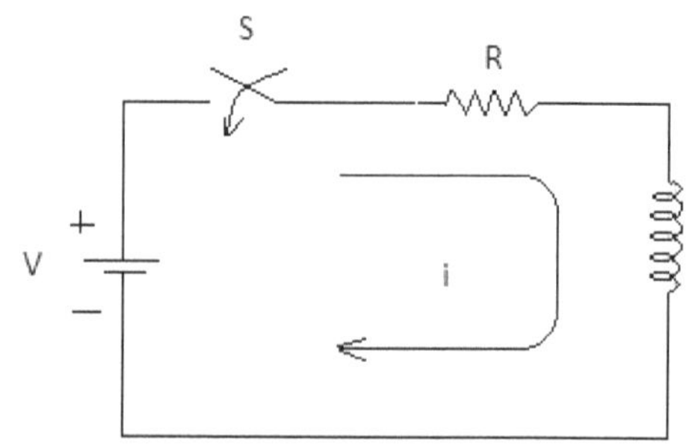

Figure 1.64 Series R-L Circuit

$$V = iR + L\frac{di}{dt}$$

Or $\frac{di}{dt} + i\frac{R}{L} = \frac{V}{L}$ which is first order linear differential equation. The current 'i' is the solution to be found.

By comparing the above equation with a non-homogeneous differential equation

$$\frac{dx}{dt} + Px = K$$

Whose solution is

$$x = e^{-Pt}\int Ke^{Pt}dt + Ce^{-Pt}$$

Where 'C' is an arbitrary constant.

Also by comparison, $x = i, P = \frac{R}{L}$ and $K = \frac{V}{L}$

By substituting these values in above equation,

$$i = e^{-\left(\frac{R}{L}\right)t}\int \frac{V}{L}e^{\left(\frac{R}{L}\right)t}dt + Ce^{-\left(\frac{R}{L}\right)t}$$

$$= e^{\left(\frac{R}{L}\right)t} \times \frac{V}{L} \times e^{\left(\frac{R}{L}\right)t} \times \frac{L}{R} + Ce^{\left(\frac{-R}{L}\right)t}$$

$$i = \frac{V}{R} + Ce^{-\left(\frac{R}{L}\right)t} \qquad\qquad .....(1)$$

To determine the value of 'C', initial conditions are used. Initial current in the circuit at t=0 is zero. ie., i = 0 at t = 0.

Writing these values in above equation (1),

$$0 = \frac{V}{R} + Ce^{-\left(\frac{R}{L}\right)\times 0}$$

Or $\qquad C = \frac{V}{R}$

Again writing the value of 'C' in above equation (1),

$$i = \frac{V}{R} - \frac{V}{R}e^{-\left(\frac{R}{L}\right)t}$$

$$i = \frac{V}{R}\left[1 - \exp\left[-\frac{R}{L}t\right]\right] \qquad\qquad .....(2)$$

The time response of the current for DC excitation in a series RL circuit is shown in the figure 1.65.

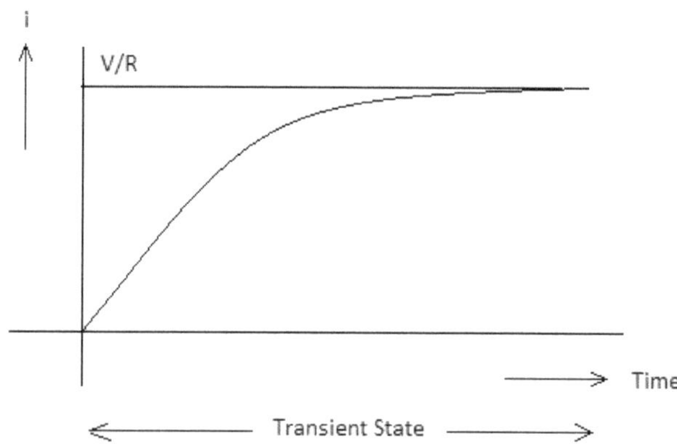

Figure 1.65 Time response of current in Series R-L Circuit

The equation (2) consists of two parts. The steady state part $\dfrac{V}{R}$ and

The transient state part $\dfrac{V}{R}e^{-\left(\frac{R}{L}\right)t}$

After the switch is closed, the response (i) takes some time to reach a steady state as shown. This transition period taken for the current to reach its final or steady state value from its initial value is called Transient period.

The time constant of the transient part of the solution $\dfrac{V}{R}e^{-\left(\frac{R}{L}\right)t}$ is the time at which the exponent of 'e' becomes unity. Where 'e' is the base of natural logarithms.

Therefore, $\left(\dfrac{R}{L}\right)t = 1$ or $t = \dfrac{L}{R} = \tau$ which is called 'Time Constant' and has the units 'seconds'.

At one Time Constant (TC), the value of 't' is $\dfrac{L}{R}$ seconds. The transient part of the solution at one TC is

$$i(\tau) = -\frac{V}{R}e^{-\left(\frac{t}{\tau}\right)} = -\frac{V}{R}e^{-1} = -0.368\frac{V}{R} \qquad\qquad \text{(since, e=2.717)}$$

That means, at one Time Constant (TC), the transient term of current reaches 36.8% of its initial value $\dfrac{V}{R}$. The total current at this one TC becomes $i = \dfrac{V}{R} - 0.368\dfrac{V}{R} = 0.632\dfrac{V}{R}$

Hence, in one Time Constant (TC), the current in the circuit increases to 63.2% of its final steady state value from its initial value as shown in figure 1.66.

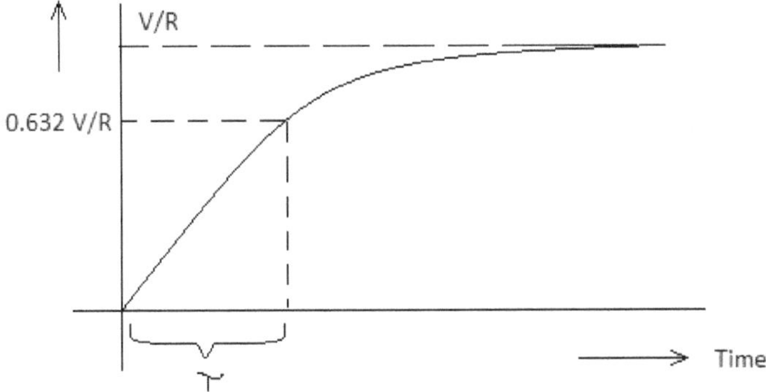

Figure 1.66 Representation of Time Constant in Series R-L Circuit

The voltage across the resistor and inductor change as follows during transient state.

$$V_R = iR = R \times \frac{V}{R}\left[1 - e^{-\left(\frac{R}{L}\right)t}\right]$$

$$= V\left[1 - e^{-\left(\frac{R}{L}\right)t}\right]$$

and, $$V_L = L\frac{di}{dt} = L \times \frac{V}{R} \times \frac{R}{L} \times e^{-\left(\frac{R}{L}\right)t} = Ve^{-\left(\frac{R}{L}\right)t}$$

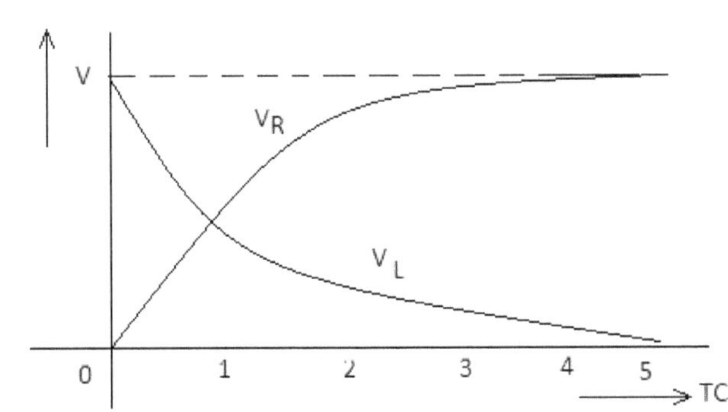

Figure 1.67 Variation of voltage and current in a Series R-L Circuit

The variations in voltages across resistance and inductor are shown in figure 1.67.

Similarly, the power consumed in the resistance during the transient state is

$$P_R = V_R i = V\left[1 - e^{-\left(\frac{R}{L}\right)t}\right]\left[1 - e^{-\left(\frac{R}{L}\right)t}\right]\frac{V}{R}$$

$$= \frac{V^2}{R}\left[1 - 2e^{-\left(\frac{R}{L}\right)t} + e^{-\left(2\frac{R}{L}\right)t}\right]$$

And the power consumed in the inductor during transient state is

$$P_L = V_L i = Ve^{-\left(\frac{R}{L}\right)t} \times \frac{V}{R}\left[1 - e^{-\left(\frac{R}{L}\right)t}\right]$$

$$= \frac{V^2}{R}\left[e^{-\left(\frac{R}{L}\right)t} - e^{-\left(\frac{2R}{L}\right)t}\right]$$

The responses of powers are shown below in the figure 1.68.

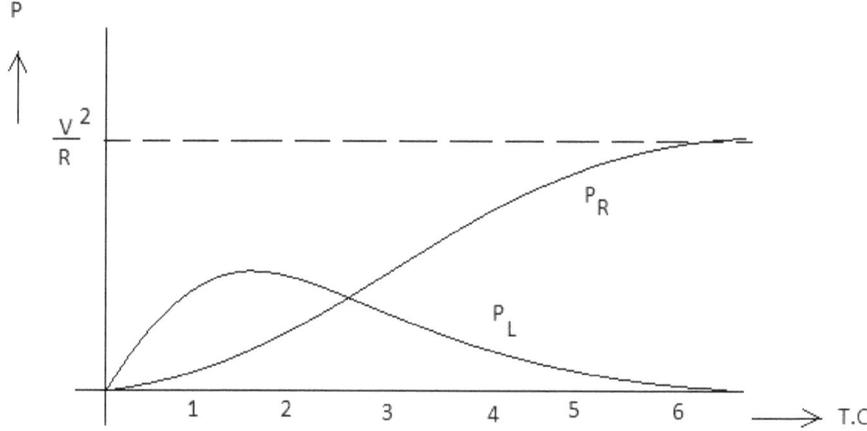

Figure 1.68 Time response of Power in Series R-L Circuit

1.10.3 DC Response of an R-C Circuit

Consider a circuit having Series Resistance- Capacitance connected across a DC source (battery) through a control switch 'S' as shown in figure 1.69. Initially, the capacitor is assumed to be uncharged.

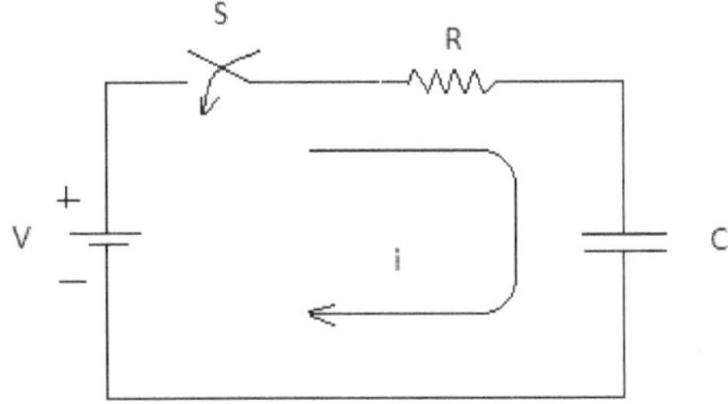

Figure 1.69 Series R-C Circuit

The switch 'S' is closed at t=0. By applying the Kirchhoff's Laws to the circuit during the transient state,

$$V = iR + \frac{1}{C}\int i.dt$$

By differentiating the above equation, we get

$$0 = \frac{di}{dt}R + \frac{1}{C}i$$

Or $$\frac{di}{dt} + \frac{1}{RC}i = 0$$

It is a linear differential equation and it is compared with a non-homogeneous differential equation

$$\frac{dx}{dt} + Px = K$$

Whose solution is

$$x = e^{-Pt}\int Ke^{Pt}dt + Ce^{-Pt}$$

Where 'C' is an arbitrary constant.

Also by comparison, $x = i$, $P = \dfrac{1}{RC}$ and $K = 0$

The solution is

$$i = e^{\frac{1}{RC}t}\int 0.e^{\frac{1}{RC}t}.dt + Ce^{\frac{1}{RC}}$$

$$= Ce^{-\frac{t}{RC}}$$

This linear differential equation has only the Complementary Function (only transient response). The particular solution (steady state response) for above equation is zero.

Initial conditions have been used to find the value of 'C'.

At t=0, the switch 'S' is closed. Since, the capacitor does not allow the sudden changes in voltages, it acts as short circuit (V=0).

Then, at $t = 0$, $i = \dfrac{V}{R}$

By writing the value of 'I' in above equation,

$$\dfrac{V}{R} = Ce^{-\frac{0}{RC}} \quad \text{or} \quad C = \dfrac{V}{R}$$

Then, the current equation becomes,

$$i = \dfrac{V}{R}e^{-\frac{t}{RC}}$$

When the switch 'S' is closed, the response (current) decays with time as shown in figure1.70.

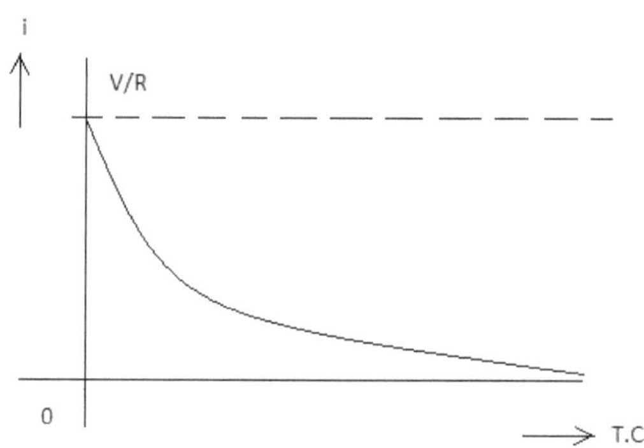

Figure 1.70 Time response of current in Series R-C Circuit

The time constant of transient is the time at which the exponent of e becomes unity.

That is, $\dfrac{t}{RC} = 1$ or $t = RC = \tau = $ Time Constant (TC)

The voltage across the resistance is

$$V_R = R \times i = R \times \dfrac{V}{R}e^{-\frac{t}{RC}}$$

$$= Ve^{-\frac{t}{RC}}$$

Similarly, the voltage across the capacitor is

$$V_C = \frac{1}{C}\int i.dt$$

$$= \frac{1}{C}\int \frac{V}{R} e^{-\frac{t}{RC}} dt$$

$$= -\left[\frac{V}{RC} \times RCe^{-\frac{t}{RC}}\right] + C$$

To find the value of C:

At t = 0, voltage across capacitor is zero.

$$0 = -Ve^{-0} + C \quad \text{or} \quad C = V$$

Therefore, $V_C = -Ve^{-\frac{t}{RC}} + V$

$$= V\left[1 - e^{-\frac{t}{RC}}\right]$$

The variations in voltages are shown in figure 1.71.

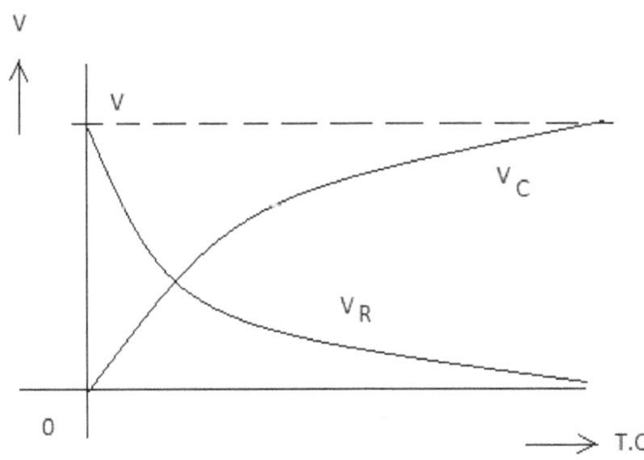

Figure 1.71 Variation of voltages across R and C

The power consumed in resistor during transient period is

$$P_R = V_R \times i = Ve^{-\frac{t}{RC}} \times \frac{V}{R} e^{-\frac{t}{RC}}$$

$$= \frac{V^2}{R} e^{-\frac{2t}{RC}}$$

Power in capacitor is

$$P_C = V_C \times i = V\left(1 - e^{-\frac{t}{RC}}\right)\frac{V}{R}e^{-\frac{t}{RC}}$$

$$= \frac{V^2}{R}\left[e^{-\frac{t}{RC}} - e^{-\frac{2t}{RC}}\right]$$

The variations in powers with respect to time are shown in the figure 1.72.

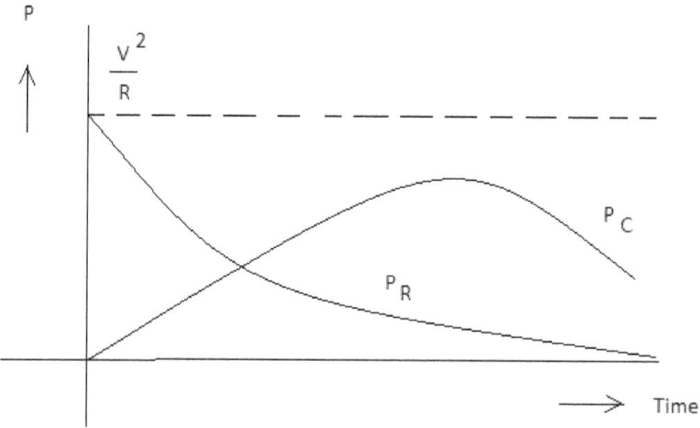

Figure 1.72 Time response of Powers in Series R-C Circuit

Review Questions

Short Answer (Two marks) Questions

1. List the advantages of electrical energy over other forms of energies
2. Define 'Electric Generator' and 'Electric Motor'
3. What is an Electric Circuit003F
4. Compare Electric and Magnetic Circuits
5. Name the basis circuit parameters
6. Differentiate between Resistance and Inductance
7. How does an inductor behave in a DC circuit?
8. Write down the expression for energy stored in a Capacitor
9. How does a capacitor behave in a DC circuit?
10. Differentiate between Independent and Dependent Voltage Sources
11. Show the V-I characteristics of ideal voltage and current sources

12. Show the equivalent circuit of practical voltage and current sources

13. Find the equivalent current source for the voltage source given in figure 1.73. (Ans: 1.92A, 25 Ohms)

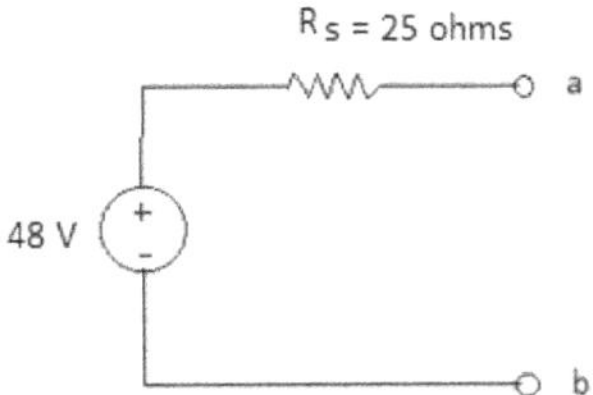

Figure 1.73

14. State KCL and KVL

15. Define Transient State

16. Write down the expression for time constant of a RC network

Essay (Six marks) Questions

1. Explain the role of Electrical Energy in modern life

2. With the help of a single line diagram, briefly explain the concept of Electrical Power Generation, Transmission and Utilization

3. Briefly discuss on the behavior of a) Inductance and b) Capacitance when placed in electrical circuits

4. What are the different types of dependent sources? Show their equivalent circuits and V-I characteristics

5. Explain the procedure for converting a given Current Source into its equivalent Voltage Source

6. Explain briefly about the Kirchhoff's Laws with examples

7. Find the current flowing through 3 ohms resistor of figure1.61 using Kirchhoff's laws (Ans: 1.736A)

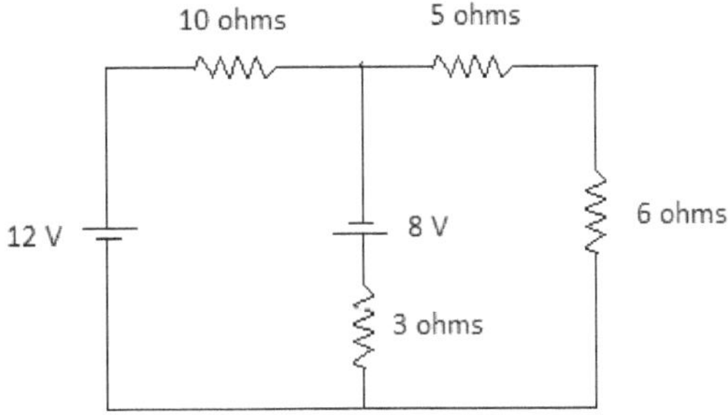

Figure 1.74

8. Find the source current using Star-Delta Transformation techniques for the circuit shown in figure 1.62. (Ans: 2.81 A)

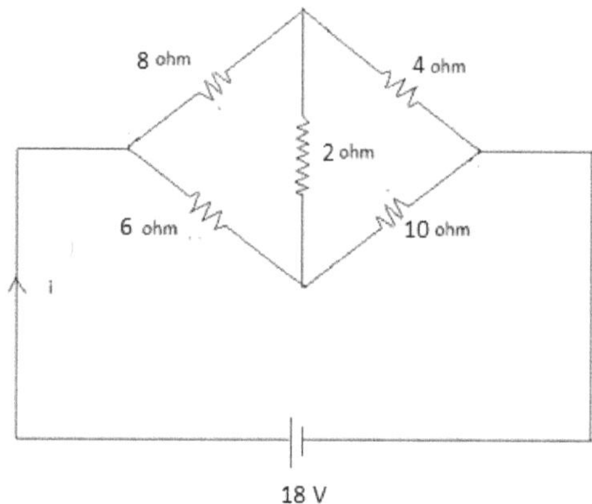

Figure 1.75

9. State and explain the Superposition theorem

10. Discuss on DC response of a RL circuit. And write down the expression for current transient with respect to time.

CHAPTER 2

Steady State A.C Circuits

2.1.1 Representation of Sinusoidal Waveforms

In Alternating Current (A.C) Circuits, both voltage and current are time varying in a periodic manner. Alternating voltage or current reverses its direction at regular time intervals and has positive and negative values. Usually, sinusoidal voltages or currents are preferred in AC circuits instead of other alternating quantities like square wave, triangular, saw-tooth wave, pulse wave etc., due to the following advantages.

1. Any other alternating quantity can be equated to the sum of many sinusoidal waveforms (Fourier Analysis)

2. Sinusoidal signals are easy to generate and transmit

3. Easy to handle mathematically. The derivative and integral of a sinusoid are themselves sinusoids.

The following figure 2.1 shows the typical Sine Wave. The magnitude of alternating quantity is represented on y-axis and the time is shown on x-axis.

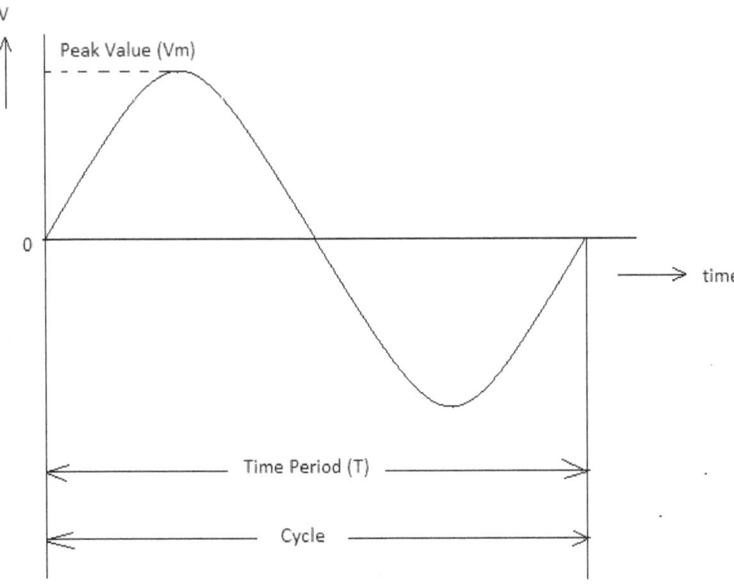

Figure 2.1 Sinusoidal Wave form

If we start from $t = 0$, the waveform goes to maximum value from zero and returns to zero, then, decreases to negative maximum and becomes zero. This pattern repeats periodically with respect to time. The terminology concerned to periodic waveforms is defined as follows.

1. ***Cycle:*** The complete set of positive and negative values formed by an alternating quantity is called a '*Cycle*'.

2. ***Time Period (T)* :** The time taken in seconds to complete one cycle is called '*Time Period*'.

3. ***Frequency (f)* :** The number of cycles completed in one second by an alternating quantity is called '*Frequency*'.

The Frequency (f) is related to the Time Period (T) as follows.

$$\text{Frequency } f = \frac{1}{T}$$

The angle (θ) can also be taken on y-axis to represent the alternating quantities.

One cycle $= 360^0$ or 2π radians.

If an alternating quantity completes 'f' number of cycles in a second, the frequency is said to be 'f' Hz. In this connection, Angular Frequency is defined as follows.

Angular Frequency ω = Total angle subtended per second

= Total angle subtended per cycle $\times$ No.of Cycles per second

$$= \left(360^0 \text{ or } 2\pi \text{ rad}\right) \times f$$

$$\omega = 2\pi f \text{ rad/sec}$$

Then, the angle subtended in the time 't' is given by

$$\theta = 2\pi ft = \omega t \text{ radians}$$

A sinusoidal quantity changes its magnitude as Sine function with respect to time or from one instant to other instant and it is shown in figure 2.2.

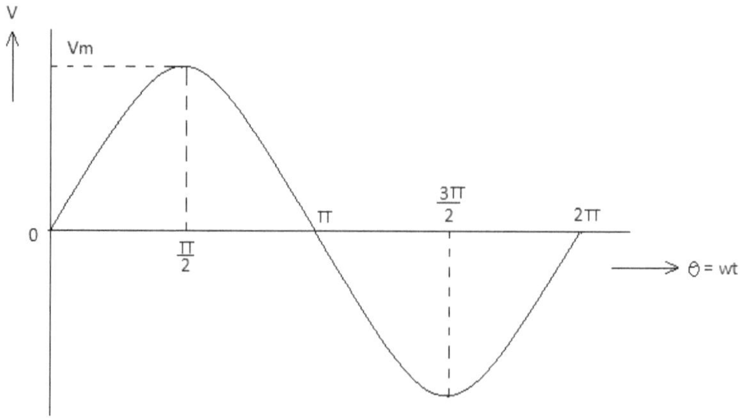

Figure 2.2 Sinusoidal Voltage wave form

Hence, the instantaneous value of a sinusoidal quantity, say voltage *(v)* at any time instant *'t'* in terms of its peak or maximum value V_m is given by

$$v = V_m \, Sin\omega t = V_m \, Sin\theta$$

Where ω is the angular velocity and $\omega = 2\pi f$ rad/sec

At $t=0$, $\omega t = \theta = 0^0$, $v = V_m \, Sin0^0 = 0$

At $t = \dfrac{T}{4}$, $\omega t = \theta = 90^0$, $v = V_m \, Sin90^0 = V_m$

Likewise, we get different values of voltages at different time instants.

The phasor representation of Sinusoidal Quantity is illustrated in figure 2.3.

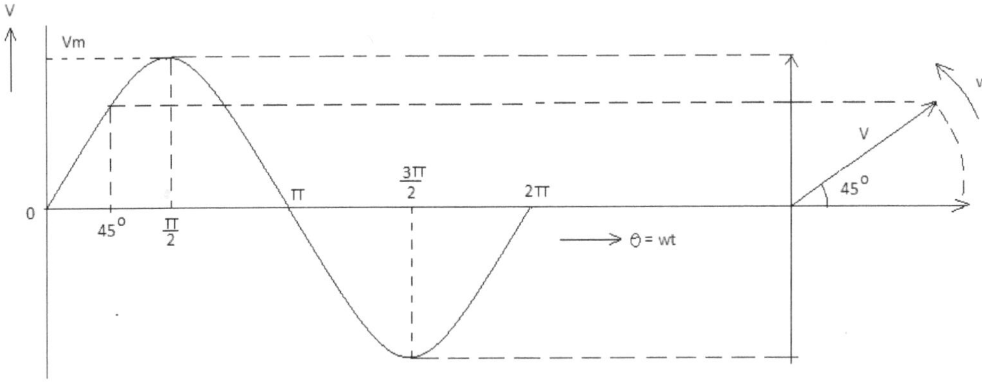

Figure 2.3 Phasor representation of sinusoidal quantity

A sinusoidal quantity can be represented by a rotating phasor. Let *'v'* be the voltage phasor rotating at an angular velocity of ω rad/sec. By the time, it completes one revolution, the corresponding sine wave completes one cycle. The vertical projection of phasor *'V'* at a given instant *'t'* is its instantaneous value.

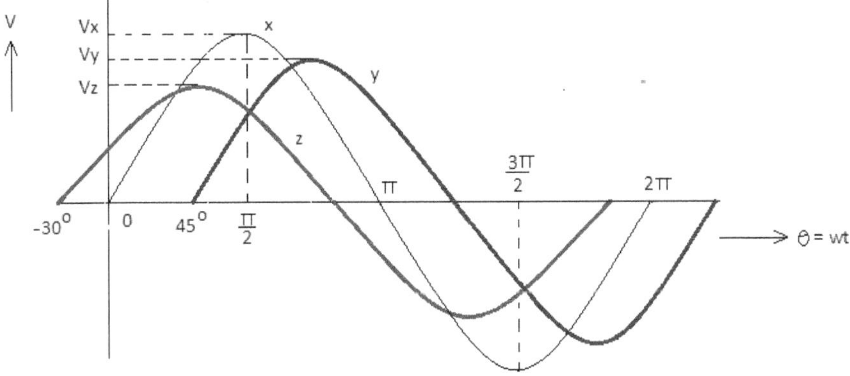

Figure 2.4 Illustration of phase differences between sine waves

The Phase of a Sine wave is an angular measurement that specifies the position of sine wave relative to a reference.

Where 'x' is the reference waveform and it is represented by

$$x = V_x Sin\omega t = V_x Sin\theta$$

The alternating quantity 'y' begun 45^0 after the start of x. Then, the 'y' is represented as

$$y = V_y Sin\left(\omega t - 45^0\right)$$

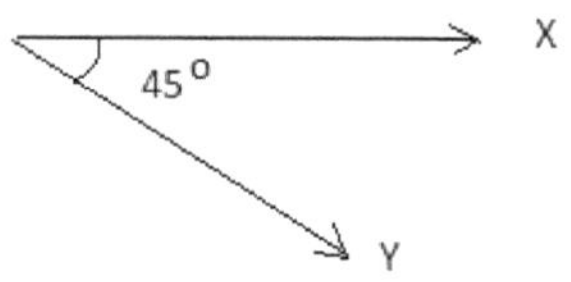

Or the phasor Y is lagging the phasor X by 45^0 and it is shown in phasor form below in figure 2.5.

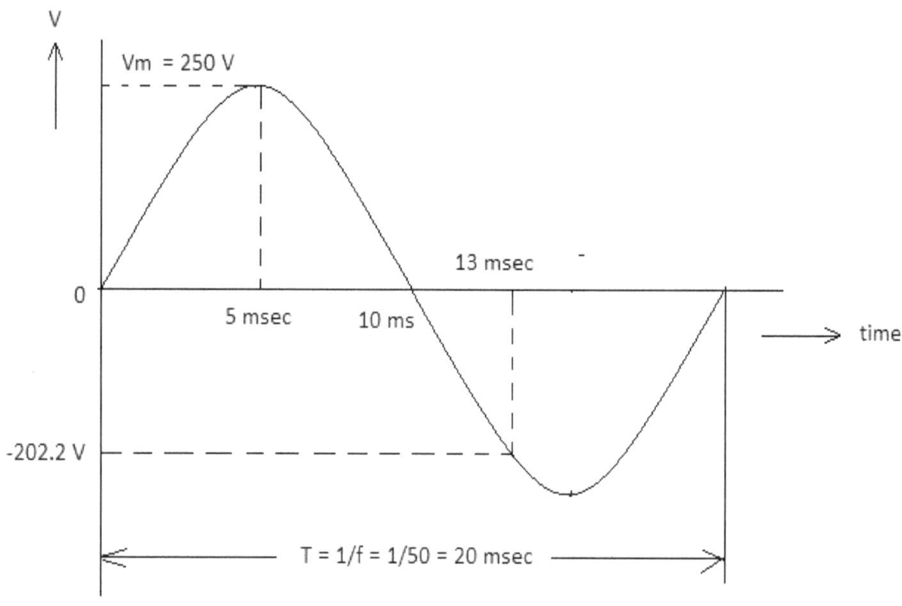

Figure 2.5

Similarly, the sinusoidal quantity 'z' starts 30^0 before the start of reference wave form 'x' and it is represented as

$$z = V_z Sin\left(\omega t + 30^0\right)$$

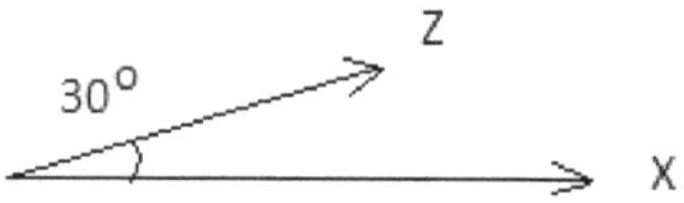

Or the phasor Z leads the phasor X by 30^0 and it is shown in figure 2.6.

Figure 2.6

Exercise 2.1.1: The maximum value of a Sinusoidal alternating voltage whose frequency is 50 Hz is 250V. Find the voltage at the instants i) t = 5 msec   ii) t = 10 m sec and iii) t = 13 msec.

Figure 2.7

Solution:

The maximum or peak value $V_m = 250\,\text{V}$

Frequency $f = 50\,\text{Hz}$

The angular frequency $\omega = 2\pi f = 2 \times \pi \times 50 = 100\pi$ rad/sec

The voltage at any instant 't' is $v = V_m\,Sin\omega t = V_m\,Sin(100\pi t)$

(i) At $t = 5\,\text{msec}$,

$$v = V_m\,Sin(100\pi t) = 250\,Sin(100\pi \times 5 \times 10^{-3}) = 250\ \text{V}$$

(ii) At $t = 10\,\text{m sec}$

$$v = V_m Sin(100\pi t) = 250 Sin(100\pi \times 10 \times 10^{-3}) = 0\ \text{V}$$

(iii) At $t = 13\,\text{m sec}$

$$v = V_m Sin(100\pi t) = 250 Sin(100\pi \times 13 \times 10^{-3}) = -202.2\ \text{V}$$

Exercise 2.1.2: Find the amplitude, phase, time period and frequency of a sinusoidal voltage

$$v = 12 Sin\left[50t + 10^0\right]$$

Solution : By comparing the above voltage equation with a standard form

$$v = V_m\,Sin\left[\omega t + \theta\right]$$

Where Θ is the phase of the alternating quantity with respect to a reference waveform.

$V_m = 12\,\text{V}$ and Phase $\theta = 10^0$ and the angular frequency $\omega = 2\pi f = 50\,\text{rad/sec}$

The frequency $f = \dfrac{50}{2\pi} = 7.958\,\text{Hz}$

Time Period $T = \dfrac{1}{f} = \dfrac{1}{7.958} = 0.1257\,\text{sec}$

Exercise 2.1.3 : Calculate the phase difference between v_1 and v_2. State which is leading.

$$v_1 = -10\,Cos\left[\omega t + 50^0\right] \quad\text{and}\quad v_2 = -12\,Sin\left[\omega t - 10^0\right]$$

Solution: $\quad v_1 = -10\,Cos\left[\omega t + 50^0\right]$

$$= 10\left(-Cos\left[\omega t + 50^0\right]\right)$$

$$= 10\left(Sin\left[270^0 + \omega t + 50^0\right]\right)$$

$$= 10 Sin\left[\omega t + 320^0\right]$$

$$v_2 = -12\,Sin\big[\omega t - 10^0\big]$$

The phase difference $= 320^0 - \left(-10^0\right) = 330^0$

The voltage v_1 leads v_2 by 330^0.

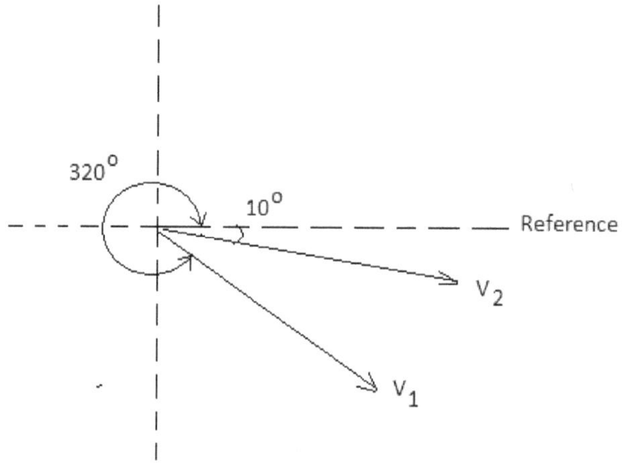

Figure 2.8

Exercise 2.1.4: Specify the Phase relations for the following Phasors shown in figure 2.9. Write down the expressions for instantaneous values. Also express these phasors in Rectangular, Polar and Exponential forms.

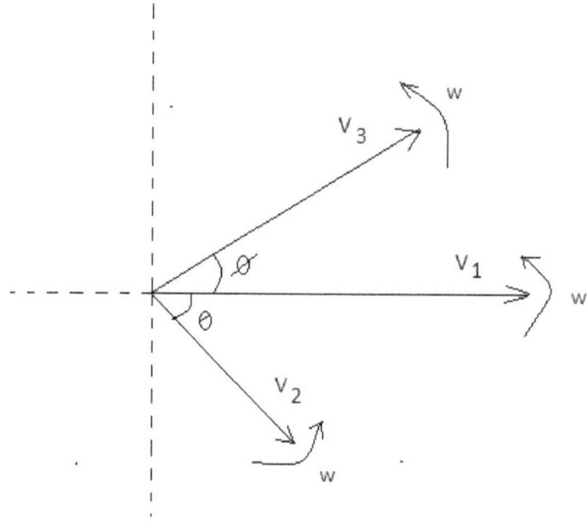

Figure 2.9

Solution:

The peak values of phasors V_1, V_2 and V_3 are V_{m1}, V_{m2} and V_{m3}.

The Phasor V_1 leads V_2 by an angle Θ and the phasor V_1 lags V_3 by angle $\varnothing$

The Phasor V_2 lags V_1 by angle Θ and the Phasor V_2 also lag V_3 by angle $(\Theta + \varnothing)$

The Phasor V_3 leads the Phasors V_1 and V_2 by an angles $\varnothing$ and $(\Theta + \varnothing)$ respectively.

The expressions for instantaneous values are written as follows by taking the phasor V_1 as reference.

$$v_1 = V_{m1} Sin\,\omega t$$

$$v_2 = V_{m2} Sin\left(\omega t - \theta\right)$$

$$v_3 = V_{m3} Sin\left(\omega t + \phi\right)$$

The phasors are represented in three forms in the following table.

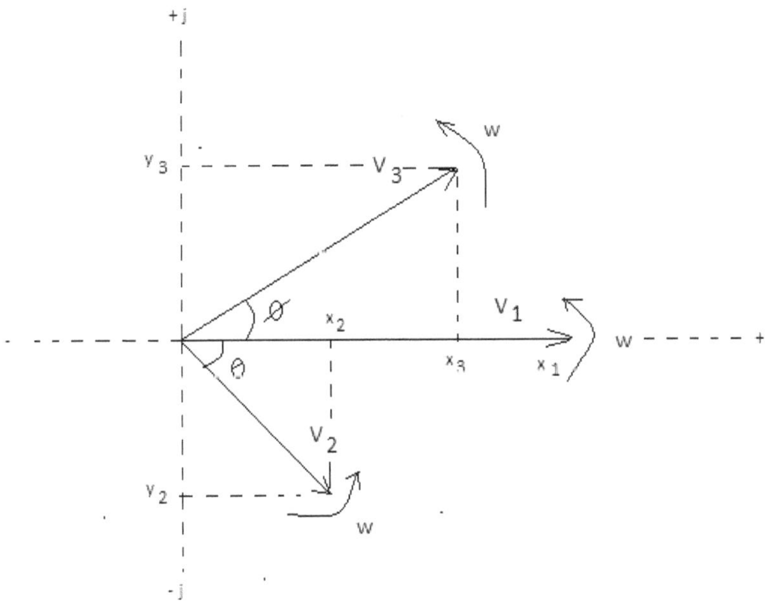

Figure 2.10

Table 2.1 Different forms of representing a Phasor

	Rectangular Form	Polar Form	Exponential Form
Phasor V_1	$x_1 + j0$	$V_{m1} \angle 0^0$	$V_{m1} e^{j0}$
Phasor V_2	$x_2 - jy_2$	$V_{m2} \angle -\theta$	$V_{m2} e^{-j\theta}$
Phasor V_3	$x_3 + jy_3$	$V_{m3} \angle \phi$	$V_{m3} e^{j\phi}$

2.1.2 Average and RMS Values

Since the value of an alternating quantity changes instant to instant, its value is specified in terms of i) Average Value and ii) Root Mean Square (RMS) Value.

A. Average Value:

The average value of a sine wave over one complete cycle is zero. Hence, for specifying the average value of sine wave, a half cycle is considered. In general, the average value of any function $f(t)$ in X-Y Plane is the total area under the curve divided by the distance of the curve.

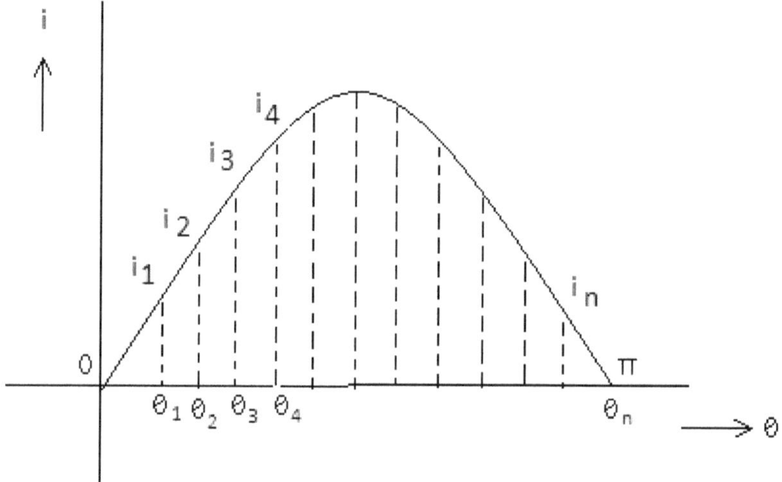

Figure 2.11 Positive half cycle of a sine wave

To take the average value of 'i', let us divide the x-axis into 'n' equal parts, $\theta_1,$ $\theta_2,$ $\theta_3, \dots \dots \theta_n$.

Also, $\theta_1 = \theta_2 = \theta_3 = \dots \dots = \theta_n = d\theta$

$$\text{Average value} = \frac{Total\ Area}{Base} = \frac{i_1\theta_1 + i_2\theta_2 + i_3\theta_3 + \dots \dots + i_n\theta_n}{\pi} = \frac{\int id\theta}{\pi}$$

Since, $i = I_m Sin\omega t$

$$\text{Average value} = \frac{1}{\pi}\left[\int I_m Sin\omega t.d\omega t\right]$$

$$= \frac{I_m}{\pi}\left[-Cos\omega t\right]_0^\pi = \frac{I_m}{\pi}\left[-\left(Cos\pi - Cos0^0\right)\right]$$

$$= \frac{I_m}{\pi}\left[-(-1-1)\right] = \frac{2I_m}{\pi} = 0.637I_m$$

The Average value of Sine wave is shown below in figure 2.12.

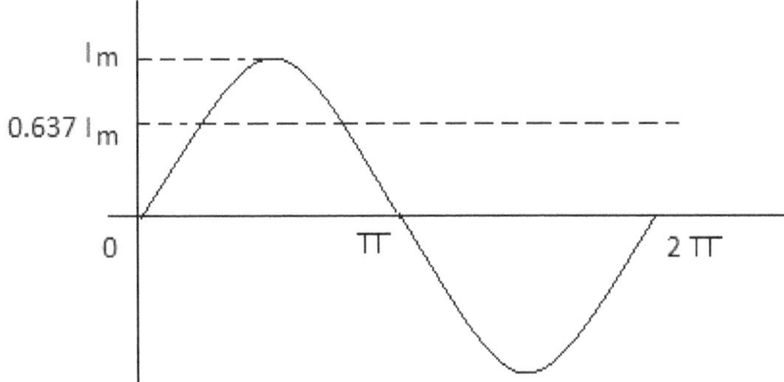

Figure 2.12 Illustration of Average value on sine wave

B. Root Mean Square (RMS) Value or Effective Value:

The need of Effective Value or RMS Value arises to measure the effectiveness of an alternating voltage or current. Consider a resistance R connected across a DC source produces a certain amount of heat energy in a given time. Now, the same resistance is connected across an AC source. If the resistance produces the same heat energy in the same time, then, the DC value will be considered as the effective value of alternating quantity.

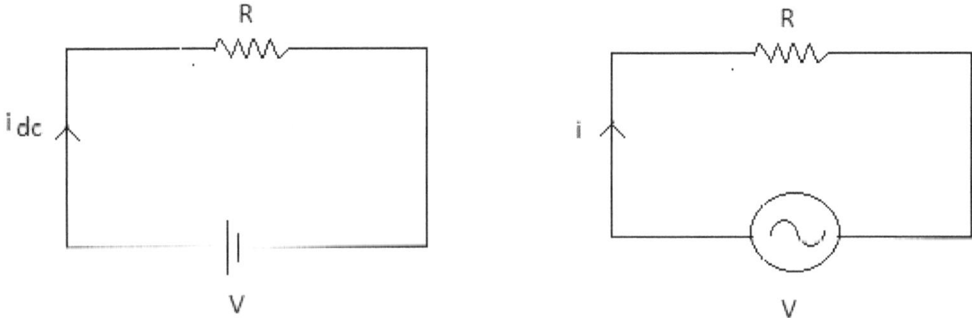

Figure 2.13 Resistance connected across a) Battery b) AC Source

Let I_{dc} be direct current which produces a certain amount of heat in the resistance R in a given time $\dfrac{T}{2}$.

Therefore, heat produced in DC circuit $\propto I_{dc}^2 R \dfrac{T}{2}$

Now, consider an alternating current having the total time period of 'T' seconds.

Let us divide the time base of half cycle $\left(\dfrac{T}{2}\right)$ into 'n' equal sections $t_1, t_2, t_3, \dots t_n$.

Also, $t_1 = t_2 = t_3 = \ldots\ldots = t_n = t$

Hence, $n \times t = \dfrac{T}{2}$

The heat produced in AC circuit $\propto I_1^2 Rt_1 + I_2^2 Rt_2 + I_3^2 Rt_3 + \ldots\ldots I_n^2 Rt_n$

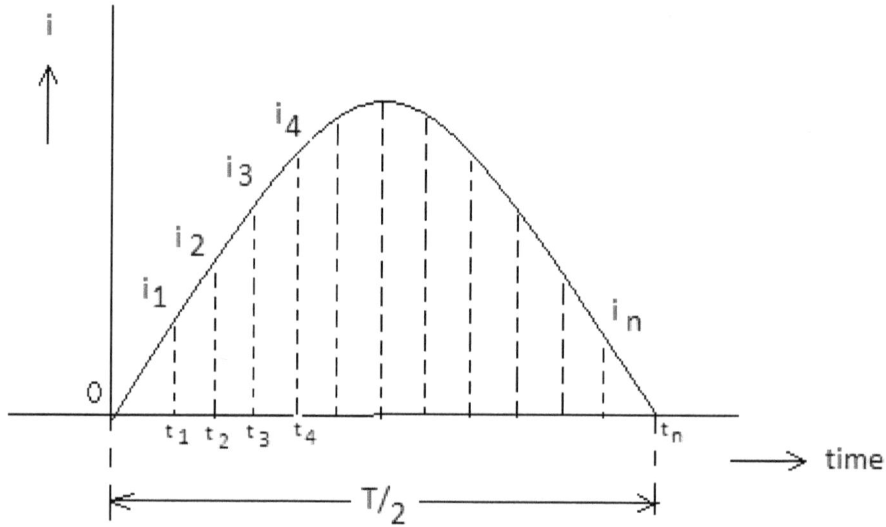

Figure 2.14 Currents at different instants

Heat produced in AC circuit $\propto Rt\left[i_1^2 + i_2^2 + i_3^2 + \ldots\ldots i_n^2 \right]$

Since, $n \times t = \dfrac{T}{2}$

Equating the two terms,

$$I_{dc}^2 R\left[n \times t\right] = R \times t\left[i_1^2 + i_2^2 + i_3^2 + \ldots\ldots i_n^2 \right]$$

$$I_{dc}^2 = \frac{i_1^2 + i_2^2 + i_3^2 + \ldots\ldots i_n^2}{n}$$

$$I_{dc} \ \text{or} \ I_{rms} = \sqrt{\frac{i_1^2 + i_2^2 + i_3^2 + \ldots\ldots i_n^2}{n}}$$

Which is the Root Mean Square (RMS) value.

To find the RMS value of Sinusoidal Wave,

$$i = I_m Sin\omega t$$

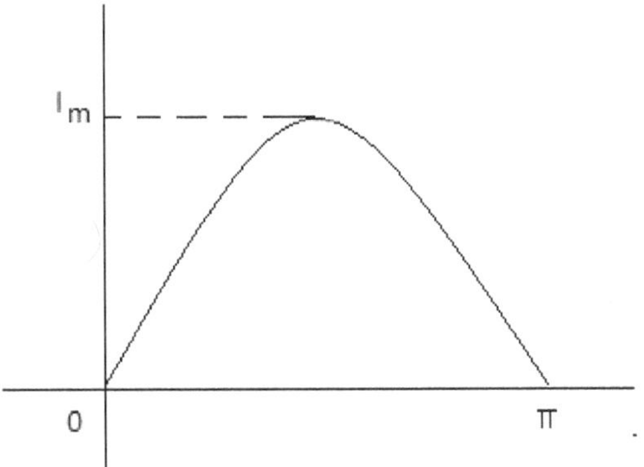

Figure 2.15

$$\text{RMS Value} = \sqrt{\frac{1}{\pi}\int\left[I_m Sin\omega t\right]^2 d\theta}$$

$$= \sqrt{\frac{I_m^2}{\pi}\int\left[Sin\omega t\right]^2 d\theta}$$

$$= \sqrt{\frac{I_m^2}{\pi}\int\left[\frac{1-Cos2\theta}{2}\right]d\theta}$$

$$= \sqrt{\frac{I_m^2}{2\pi}\int\left[1-Cos2\theta\right]d\theta}$$

$$= \sqrt{\frac{I_m^2}{2\pi}\left[\theta-\frac{Sin2\theta}{2}\right]_0^\pi}$$

$$= \sqrt{\frac{I_m^2}{2\pi}\left[\pi-\frac{Sin2\pi}{2}+0+0\right]}$$

$$= \sqrt{\frac{I_m^2}{2\pi}\left[\pi-0\right]}$$

$$= \frac{I_m}{\sqrt{2}}$$

$$= 0.707 I_m$$

The Form Factor of an alternating quantity is ratio of it RMs value to its Average value.

$$\text{Form factor} = \frac{RMS\ Value}{Ave\ Value}$$

$$= \frac{0.707 I_m}{0.637 I_m} = 1.11 \quad \text{for Sine waves}$$

Similarly, the Peak factor of an alternating quantity is the ratio of Peak value to its RMS value.

$$\text{Peak factor} = \frac{Peak\ Value}{RMS\ Value}$$

$$= \frac{I_m}{0.707 I_m} = 1.414 \ \text{ for Sine Waves.}$$

Exercise 2.1.5: Find the Average and RMS values for the periodic wave shown in figure 2.16.

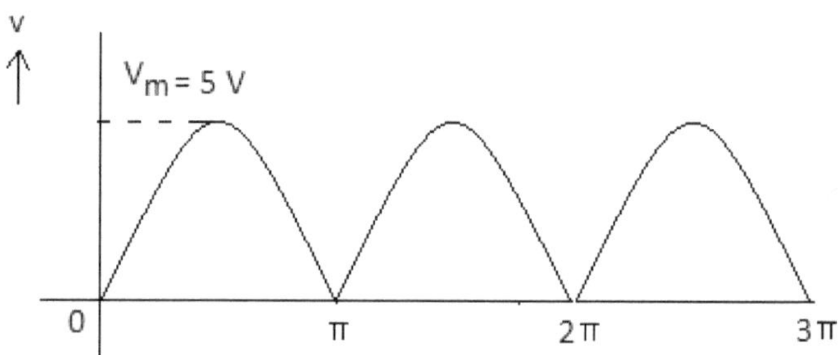

Figure 2.16

Solution:

The instantaneous value of the voltage is given by $v = V_m \, Sin\omega t$

$$\text{The Average value} = \frac{1}{\pi} \int_0^\pi V_m \, Sin\omega t . d\theta$$

$$= \frac{1}{\pi} \int_0^\pi 5 \, Sin\omega t . d\theta$$

$$= \frac{5}{\pi} \int_0^\pi Sin\omega t . d\theta = \frac{5}{\pi} \left[-Cos\omega t \right]_0^\pi$$

$$= \frac{5}{\pi} \left[-(-1-1) \right] = \frac{10}{\pi} = 3.185 \text{ V}$$

$$\text{RMS value} = \sqrt{\frac{1}{\pi} \int_0^\pi v^2 . d\theta} = \sqrt{\frac{1}{\pi} \int_0^\pi \left(V_m Sin\omega t \right)^2 . d\theta}$$

$$= \sqrt{\frac{V_m^2}{\pi} \int_0^\pi \left[\frac{1-Cos2\theta}{2} \right] . d\theta}$$

$$= \sqrt{\frac{V_m^2}{2\pi} \int_0^\pi [1 - Cos2\theta].d\theta}$$

$$= \sqrt{\frac{V_m^2}{2\pi} \left[\theta - \frac{Sin2\theta}{2} \right]_0^\pi}$$

$$= \sqrt{\frac{V_m^2}{2\pi} [\pi - 0 + 0 - 0]}$$

$$= \sqrt{\frac{V_m^2}{2\pi}} = \frac{V_m}{\sqrt{2}} = 0.707V_m = 0.707 \times 5 = 3.54 \text{ V}$$

Exercise 2.1.6: Find the Average and Effective values of saw-tooth wave form shown in figure 2.17.

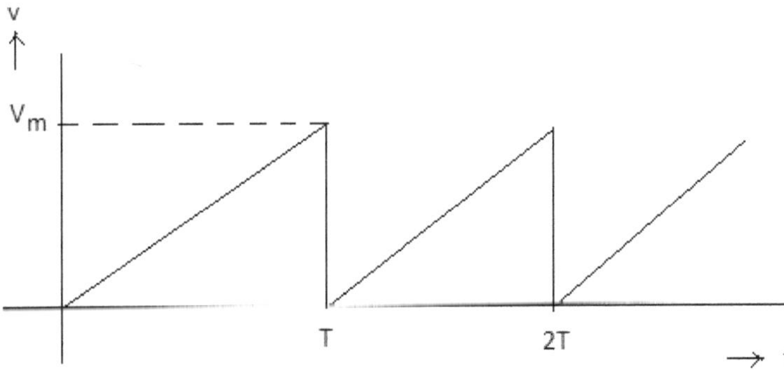

Figure 2.17

Solution:

The equation for voltage as a function of time $= v = m \times t$

Where m is the slope, $m = \dfrac{V_m}{T}$

$$\text{Average value} = = \frac{1}{T}\int_0^T v.dt = \frac{1}{T}\int_0^T \frac{V_m}{T}.t.dt = \frac{V_m}{T^2}\left[\frac{t^2}{2}\right]_0^T = \frac{V_m}{T^2} \times \frac{T^2}{2} = \frac{V_m}{2} = 0.5V_m$$

$$\text{Effective value or RMS value} = \sqrt{\frac{1}{T}\int_0^T v^2 dt} = \sqrt{\frac{1}{T}\int_0^T m^2 t^2 dt} = \sqrt{\frac{1}{T} \times \frac{V_m^2}{T^2} \int_0^T t^2 dt}$$

$$= \sqrt{\frac{V_m^2}{T^3}\left[\frac{t^3}{3}\right]_0^T} = \sqrt{\frac{V_m^2}{T^3} \times \frac{T^3}{3}} = \sqrt{\frac{V_m^2}{3}} = \frac{V_m}{\sqrt{3}}$$

Exercise 2.1.7: Find the Form factor for the following periodic wave form shown in figure 2.18.

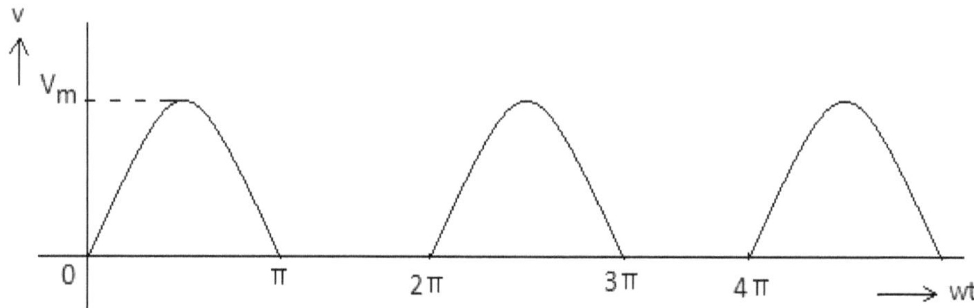

Figure 2.18

Solution:

Since, the wave form is not symmetric about x-axis, for calculating Average and RMS values, the base 2π is taken.

$$\text{Average value} = \frac{1}{2\pi}\int_0^{2\pi} v.dt$$

But, 0 to π : $v = V_m \; Sin \; \omega t$

π to 2π: $v = 0$

Then, Average value $= \dfrac{1}{2\pi}\int_0^{2\pi} v.d\theta = \dfrac{1}{2\pi}\left[\int_0^{\pi} V_m Sin\omega t.d\omega t + \int_{\pi}^{2\pi} 0.d\omega t\right]$

$$= \frac{V_m}{2\pi}\left[-Cos\omega t\right]_0^{\pi} = \frac{V_m}{2\pi}\left[-(Cos\pi - Cos0)\right] = \frac{V_m}{2\pi}\left[-(-1-1)\right] = \frac{V_m}{\pi}$$

The RMS value $= \sqrt{\dfrac{1}{2\pi}\int_0^{2\pi} v^2.d\theta}$

$$= \sqrt{\frac{1}{2\pi}\left[\int_0^{\pi} v^2 d\theta + \int_{\pi}^{2\pi} 0.d\theta\right]} = \sqrt{\frac{1}{2\pi}\int_0^{\pi} V_m^2 Sin^2\theta.d\theta} = \sqrt{\frac{V_m^2}{2\pi}\int_0^{\pi}\left(\frac{1-Cos2\theta}{2}\right)d\theta}$$

$$= \sqrt{\frac{V_m^2}{4\pi}\left[\theta - \frac{Sin2\theta}{2}\right]_0^{\pi}} = \sqrt{\frac{V_m^2}{4\pi}\left[\pi - 0 + 0 + 0\right]} = \sqrt{\frac{V_m^2}{4\pi}\left[\pi\right]} = \frac{V_m}{2}$$

Therefore, the Form Factor $= \dfrac{V_{RMS}}{V_{AVE}} = \dfrac{V_m}{2} \times \dfrac{\pi}{V_m} = \dfrac{\pi}{2} = 1.571$

Exercise 2.1.8: Find the Form Factor for the following wave form shown in figure 2.19.

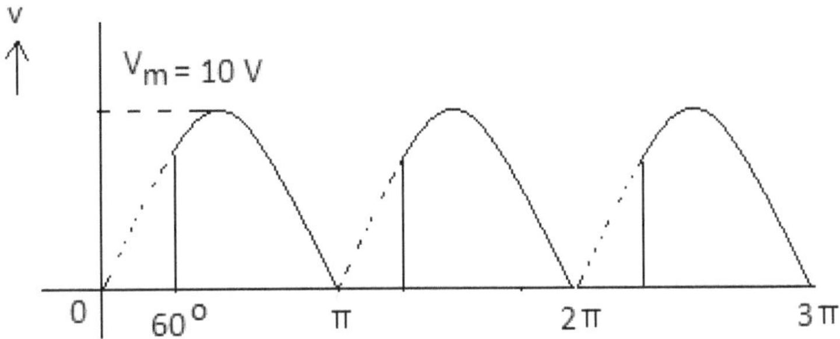

Figure 2.19

Solution: The Average Value $=\dfrac{1}{\pi}\int\limits_0^\pi v.d\theta$

But, 0 to 60^0, $v=0$:

60^0 to π, $v=V_m\,Sin\omega t$

The Average Value $=\dfrac{1}{\pi}\int\limits_0^\pi v.d\theta$

$$=\dfrac{1}{\pi}\left[\int\limits_0^{60^0}0.d\theta+\int\limits_{60^0}^\pi V_m\,Sin\omega t.d\theta\right]=\dfrac{V_m}{\pi}\left[-Cos\theta\right]_{60^0}^\pi$$

$$=\dfrac{V_m}{\pi}\left[-(-1-0.5)\right]=1.5\times\dfrac{V_m}{\pi}=0.477\times10=4.77\,V$$

The RMS value $=\sqrt{\dfrac{1}{\pi}\int\limits_0^\pi v^2 d\theta}$

$$=\sqrt{\dfrac{1}{\pi}\left[\int\limits_0^{60^0}0.d\theta+\int\limits_{60^0}^\pi V_m^2\,Sin^2\theta.d\theta\right]}=\sqrt{\dfrac{V_m^2}{2\pi}\left[\theta-\dfrac{Sin2\theta}{2}\right]_{60^0}^\pi}$$

$$=\sqrt{\dfrac{V_m^2}{2\pi}\left[\pi-0-\dfrac{\pi}{3}+0.29\right]}=\sqrt{\dfrac{V_m^2}{2\pi}\left[2.39\right]}$$

$$=0.6168\times V_m=0.6168\times10=6.168\,V$$

Therefore, Form factor $=\dfrac{6.168}{4.77}=1.293$

Exercise 2.1.9: Find the Average and RMS values of the following triangular waveform shown in figure 2.20.

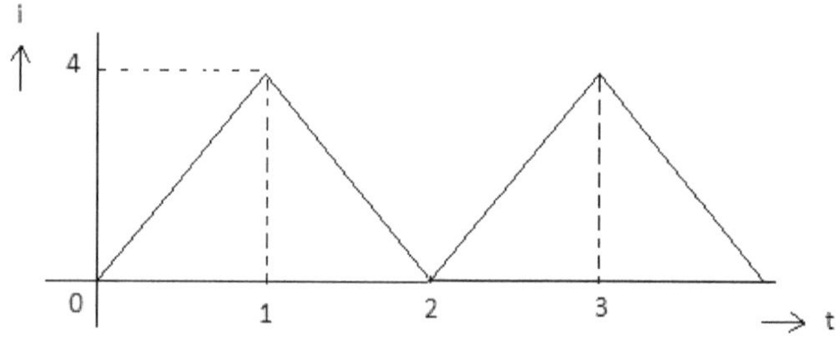

Figure 2.20

Given that: $i = 4t$ for t value between 0 and 1

$i = -4t + 8$ for t value between 1 and 2

Solution :

$$\text{Average value} = \frac{1}{2}\left[\int_0^1 4t.dt + \int_1^2 (-4t + 8)dt\right] = \frac{1}{2}\left[\left(\frac{4t^2}{2}\right)_0^1 + \left(\frac{-4t^2}{2} + 8t\right)_1^2\right]$$

$$= \frac{1}{2}\left[\left(2t^2\right)_0^1 + \left(-2t^2 + 8t\right)_1^2\right] = \frac{1}{2}\left[2 \times 1 - 0 + \left(-2 \times 2^2 + 8 \times 2\right) - \left(-2 \times 1 + 8\right)\right]$$

$$= \frac{1}{2}\left[2 + (-8 + 16) - (-2 + 8)\right] = \frac{1}{2}[2 + 8 - 6] = 2 \text{ A}$$

$$\text{RMS value} = \sqrt{\frac{1}{2}\left[\int_0^1 (4t)^2 + \int_1^2 (-4t + 8)^2\right]}$$

$$= \sqrt{\frac{1}{2}\left[\left(\frac{16t^3}{3}\right)_0^1 + \int_1^2 (16t^2 + 64 - 64t)\right]}$$

$$= \sqrt{\frac{1}{2}\left[\left(\frac{16t^3}{3}\right)_0^1 + \left(\frac{16t^3}{3} + 64t - \frac{64t^2}{2}\right)_1^2\right]}$$

$$= \sqrt{\frac{1}{2}\left[\left(\frac{16 \times 1}{3}\right) + \left(\frac{16 \times 8}{3} + 132 - \frac{64 \times 4}{2}\right) - \left(\frac{16 \times 1}{3} + 64 - \frac{64}{2}\right)\right]}$$

$$= \sqrt{\frac{1}{2} \times [5.333 + 46.66 - 37.33]}$$

$$= 2.707 \text{ A}$$

2.2   Analysis of Single Phase AC Circuits

2.2.1   A.C through Pure Resistance

Consider a pure resistance connected across a sinusoidal voltage as shown in figure 2.21. Let 'v' be the instantaneous value of voltage.

$$v = V_m \, Sin\omega t$$

This time varying voltage is dropped across the resistance. At any instant, the voltage drop across the resistance is given by

$$v = i \times R \quad and \quad i = \frac{v}{R}$$

$$i = \frac{V_m \, Sin\omega t}{R} = \frac{V_m}{R} Sin\omega t = I_m \, Sin\omega t$$

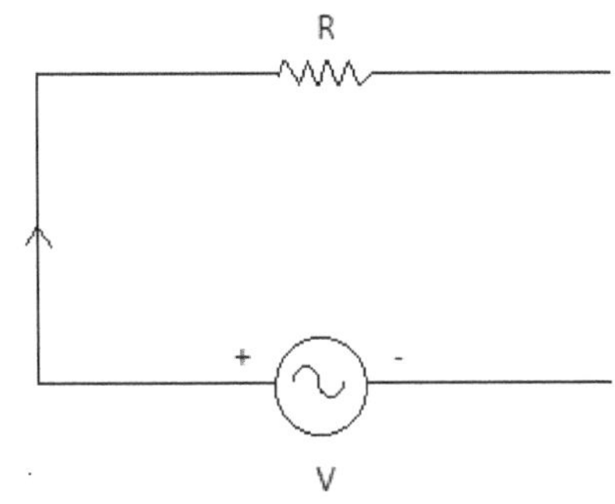

Figure 2.21 Resistance across AC source

Where I_m is the maximum value of current. It is also cleared from above equations that the phase of current is same as that of voltage. That means, both voltage and current are in-phase quantities. The voltage and current are shown figure 2.22 as waveforms.

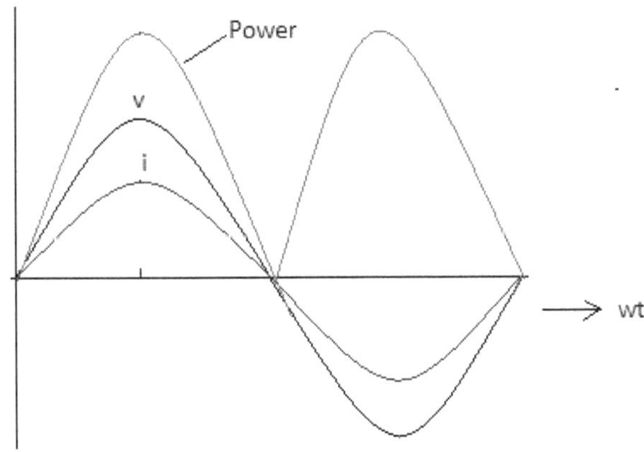

Figure 2.22 voltage and current wave forms in a Resistive network

The voltage and current phasors are shown in figure 2.23.

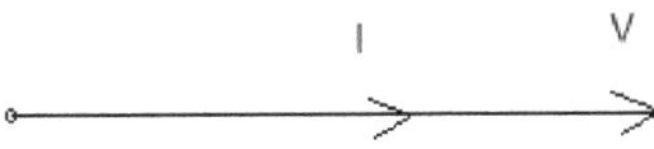

Figure 2.23 voltage and current phasors in a Resistive network

The power factor is defined as the Cosine of angle between the voltage and current phasors.

The Power Factor $= Cos\phi$

Since, V and I are in same phase, $\phi = 0$, hence, the power factor is 1 (unity) in resistive circuits.

In AC circuits, the current opposing property is termed as *'Impedance'* which includes the resistance and the reactive effect of elements like Inductance and Capacitance. This will be dealt in detail in subsequent sections. The impedance Z is given by

$$Z = \frac{\overline{V}}{\overline{I}} = \frac{V_m \angle 0^0}{I_m \angle 0^0} = R \angle 0^0 = R$$

The instantaneous Power (p) is given by

$$p = v \times i$$

$$= V_m \, Sin\omega t \times I_m \, Sin\omega t$$

$$= V_m I_m \, Sin^2 \omega t$$

$$= V_m I_m \left[\frac{(1 - Cos2\omega t)}{2} \right]$$

$$= \frac{V_m I_m}{2} - \frac{V_m I_m}{2} Cos2\omega t$$

The instantaneous power has two terms: the first term is independent of time and it is constant. The second term is sinusoid whose frequency is twice the frequency of voltage or current and its average value is zero. Since, the value of second term cannot become more than the first term at any value of 't', the instantaneous power cannot become negative as shown.

Therefore, the average power $= P = \dfrac{V_m I_m}{2} = \dfrac{V_m}{\sqrt{2}} \times \dfrac{I_m}{\sqrt{2}} = V_{RMS} \times I_{RMS}$

The instantaneous power is always positive and unidirectional. It means that the power always flows from source to load and the resistance continuously consumes it. Hence, it gets heated up after some time. This power is called *'Active Power'* or *'Real Power'*.

2.2.2 AC through Pure Inductance

Consider a pure inductance L whose internal resistance is zero connected across a sinusoidal voltage source as shown in figure 2.24.

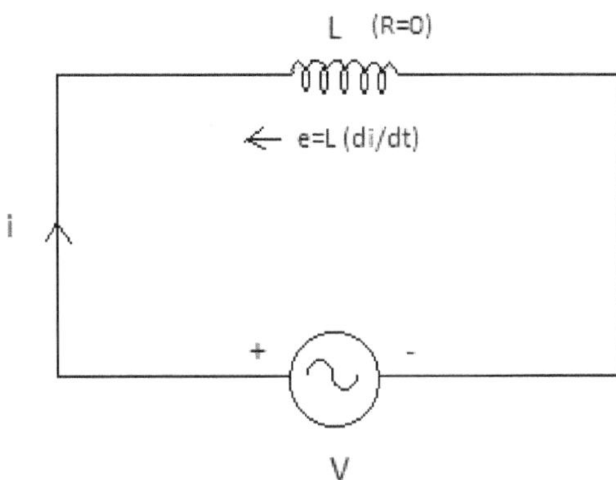

Figure 2.24 Inductance connected across an AC source

The instantaneous voltage is given by $v = V_m\ Sin\omega t$

This voltage will be opposed by the emf generated in the coil due to the change of its own flux linkages. Hence,

$$v = e + L\frac{di}{dt}$$

or
$$\frac{di}{dt} = \frac{1}{L} \times v$$

or
$$di = \frac{1}{L} \times v.dt$$

$$i = \frac{1}{L}\int v.dt = \frac{1}{L}\int (V_m\ Sin\omega t).dt$$

$$= \frac{V_m}{\omega L}(-Cos\omega t)$$

$$= \frac{V_m}{\omega L}\left(-Sin\left(90^0 - \omega t\right)\right)$$

$$= \frac{V_m}{\omega L}\left(Sin\left(\omega t - 90^0\right)\right)$$

$$= I_m Sin\left(\omega t - 90^0\right)$$

It is cleared from above equation that *the current lags the voltage by 90^0* and it is depicted in figure 2.25.

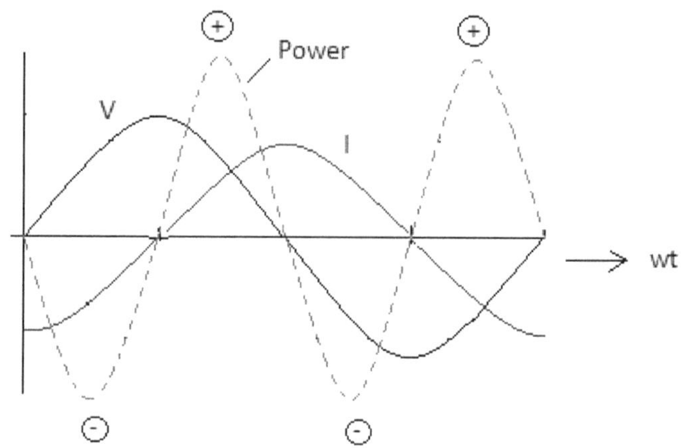

Figure 2.25 Voltage and Current wave forms in a Pure Inductive network

The phasor relation between voltage and current is shown below in figure 2.26.

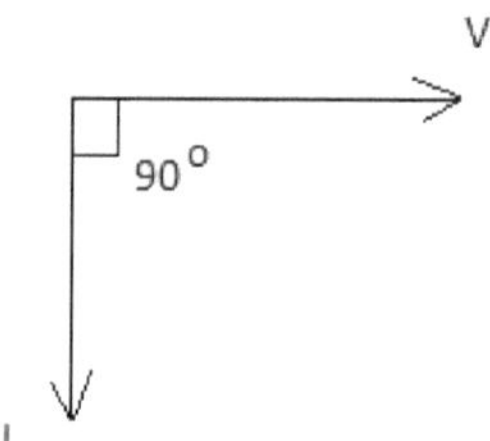

Figure 2.26 Voltage and Current phasors in a Pure Inductive network

It is also cleared from above equation of instantaneous current, $I_m = \dfrac{V_m}{\omega L} = \dfrac{V_m}{X_L}$

Where X_L is called 'Inductive Reactance' which opposes the alternating current.

$$X_L = 2\pi f L \quad \text{since } \omega = 2\pi f$$

The inductive reactance is proportional to the frequency. It becomes zero for DC circuits as the frequency $f = 0$ in DC circuits. *Hence, it behaves as short circuit in DC circuits.*

The impedance value for pure inductive circuit is equal to its inductive reactance as the resistance is zero. Then, the Impedance is given by

$$Z = \frac{\overline{V}}{\overline{I}} = \frac{V_m \angle 0^0}{I_m \angle -90^0} = X_L \angle 90^0 = jX_L$$

The power factor = Cosine of angle between V and I

$$= Cos\left(90^0\right) = 0 \text{ for pure inductor}$$

It is understood from above wave form that the frequency of power is twice the frequency of voltage or current. The power is bidirectional. It flows from source to load and load to source back repeatedly. Hence, the net power delivered to load or inductance is zero. The pure inductance does not consume any *'Active or Real Power'* and it does not get heated up. But, the power which got exchanged or swapped between source and inductance is called *'Reactive Power'* or *'Imaginary Power'*.

It is also illustrated analytically as follows.

Instantaneous Power $p = v \times i$

$$= V_m \, Sin\omega t \times I_m \, Sin\left(\omega t - 90^0\right)$$

$$= V_m I_m \, Sin\omega t \times Sin\left(\omega t - 90^0\right)$$

$$= V_m I_m \left[Cos\left(\omega t - \omega t + 90^0\right) - Cos\left(\omega t + \omega t - 90^0\right)\right]$$

$$= \frac{V_m I_m}{2}\left[0 - Cos\left(2\omega t - 90^0\right)\right]$$

$$= \frac{V_m I_m}{2} Cos\left(2\omega t - 90^0\right)$$

Its average value is zero.

2.2.3   A.C through Pure Capacitance

Consider a pure capacitor connected across a sinusoidal voltage source as shown in figure 2.27. The instantaneous value of voltage is given by

$$v = V_m \, Sin\omega t$$

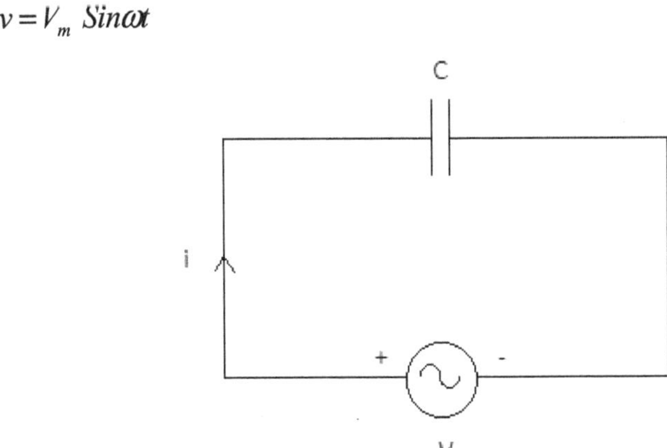

Figure 2.27 Pure Capacitor connected across an AC source

The charge accumulated on the capacitor plates is given by

$$Q = Cv$$

The rate of change of charge is the current $i = \dfrac{dQ}{dt} = \dfrac{d(Cv)}{dt} = C\dfrac{dv}{dt}$

$$= C\frac{d(V_m\,Sin\omega t)}{dt} = C\omega V_m\left(Cos\omega t\right)$$

$$= C\omega V_m\left(Sin\left(\omega t + 90^0\right)\right)$$

$$= I_m\,Sin\left(\omega t + 90^0\right)$$

It is cleared that the phase of the current with respect to voltage is $+90^0$. *Hence, the current leads the voltage by 90^0.*

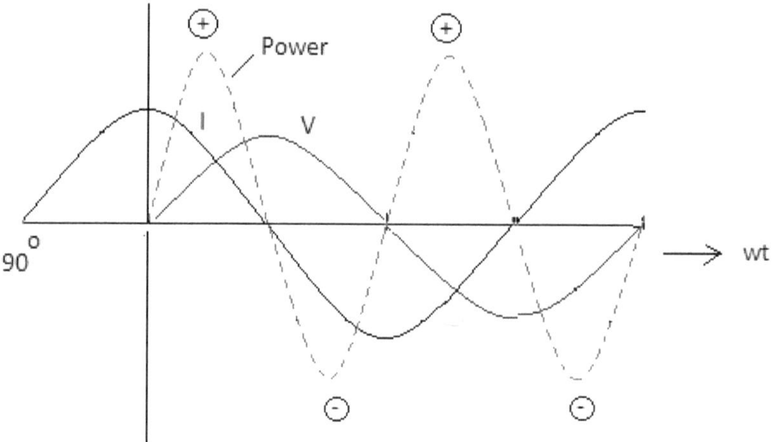

Figure 2.28 Voltage, Current and Power wave forms in a Pure capacitive network

The phasor representation of voltage and current is shown in figure 2.29.

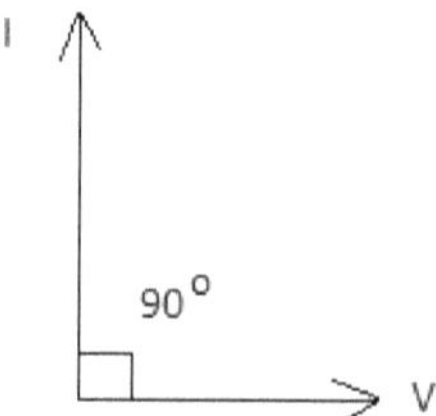

Figure 2.29 Voltage and Current Phasors in a Pure Capacitive network

And also $I_m = C\omega V_m$

Or $\qquad \dfrac{V_m}{I_m} = \dfrac{1}{C\omega} = X_C$ which is called Capacitive Reactance

$$X_C = \dfrac{1}{2\pi fC} \qquad \text{since } \omega = 2\pi f$$

The capacitive reactance is inversely proportional to the frequency. It becomes infinite for DC circuits as the frequency $f = 0$ in DC circuits. *Hence, it behaves as Open Circuit in DC circuits.*

The impedance value for pure capacitive circuit is equal to its capacitive reactance as the resistance is zero. Then, the Impedance is given by

$$Z = \dfrac{\overline{V}}{\overline{I}} = \dfrac{V_m \angle 0^0}{I_m \angle +90^0} = X_C \angle -90^0 = -jX_C \text{ since } -j = \angle -90^0$$

The power factor = Cosine of angle between V and I

$$= Cos\left(90^0\right) = 0 \quad \text{for pure Capacitor.}$$

It is evident from above wave form that the frequency of power is twice the frequency of voltage or current. The power is bidirectional. It flows from source to load and load to source back repeatedly. Hence, the net power delivered to load or capacitance is zero. The pure capacitance does not consume any *'Active or Real Power'* and it does not get heated up. But, the power which exchanged between source and inductance is called *'Reactive Power 'or 'Imaginary power'.*

It is also illustrated analytically as follows.

Instantaneous Power $\quad p = v \times i$

$$= V_m Sin\omega t \times I_m Sin\left(\omega t + 90^0\right)$$

$$= V_m I_m Sin\omega t \times Sin\left(\omega t + 90^0\right)$$

$$= \dfrac{V_m I_m}{2}\left[Cos\left(\omega t - \omega t - 90^0\right) - Cos\left(\omega t + \omega t + 90^0\right)\right]$$

$$= \dfrac{V_m I_m}{2}\left[Cos 90^0 - Cos\left(2\omega t + 90^0\right)\right]$$

$$= \dfrac{V_m I_m}{2} Cos\left(2\omega t + 90^0\right)$$

Its average value is zero.

The important points from above discussions are summarized in the following table 2.2.

Table 2.2 Different quantities in AC circuits having Resistance, Inductance and Capacitance

	Resistance	**Inductance**	**Capacitance**
Voltage	$v = V_m\ Sin\ \omega t$	$v = V_m\ Sin\ \omega t$	$v = V_m\ Sin\ \omega t$
Current	$i = I_m\ Sin\ \omega t$	$i = I_m\ Sin\left(\omega t - 90^0\right)$	$i = I_m\ Sin\left(\omega t + 90^0\right)$
Reactance	0	$X_L - 2\pi fL$	$X_C = \dfrac{1}{2\pi fC}$
Impedance	$Z = R$	$Z = jX_L$	$Z = -jX_C$
Power Factor	1	0	0
Active Power	*Consumes*	*Does not consume*	*Does not consume*
Reactive Power	*Does not consume*	*Consumes*	*Consumes*
Phasor Relation	*'V' and 'I' are in phase*	*'I' lags 'V' by 90^0*	*'I' leads 'V' by 90^0*

2.2.4 AC through Series R-L Circuit

Consider an AC circuit having Resistance (R) and Inductance (L) in series as shown in figure 2.30.

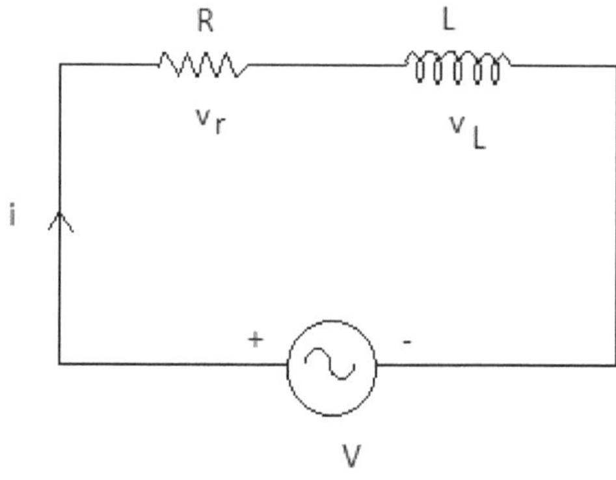

Figure 2.30 Series R-L Circuit

As the circuit consists of resistance and inductance, it consumes both active and reactive powers. Also, the current does not lag voltage by 90^0 or not in phase with the voltage. It lags the voltage by an angle Ø which lies between 0 and 90^0.

$$v = V_m \; Sin \; \omega t$$

$$i = I_m \; Sin \left(\omega t - \phi^0 \right)$$

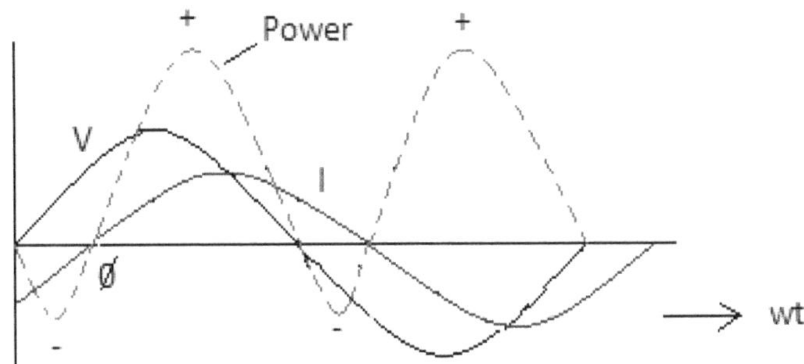

Figure 2.31 Voltage, Current and Power wave forms in Series R-L Circuit

Since, it is a series circuit, the current is taken as reference for drawing the phasor diagram. The phasor diagram is illustrated in figure 2.32.

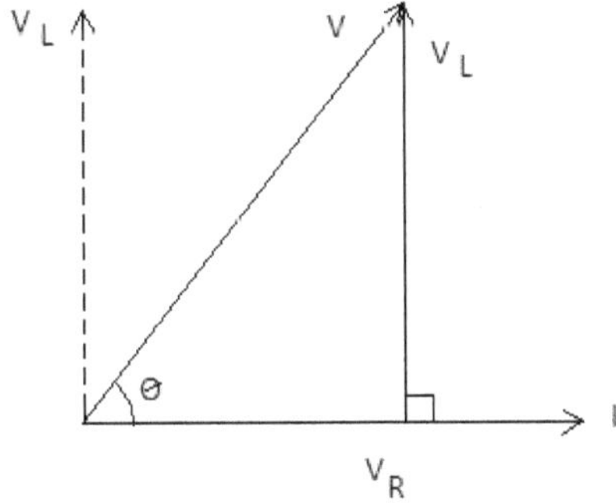

Figure 2.32 Series R-L Circuit Phasor diagram

V_R and V_L are the voltage drops in the resistance and inductance. The voltage V_R is in phase with current in resistance and V_L leads the current by 90^0.

From the above phasor diagram, it is cleared that the supply voltage V is the phasor sum of these two voltages.

$$V = V_R + jV_L$$

$$IZ = IR + jIX_L$$

Or $\quad Z = R + jX_L$

Where Z is the impedance of the circuit and the impedance triangle shown below in figure 2.33.

$$R = Z \times Cos\phi$$

$$X_L = Z \times Sin\phi$$

And also $Z = \sqrt{R^2 + X_L^2}$

The power factor of the circuit, $Cos\phi = \dfrac{R}{Z}$

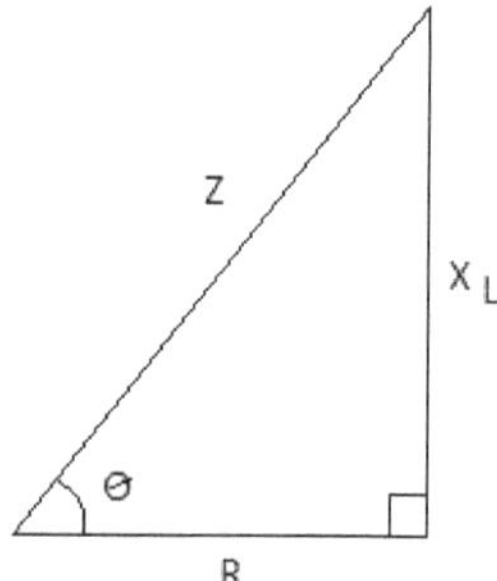

Figure 2.33 Impedance Triangle in Series R-L Circuit

The Real Power or Active Power consumed is the product of voltage and in-phase component of current with voltage.

The Real Power or Active Power $P = V \times I \times Cos\phi$

$$P = V \times I \times Power\ Factor$$

Where Ø is the angle between Voltage and Current

Similarly, the Reactive Power or Imaginary Power is the product of voltage and quadrature component of current with it.

The Reactive Power $jQ = V \times I \times Sin\phi$

Where 'S' is called 'Apparent Power'.

$$S = P + jQ$$

And $\quad S = \sqrt{P^2 + Q^2}$

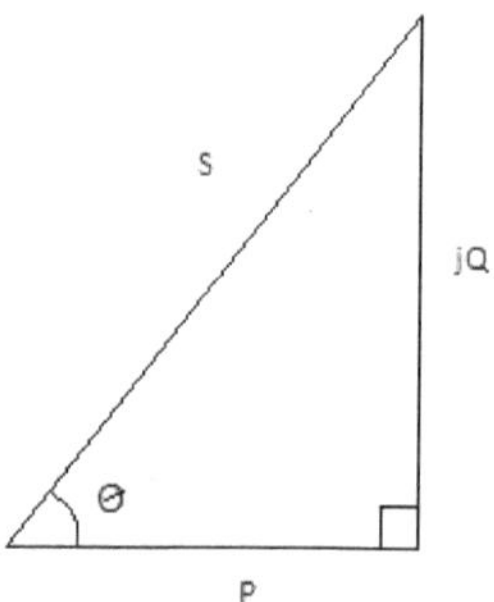

Figure 2.34 Different Powers in Series R-L Circuit

The units for Active, Reactive and Apparent powers are KW, KVAR and KVA respectively.

2.2.5 AC through Series R-C Circuit

Consider an AC circuit having Resistance (R) and Capacitance (C) in series as shown in figure 2.35.

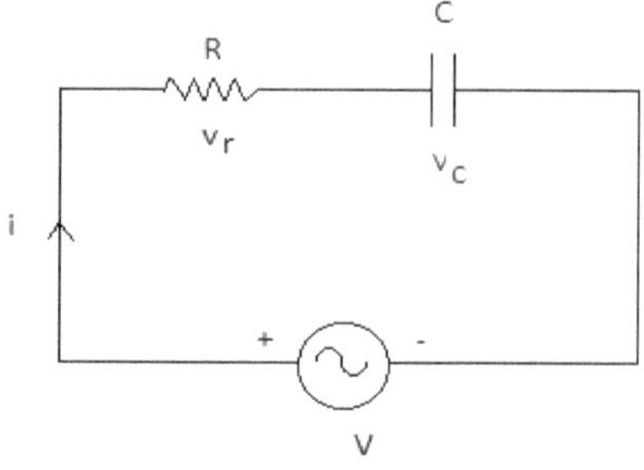

Figure 2.35 Series R-C Circuit

As the circuit consists of resistance and capacitance, it consumes both active and reactive powers. Also, the current does not lead voltage by 90^0 or not in phase with the voltage. It leads the voltage by an angle Ø which lies between 0 and 90^0.

$$v = V_m \ Sin \ \omega t$$

$$i = I_m \ Sin \ \left(\omega t + \phi^0 \right)$$

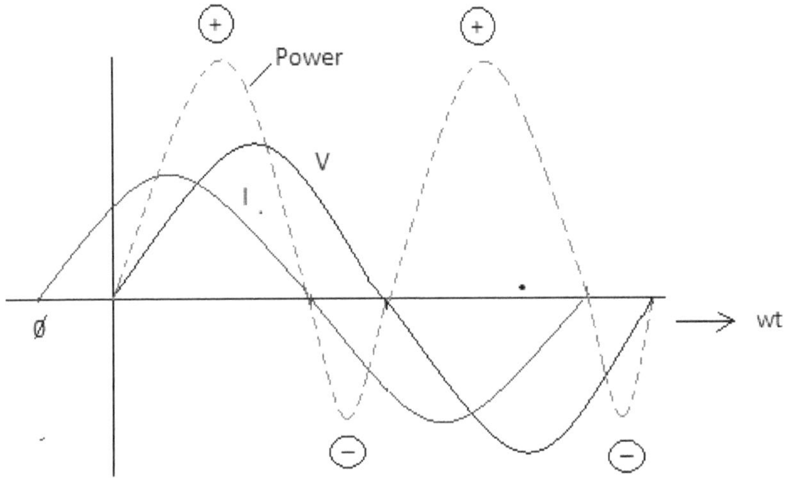

Figure 2.36 Voltage, Current and Power wave forms in Series R-C Circuit

Since, it is a series circuit, the current is taken as reference for drawing the phasor diagram. The phasor diagram is shown in figure 2.37.

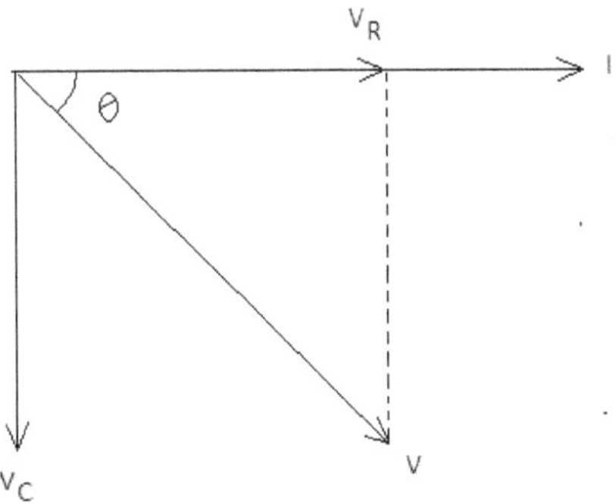

Figure 2.37 Series R-C Circuit Phasor diagram

V_R and V_C are the voltage drops in the resistance and capacitance. The voltage V_R is in phase with current in resistance and V_C lags the current by 90^0.

From the above phasor diagram, it is cleared that the supply voltage V is the phasor sum of these two voltages.

$$V = V_R - jV_C$$

$$Z = R - jX_C$$

Or $\quad Z = R - jX_C$

Where Z is the impedance of the circuit and the impedance triangle shown below.

$$R = Z \times Cos\phi$$

$$X_C = Z \times Sin\phi$$

And also $Z = \sqrt{R^2 + X_C^2}$

The power factor of the circuit, $Cos\phi = \dfrac{R}{Z}$

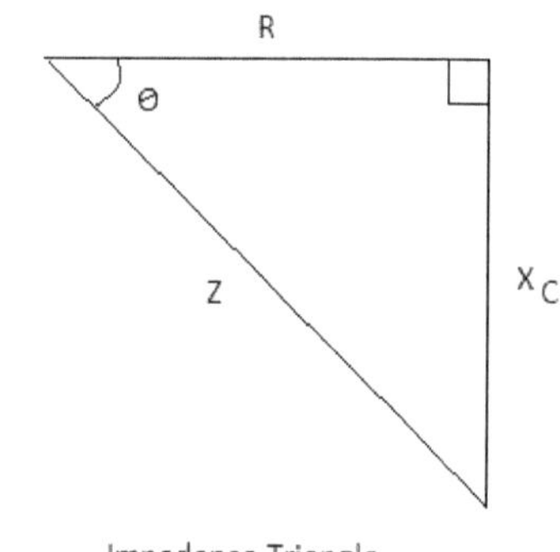

Figure 2.38 Impedance Triangle of Series R-C Circuit

The Real Power or Active Power consumed is the product of voltage and in-phase component of current with voltage.

The Real Power or Active Power $P = V \times I \times Cos\phi$

$$P = V \times I \times Power\ Factor$$

Where Ø is the angle between Voltage and Current

Similarly, the Reactive Power or Imaginary Power is the product of voltage and quadrature component of current with it.

The Reactive Power $-jQ = V \times I \times Sin\phi$

Where 'S' is called 'Apparent Power'.

$$S = P - jQ$$

And $\quad S = \sqrt{P^2 + Q^2}$

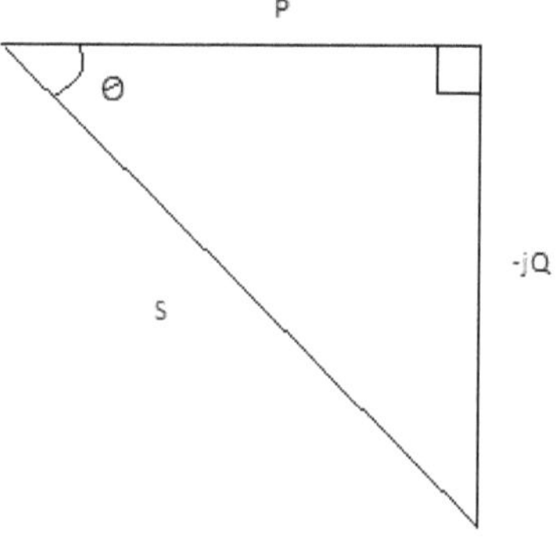

Figure 2.39 Different Powers in Series R-C Circuit

2.2.6 AC through Series R-L-C Circuit

Consider an AC circuit having Resistance (R), Inductance (L) and Capacitance (C) in series as shown in figure 2.40.

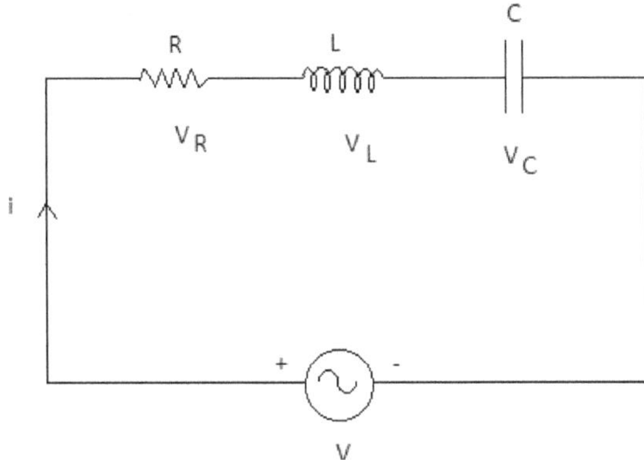

Figure 2.40 Series R-L-C Circuit

The instantaneous values of voltage and current are

$$v = V_m \ Sin \ \omega t$$

$$i = I_m \ Sin \ \left(\omega t \pm \phi^0\right)$$

The phase difference (Ø) between voltage(V) and current (I) depends up on magnitudes of X_L and X_C.

Ø is positive if $X_L > X_C$

Ø is negative if $X_C > X_L$

Ø is zero if $X_L = X_C$

By taking current as reference, the phasor diagram is shown below in figure 2.41.

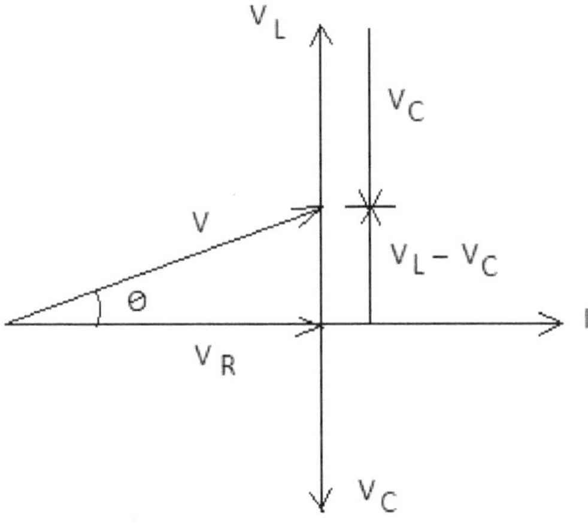

Figure 2.41 phasor diagram of Series R-L-C Circuit

The supply voltage is the phasor sum of voltages across all the elements.

$$V = V_R + j(V_L - V_C)$$

$$IZ = IR + jI(X_L - X_C)$$

$$Z = R + j(X_L - X_C)$$

If $X_L > X_C$, the circuit is inductive, the current 'I' lags the voltage 'V' by angle Ø. The power factor is lagging.

If $X_C > X_L$, the circuit is capacitive, the current 'I' leads the voltage 'V' by angle Ø. The power factor is leading.

If $X_L = X_C$, the circuit is resistive, the current 'I' is in phase with the voltage 'V'. The power factor is unity. Under this condition, the impedance of the circuit becomes minimum($Z= R$ only) and current becomes maximum. This condition is called '*Resonance*'.

$$R = Z \times Cos\phi$$

Or the power factor $Cos\phi = \dfrac{R}{Z}$

Active Power consumed $P = V \times I \times Cos\phi$

Reactive power consumed $Q = V \times I \times Sin\phi$

The Apparent Power $S = V \times I$

And also $S = P + jQ$ or $S = \sqrt{P^2 + Q^2}$

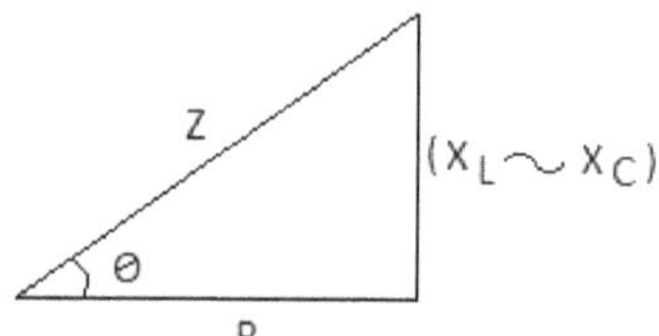

Figure 2.42 Impedance Triangle of Series R-L-C Circuit

Exercise 2.2.1 : A Sinusoidal voltage of $v = 311\,Sin314t$ is applied to a circuit having Resistance of 5 ohms and Inductance of 0.2H in series. Find a) frequency of the voltage and current b) impedance c) power factor of the circuit d) expression for instantaneous current e) voltage across Resistance f) voltage across inductance g) Real Power and h) Reactive Power consumed.

Solution:

The instantaneous value of voltage is given by $v = 311\,Sin314t$

By comparing this expression with the standard form $v = V_m\,Sin\,\omega t$

$$V_m = 311\,V, \omega = 314\,\text{rad/sec} \ \text{ or } \omega = 2\pi f \ \text{ or } f = \dfrac{\omega}{2\pi} = \dfrac{314}{2\pi} = 50\,\text{Hz}$$

(a) The frequency of voltage and current $f = 50\,\text{Hz}$

(b) Resistance of the circuit $= R = 5\ ohms$

Inductance of the circuit $= L = 0.02\ H$

Inductive Reactance $X_L = 2\pi fL = 2\pi \times 50 \times 0.02 = 6.28\,\text{Ohms}$

Impedance of the circuit $Z = R + jX_L$

And the magnitude of impedance $Z = \sqrt{R^2 + X_L^2} = \sqrt{5^2 + 6.28^2} = 8.027\,\text{Ohms}$

(c) The power factor $Cos\phi = \dfrac{R}{Z} = \dfrac{5}{8.027} = 0.6229$ lagging

The power factor angle $\phi = Cos^{-1}(0.6229) = 51.47^0$ lagging

(d) Expression for the instantaneous value of current is given by

$$i = I_m Sin(\omega t - \phi) \qquad (\text{\O\ is negative as the power factor is lagging})$$

$$I_m = \frac{V_m}{Z} = \frac{311}{8.027} = 38.74\,\text{A}$$

$$i = I_m Sin(\omega t - \phi^0) = 38.74 Sin(314t - 51.47^0)$$

(e) RMS value of current $I_{rms} = 0.707 \times I_m = 0.707 \times 38.74 = 27.39\,\text{A}$

RMS value of voltage $V_{rms} = 0.707 \times V_m = 0.707 \times 311 = 219.877\,\text{V}$

Voltage across Resistance $V_R = I_{rms} \times R = 27.39 \times 5 = 136.94\,\text{V}$

Voltage across the inductance $V_L = I_{rms} \times X_L = 27.39 \times 6.28 = 172\,\text{V}$

(f) Real Power or Active Power consumed $P = V_{rms} \times I_{rms} \times Cos\phi$

$$= 219.87 \times 27.39 \times 0.6229 = 3751.37 = 3.751\,\text{KW}$$

(g) Reactive power or Imaginary power consumed $Q = V_{rms} \times I_{rms} \times Sin\phi$

$$= 219.87 \times 27.39 \times Sin51.47^0 = 4710.74\,\text{VAR or }\ 4.71\,\text{KVAR}$$

Exercise 2.2.2 : In a series R-L circuit, the current and voltages are given as

$i = 1\,Cos(314t - 20^0)$ and $v = 10\,Cos(314t + 10^0)$. Find the values of R and L.

Solution:

Since both voltage and currents are expressed as Cosine functions, they need not be converted into Sine functions to find the phase difference between them. These voltage and current phasors are shown below to a reference.

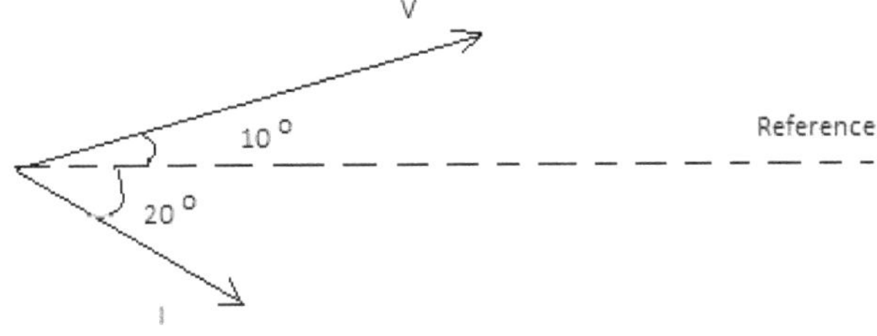

Figure 2.43 Phasor diagram

The phase difference $=10^0 - (-20^0) = 30^0$

Since $\omega = 314$, and $2\pi f = 314$, or $f = \dfrac{314}{2\pi} = 50$ Hz

The maximum values of current and voltages from above equations are 1A and 10 V respectively.

Then, the impedance of the circuit $Z = \dfrac{V_m}{I_m} = \dfrac{10}{1} = 10$ ohms

Resistance of the circuit $R = Z \times Cos\phi = 10 \times Cos30^0 = 8.68$ ohms

Since, the current is lagging the voltage, the circuit is inductive.

Inductive Reactance of the circuit $X_L = Z \times Sin\phi = 10 \times Sin30^0 = 5$ ohms

Inductance of the circuit $L = \dfrac{X_L}{2\pi f} = \dfrac{5}{2\pi \times 50} = 15.9$ mH

Exercise 2.2.3 : In a series R-L-C circuit, R = 4.2 ohms, L = 0.03 H and C = 450 μF. Find i) the supply voltage ii) the voltage across each element and iii) power factor angle if I = 10 A. Also draw the phasor diagram. Assume the supply frequency as 50 Hz.

Solution:

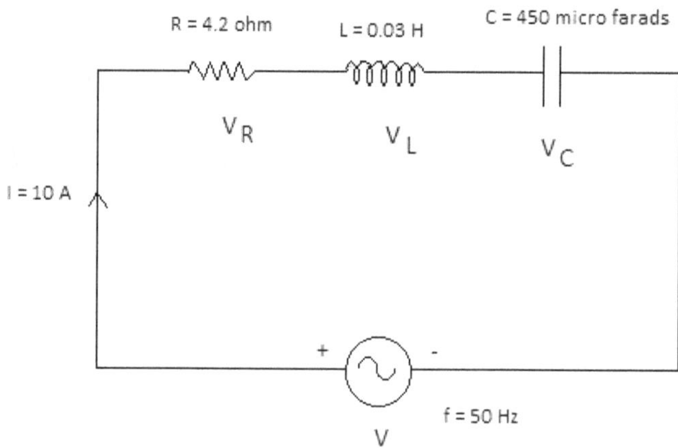

Figure 2.44 Series R-L-C Circuit

R = 4.2 ohm

L = 0.03 H

C = 450 μF

Inductive Reactance $X_L = 2\pi fL = 2\pi \times 50 \times 0.03 = 9.42$ ohms

Capacitive Reactance $X_C = \dfrac{1}{2\pi fC} = \dfrac{1}{2\pi \times 50 \times 450 \times 10^{-6}} = 7.073$ ohms

Total impedance of the circuit,

$$Z = R + j(X_L - X_C) = 4.2 + j(9.42 - 7.073) = 4.2 + j2.351 \, \text{ohms}$$

The magnitude of the impedance $Z = \sqrt{R^2 + X^2} = \sqrt{4.2^2 + 2.351^2} = 4.813 \, \text{ohms}$

The power factor $Cos\phi = \dfrac{R}{Z} = \dfrac{4.2}{4.813} = 0.872$

Or the power factor angle $\phi = Cos^{-1}(0.872) = 29.3^0$

The supply voltage $V = I \times Z = 10 \times 4.83 = 48.3 \, \text{V}$

The voltage across the Resistance $I \times R = 10 \times 4.2 = 42 \, \text{V}$

The voltage across the Inductance $I \times X_L \times 9.42 = 94.2 \, \text{V}$

The voltage across the Capacitance $I \times X_C \times 7.07 = 70.7 \, \text{V}$

The phasor diagram for the above circuit by taking current as reference is shown in figure 2.45. The current which is common to all elements, it is taken as reference in series circuits.

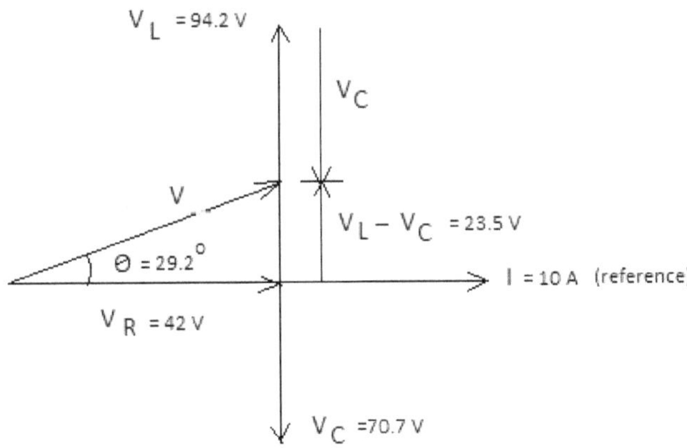

Figure 2.45 Phasor diagram of Series R-L-C Circuit

The voltage across inductor V_L leads the current 'I' by 90^0 and the voltage across capacitance V_C lags the current 'I' by 90^0 as shown. The V_L and V_C are in phase opposition, hence, the resultant is $(V_L - V_C)$. It is cleared from above phasor diagram that the supply voltage is the phasor sum of V_R and $(V_L \sim V_C)$.

$$V = \sqrt{V_R^2 + (V_L - V_C)^2} = \sqrt{42^2 + 23.5^2} = 48.12 \, \text{V}$$

Exercise 2.2.4: A Series circuit having resistance and inductance draws 11 A and consumes 0.8 Kw from a 220 V, 50 Hz supply. Determine (i) Impedance of the circuit (ii) Power factor (iii) Reactive and Apparent Powers

Solution :

Current $I_{rms} = I = 11\,A$

Voltage $V_{rms} = V = 220\,V$

The given values are considered as RMS values unless otherwise specified.

(a) Impedance $Z = \dfrac{V}{I} = \dfrac{220}{11} = 20$ ohms

(b) Since the pure inductance does not consume any active power, the entire active power of 0.8 KW has been consumed by resistance only.

$$P = I^2 R$$

$$0.8 \times 10^3 = 11^2 \times R$$

$$R = \frac{0.8 \times 10^3}{11^2} = 6.61\,\text{ohms}$$

Power factor $Cos\phi = \dfrac{R}{Z} = \dfrac{6.61}{20} = 0.3305$ lagging

(c) Since, $Cos\phi = 0.3305$, $Sin\phi = 0.9432$

Reactive power $Q = V \times I \times Sin\phi = 220 \times 11 \times 0.9432 = 2282$ VAR or $2.282\,kVAR$

Apparent Power $S = V \times I = 220 \times 11 = 2420$ VAor $2.42\,kVA$

Exercise 2.2.5: A Series R-L circuit having R = 4 ohms and L= 0.09 H is connected across a sinusoidal voltage $v = 325\ Sin(314t + 15^0)$. Find i) Expression for instantaneous value of current ii) RMS value of current iii) Form factor iv) Active power v) Reactive Power and vi) Apparent Power.

Solution:

R = 4 ohms   and L = 0.09 H

$$v = 325\ Sin(314t + 15^0)$$

By comparing the above voltage expression with a standard form

$$v = V_m Sin(\omega t + \phi^0)$$

$V_m = 325$, $\omega = 314\,\text{rad/sec}$ and Ø is the initial phase of the voltage with respect to a reference.

$$\omega = 2\pi f$$

$$314 = 2\pi \times 50 \text{ or } f = \frac{314}{2\pi} = 50\,\text{Hz}$$

Inductive Reactance $X_L = 2\pi fL = 2\pi \times 50 \times 0.09 = 28.26$ Ohms

Impedance $Z = \sqrt{R^2 + X_L^2} = \sqrt{4^2 + 28.26^2} = 28.54$ Ohms

(a) Maximum value of current $I_m = \dfrac{V_m}{Z} = \dfrac{325}{28.54} = 11.38\,\text{A}$

Power factor $Cos\phi = \dfrac{R}{Z} = \dfrac{4}{28.54} = 0.1401$ lagging

And the phase angle of current with respect to voltage is $\phi = Cos^{-1}(0.1401) = 81.94^0$. It means current lags the voltage by 81.94^0.

Therefore, the instantaneous value of current is given by

$$i = 11.38\ Sin\left(314t + 15^0 - 81.94^0\right)$$

Or $\qquad i = 11.38\ Sin\left(314t - 66.94^0\right)$

(b) RMS value of the current $I_{rms} = 0.707 \times I_m = 0.707 \times 11.38 = 8.04\,\text{A}$

(c) Power factor $Cos\phi = \dfrac{R}{Z} = \dfrac{4}{28.54} = 0.1401$ lagging

(d) RMS value of voltage $V_{rms} = 0.707 \times V_m = 0.707 \times 325 = 229.7\,\text{V}$

Active Power $P = V \times I \times Cos\phi = 229.7 \times 8.04 \times 0.1401 = 258.73\,watts$

(e) Since, $Cos\phi = 0.1401$ and $Sin\phi = 0.990$

Reactive power or Imaginary power $Q = V.I.Sin\phi = 229.7 \times 8.04 \times 0.99 = 1828.32$ VAR

(f) Apparent power $S = V \times I = 229.7 \times 8.04 = 1846.78$ VA

Exercise 2.2.6: A Pure 100 ohm resistor is placed in series with a pure capacitance of 120μF. This combination is connected across 220V, 50 Hz supply. Find the a) impedance of the circuit b) current c) power factor d) power factor angle or phase angle e) voltage across resistance and f) Voltage across capacitance.

Solution:

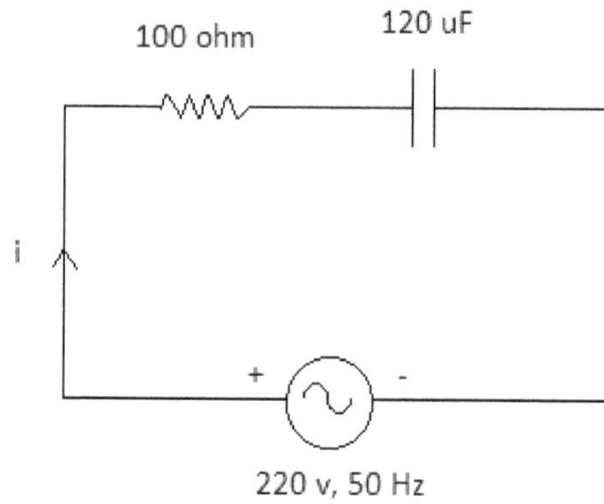

Figure 2.46 Series R-C Circuit

Resistance $R = 100\,\text{ohm}$

Capacitance $C = 120\,\mu F$

Capacitive Reactance $X_C = \dfrac{1}{2\pi fC} = \dfrac{1}{2\pi \times 50 \times 120 \times 10^{-6}} = 26.54\ \text{ohms}$

(a) Impedance $Z = R - jX_C$

$$Z = \sqrt{R^2 + X_C^2} = \sqrt{100^2 + 26.54^2} = 103.54\ \text{ohms}$$

(b) Current $I = \dfrac{V}{Z} = \dfrac{220}{103.54} = 2.127\ \text{A}$

(c) Power Factor $Cos\phi = \dfrac{R}{Z} = \dfrac{100}{103.54} = 0.967$

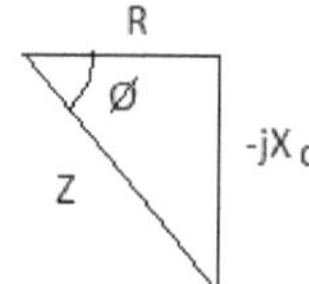

Figure 2.47
Impedance
Triangle of Series
R-C Circuit

(d) Phase angle or power factor angle $\phi = Cos^{-1}(0.967) = 14.76^{0}$

(e) Voltage across the resistance $V_R = I \times R = 2.127 \times 100 = 212.7\ \text{V}$

(f) Voltage across the capacitance $V_C = I \times X_C = 2.127 \times 26.54 = 56.45\ \text{V}$

Exercise 2.2.7: A voltage $v = 200 Sin314t$ is applied to a circuit consisting of a 50 ohm resistor and a 40 μF capacitor in series. Determine a) expression for current at a given time instant 't' and b) power consumed.

Solution:

$$v = 200 Sin314t$$

Comparing this expression with standard form $v = V_m\ Sin\omega t$

$$V_m = 200\,\text{V}, \quad \omega = 314\,\text{rad/sec}$$

$$f = \dfrac{\omega}{2\pi} = \dfrac{314}{2\pi} = 50\,\text{Hz}$$

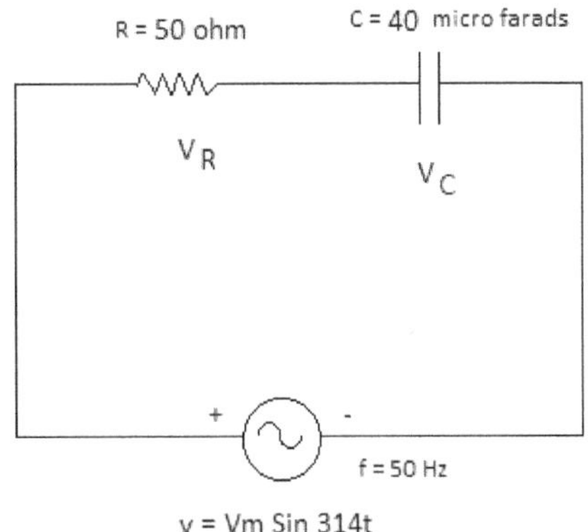

Figure 2.48 Series R-C Circuit

(a) Capacitive Reactance $X_C = \dfrac{1}{2\pi fC} = \dfrac{1}{2\pi \times 50 \times 40 \times 10^{-6}} = 79.6$ ohms

Impedance $Z = R - jX_C$

$$Z = \sqrt{R^2 + X_C^2} = \sqrt{50^2 + 79.6^2} = 94 \text{ ohms}$$

Power Factor $Cos\phi = \dfrac{R}{Z} = \dfrac{50}{94} = 0.5319$ leading

$Cos\phi = 0.5319$ and hence, $\phi = Cos^{-1}(0.5319) = 57.86^0$

$$I_m = \frac{V_m}{Z} = \frac{200}{94} = 2.127 \text{ A}$$

$$i = I_m \ Sin\left(\omega t + \phi^0\right) = 2.127 \ Sin\left(314t + 57.86^0\right)$$

In capacitive circuits, the current leads the voltage.

(b) The power consumed $P = V \times I \times Cos\phi$

$$= \left(0.707 \times V_m\right) \times \left(0.707 \times I_m\right) \times Cos \ \phi$$

$$= \left(0.707 \times 200\right) \times \left(0.707 \times 2.127\right) \times 0.5319 = 113.1 \text{ watts}$$

Exercise 2.2.8 : A 220 V, 50 Hz supply sends a current of 5A through a series RC circuit. The maximum value of voltage occurs 1/1200 seconds after the maximum current. Calculate a) Power factor b) Active Power or Average Power c) elements of the circuit

Solution :

Supply Voltage $V = 220$ V

Frequency $f = 50 \, Hz$

Current $I = 5 \, A$

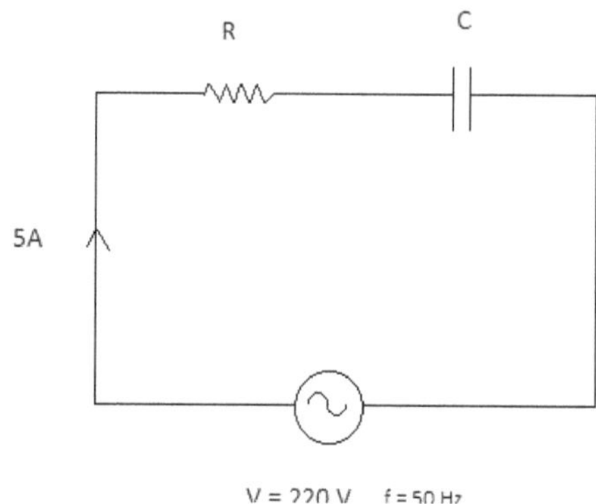

Figure 2.49 Series R-C Circuit

The voltage lags the current by 1/1200 seconds and it is cleared from following figure.

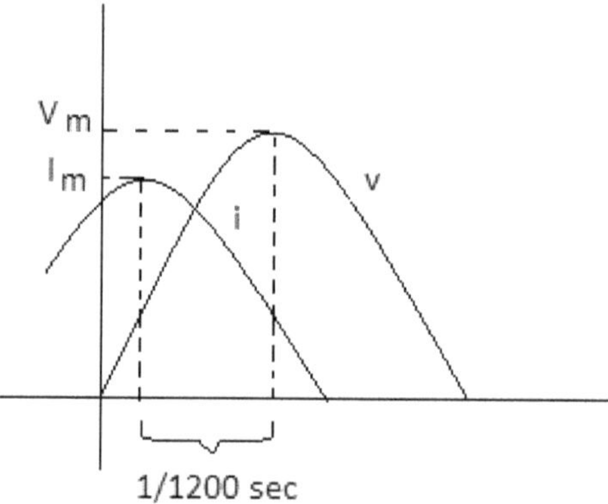

Figure 2.50 Voltage and Current wave forms in Series R-C Circuit

Time Period $T = \dfrac{1}{f} = \dfrac{1}{50} = 0.02$ sec or 20 msec

One Time period T corresponds to 360^0. That means, 20 msec corresponds to 360^0.

Therefore $\dfrac{1}{1200} = 0.833$ corresponds to $= \dfrac{0.833 \times 360^0}{20} = 15^0 = \phi$

(a) Power factor $Cos\ \phi^0 = Cos\ 15^0 = 0.966$ leading

(b) Active Power or Average Power $P = V \times I \times Cos\phi$

$$= 220 \times 5 = 0.966 = 1207.5 \text{ watts}$$

(c) Impedance $Z = \dfrac{V}{I} = \dfrac{220}{5} = 44$ ohms

Resistance $R = Z \times Cos\phi = 44 \times 0.966 = 42.5$ ohms

Capacitive Reactance $X_c = Z \times Sin\phi = 44 \times Sin15^0 = 44 \times 0.2588 = 11.38$ ohms

$$X_C = \frac{1}{2\pi fC}$$

$$11.38 = \frac{1}{2\pi \times 50 \times C}$$

Or $\quad C = \dfrac{1}{2\pi \times 50 \times 11.38} = 279.8\,\mu F$

Exercise 2.2.9: A coil of resistance 15 ohm and inductance 0.2H is connected in series with a capacitor of 100 µF across a 220 V, 50 Hzsupply. Calculate a) Inductive Reactance b) capacitive Reactance c) Impedance of the circuit d) Current e) Power factor f) Voltage across the coil g) Voltage across the capacitor.

Solution:

Resistance $R = 15$ ohms

Inductance $L = 0.2\,\mathrm{H}$

(a) Inductive Reactance $X_L = 2\pi fL = 2\pi \times 50 \times 0.2 = 62.8$ ohms

(b) Capacitive Reactance $X_C = \dfrac{1}{2\pi fC} = \dfrac{1}{2\pi \times 50 \times 100 \times 10^{-6}} = 31.84$ ohms

(c) Impedance of the circuit $Z = R + j(X_L \sim X_C) = 15 + j(62.8 \sim 31.84) == 15 + j30.95$ ohms

Or $\quad Z = \sqrt{R^2 + X^2} = \sqrt{15^2 + 30.95^2} = 34.39$ ohms

(d) Current $I = \dfrac{V}{Z} = \dfrac{220}{34.39} = 6.397$ A

(e) Power factor $Cos\phi = \dfrac{R}{Z} = \dfrac{15}{34.39} = 0.436$ lagging since X_L is greater than X_C

(f) Impedance of the coil $Z_L = R + jX_L$

$$Z_L = \sqrt{R^2 + X^2} = \sqrt{15^2 + 62.8^2} = 64.56 \text{ ohms}$$

Voltage drop across the coil $I \times Z_L = 6.397 \times 64.56 = 413\,\mathrm{V}$

(g) Voltage across the capacitance $I \times X_C = 6.397 \times 31.84 = 203.68$ V

Exercise 2.2.10 : A series RLC circuit having R = 10 ohms, L = 0.2 H and C = 100 µF is connected across 220 V AC supply. Find the frequency at which the inductive reactance becomes equal to capacitive reactance. Also find the current and power factor at this condition.

Solution:

Inductive Reactance = Capacitive Reactance

$$X_L = X_C$$

$$2\pi fL = \frac{1}{2\pi fC}$$

$$\text{or} \quad \left(2\pi f\right)^2 = \frac{1}{LC}$$

$$\text{or} \quad 2\pi f = \frac{1}{\sqrt{LC}}$$

$$\text{or} \quad f = \frac{1}{2\pi\sqrt{LC}}$$

$$f = \frac{1}{2\pi\sqrt{0.2\times100\times10^{-6}}} = 35.6 \text{ Hz}$$

This condition is called *'Resonance'* and the frequency is called *'Resonant Frequency'*. At this condition,

$$Z = R + j\left(X_L \sim X_C\right) = R = 10 \text{ ohms(since } X_L = X_C)$$

$$\text{Current } I = \frac{V}{Z} = \frac{V}{R} = \frac{220}{10} = 22 \text{ A}$$

Since, net reactance is zero, the circuit is resistive. Hence, the power factor is 1 or Power factor $Cos\phi = \dfrac{R}{Z} = \dfrac{R}{R} = 1$

2.3 Three Phase Balanced Circuits

Three Phase circuits are used to consume or generate the power in huge quantities. The usage of three phase system instead of using three single phase systems provides many economic and operational advantages. The three phase systems are cheap, compact and efficient for a given power rating. Hence, three phase systems are universally adopted.

A 3-Phase system is having three voltages or currents of same amplitude and frequency, but mutually displaced in phase by 120^0. These 3 phase quantities say, voltages are represented by V_A, V_B and V_C. The instantaneous values of voltages are presented below by taking phase 'A' as reference.

$$v_a = V_m \, Sin\omega t$$

$$v_b = V_m \, Sin\left(\omega t - 120^0\right)$$

$$v_c = V_m \, Sin\left(\omega t - 240^0\right)$$

It is cleared from above expressions that all three phases have same frequency ω and the peak value V_m. The above alternating quantities are represented in polar form as follows.

$$V_a = V_A \angle 0^0$$

$$V_b = V_B \angle -120^0$$

$$V_C = V_C \angle -240^0$$

The wave forms of three phase voltages are shown below in figure 2.51.

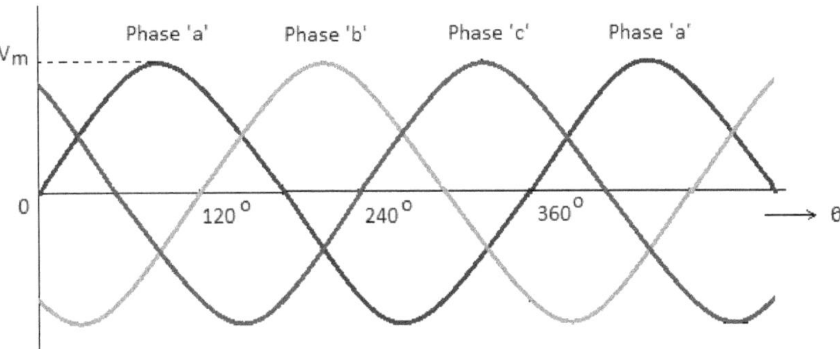

Figure 2.51 Three Phase Wave Forms

The phasor representation is shown in figure 2.52.

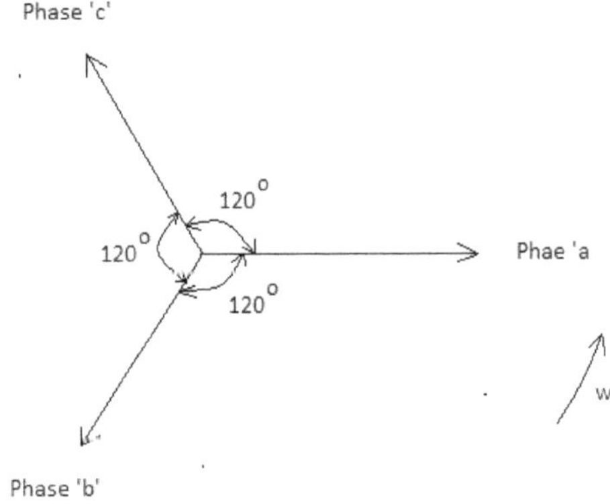

Figure 2.52 Phasors in a three phase system

Phase Sequence: The Phase sequence is the order in which individual phases occur or exist with respect to time. In the above case considered, phase 'b' voltage becomes maximum, 120^0 after the phase 'a' becomes maximum. Similarly, phase 'c' gets its peak value, 120^0 after the phase 'b' attains its maximum value. Hence, the phase sequence for a given 3-phase system is ABC.

Balanced 3-Phase system: If the magnitudes of all 3 phases are same and the mutual phase difference between them is 120^0, the system is called *'Balanced System'*. Otherwise, it is called *'Unbalanced System.'*

Let V_A, V_B and V_C be the voltages across three phases or windings of load or generator. These phases or windings are represented as shown in figure 2.53.

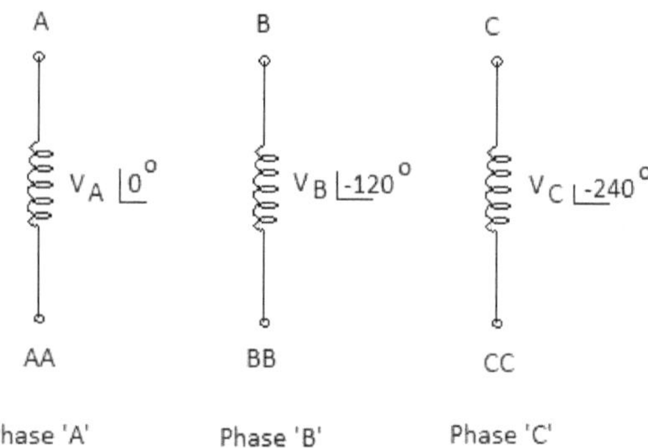

Figure 2.53 Three Phase System

There are two types of interconnecting these phases. 1. Star Connection and 2. Delta Connection.

2.3.1   Star Connection

In this Star connection, Similar polarity ends (say, either A, B, C or AA, BB, CC) of three phases are connected together to form a neutral point or Star point 'n'. It has been shown in figure 2.54. The other three terminals or lines are taken out from the 3- phase system.

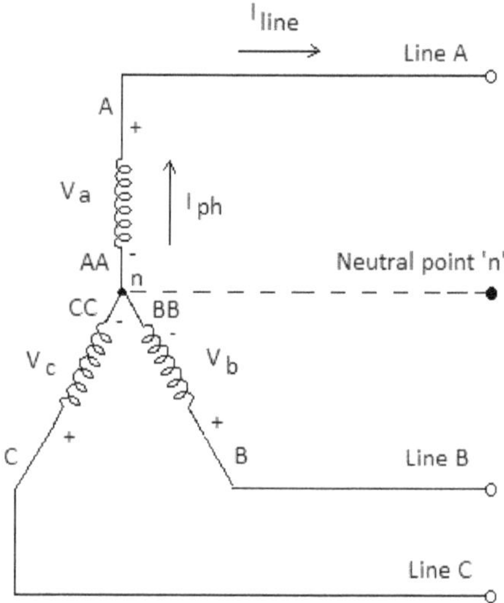

Figure 2.54 Star Connection

The voltage across each phase is called 'Phase Voltage'.

$$|V_a| = |V_b| = |V_c| = V_{ph} = \text{Phase Voltage}$$

Or, the phase voltage is the voltage between any line and neutral in a star connected system.

$$V_{Ph} = |V_{an}| = |V_{bn}| = |V_{cn}|$$

The line voltage between any two lines is called 'Line Voltage' V_L. The line voltage between the lines A and B is given by $V_{ab} = V_a - V_b$ (as we move from the point 'b' to point 'a' to obtain the voltage V_{ab}, the potential across the phase 'b' decreases by V_b. Hence, V_b is considered negative. The potential across phase 'a' rises by V_a. So, it taken as positive)

Similarly, $V_{ab} = V_a - V_b$

$$V_{bc} = V_b - V_c$$

$$V_{ca} = V_c - V_a$$

The line voltages are obtained from the phasor diagram.

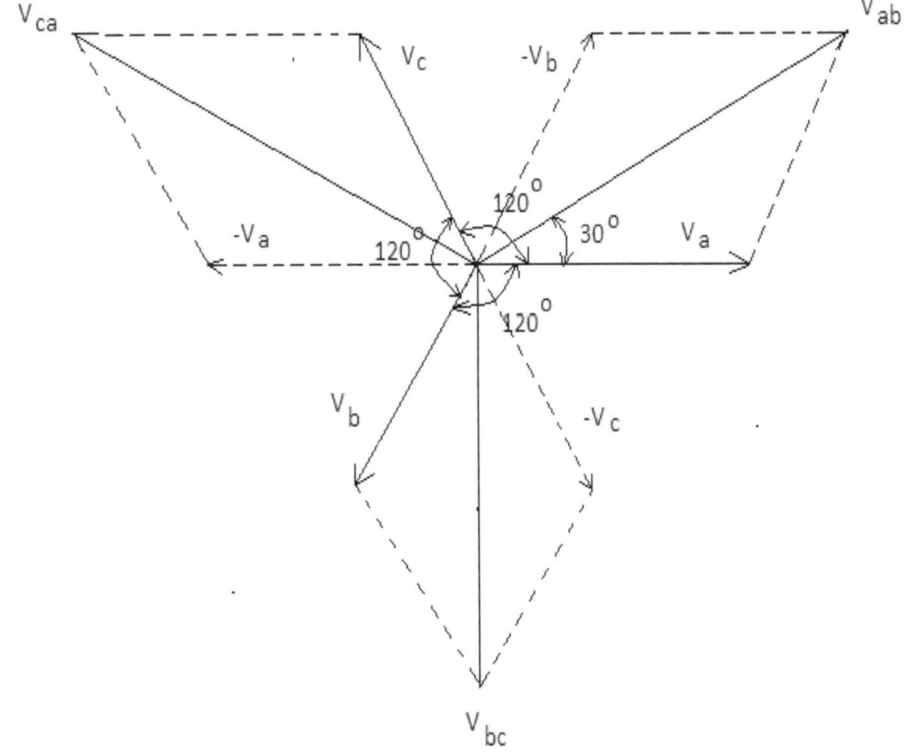

Figure 2.55 Phasor diagram of a three-phase star connected system

The line voltage V_{ab} is obtained using parallelogram law from above phase diagram (figure 2.55).

$$V_{ab} = \sqrt{V_a^2 + V_b^2 + 2V_a V_b Cos60^0}$$

$$= \sqrt{V_{ph}^2 + V_{ph}^2 + 2V_{ph}V_{ph}Cos60^0} \quad (\text{ since,} |V_a| = |V_b| = |V_c| = V_{ph})$$

$$= \sqrt{2V_{ph}^2 + 2V_{ph}^2 Cos60^0}$$

$$= \sqrt{2V_{ph}^2 \left(1 + Cos60^0\right)}$$

$$= \sqrt{2V_{ph}^2 \left(1 + \frac{1}{2}\right)}$$

$$= \sqrt{2V_{ph}^2 \left(\frac{3}{2}\right)}$$

$$= \sqrt{3} \times V_{ph}$$

Similarly, the other line voltages are given by $V_{bc} = V_{ca} = \sqrt{3} \times V_{ph}$

And also, the line voltages are balanced and displaced by 120^0. The line voltages lead the corresponding phase voltages by 30^0.

It is cleared from the star connection that the phase and line currents are same.

$$I_L = I_{Ph}$$

The following phasor diagram (figure 2.56) shows phase voltages and corresponding phase currents.

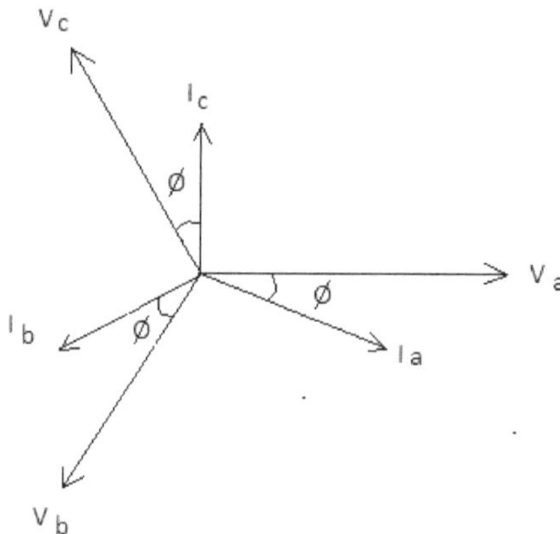

Figure 2.56 Phasor diagram of Phase Voltage and phase currents

The power factor of three phase circuit is $Cos \phi$. Where $\emptyset$ is the angle between V_{ph} and I_{ph}.

The average power consumed in each phase $P_{ph} = V_{ph} I_{ph} Cos\phi$

The total active power consumed by all three phases $P = 3V_{ph} I_{ph} Cos\phi$

$$= \sqrt{3} \times \sqrt{3} V_{ph} \times I_{ph} \times Cos\phi$$

$$= \sqrt{3} \times V_L \times I_L \times Cos\phi \text{ (Since } V_L = V_{ph} \text{ in Star connection)}$$

Similarly, the total reactive power consumed in the three phase circuit,

$$Q = \sqrt{3} \times V_L \times I_L \times Sin\phi$$

And the Apparent Power $S = \sqrt{3} \times V_L \times I_L$

2.3.2 Delta Connection

In this **D**elta connection, **D**issimilar polarity ends of phases are joined as shown in figure 2.57 and hence, a closed loop or Delta is formed. Say, AA is connected to B, BB to C and CC to A. It is evident from the diagram that no neutral is available.

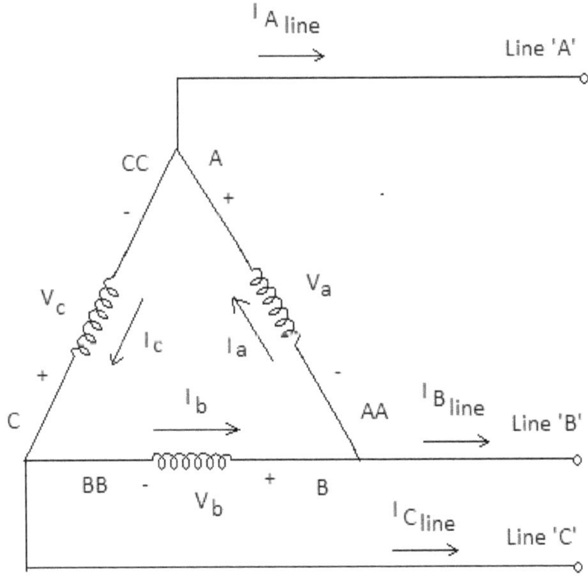

Figure 2.57 Delta Connection

It is also cleared that the voltage between any two lines is equal to voltage across the phase or winding which is placed between those lines.

Therefore, $V_L = V_{Ph}$

And $\quad V_{ab} = V_a$, $V_{bc} = V_b$ and $V_{ca} = V_c$

The relation between line currents and phase currents are obtained as follows.

$$I_{A/Line} = I_A - I_C$$

$$I_{B/Line} = I_B - I_A$$

$$I_{C/Line} = I_C - I_B$$

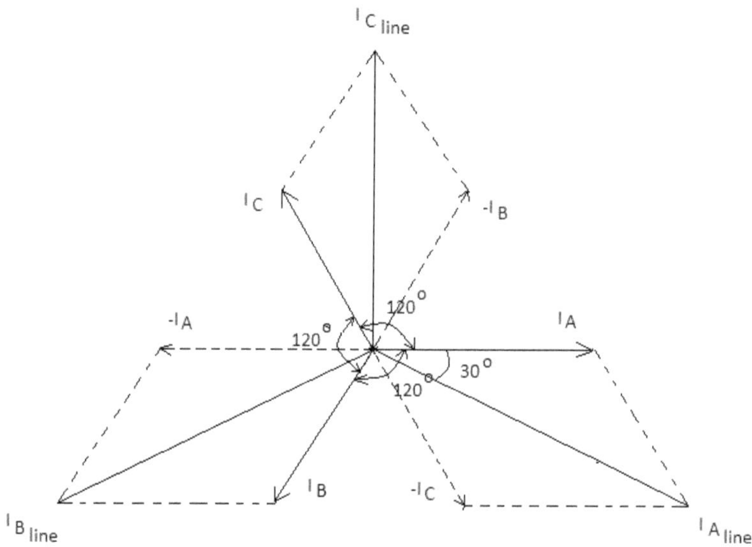

Figure 2.58 Phasor diagram of Delta Connected three phase system

By applying parallelogram law, the magnitude of line current

$$I_{A/line} = \sqrt{I_A^2 + I_B^2 + 2I_A I_B Cos60^0}$$

But $\qquad |I_A| = |I_B| = |I_C| = I_{ph}$

Then, $\qquad I_{A/line} = \sqrt{I_{ph}^2 + I_{ph}^2 + 2I_{ph}I_{ph}Cos60^0}$

$$= \sqrt{2I_{ph}^2 + 2I_{ph}^2 Cos60^0}$$

$$= \sqrt{2I_{ph}^2\left(1 + Cos60^0\right)}$$

$$= \sqrt{2I_{ph}^2\left(1 + \frac{1}{2}\right)}$$

$$= \sqrt{2I_{ph}^2\left(\frac{3}{2}\right)}$$

$$= \sqrt{3} \times I_{ph}$$

Therefore, $\left|I_{A/line}\right| = \left|I_{B/line}\right| = \left|I_{C/line}\right| = I_L$

In Delta Connection, $I_L = \sqrt{3} \times I_{ph}$

The line currents are also mutually displaced in phase by 120^0. The line currents lag the corresponding phase currents by 30^0.

If Ø is the phase angle between phase voltage and phase current, the active power consumed in a phase is given by

Active power/phase $P_{ph} = V_{ph}I_{ph}Cos\phi$

The total active power consumed by all three phases $P = 3V_{ph}I_{ph}Cos\phi$

$$= \sqrt{3}V_{ph} \times \sqrt{3} \times I_{ph} \times Cos\phi$$

$$= \sqrt{3} \times V_L \times I_L \times Cos\phi \quad (\text{since, } V_L = V_{Ph} \text{ and } I_L = \sqrt{3} \times I_{ph})$$

Similarly, the total Reactive Power consumed $Q = \sqrt{3} \times V_L \times I_L \times Sin\phi$

The Apparent Power $S = \sqrt{3} \times V_L \times I_L$

Exercise 2.3.1: A balanced star connected load of $(4 + j3)$ ohms per phase is connected to a balanced three phase 440V supply. Find i) Phase voltage ii) Phase current iii) Line current iv) Power factor v) Total Active Power vi) Total Reactive Power and vii) Total Apparent Power

Solution:

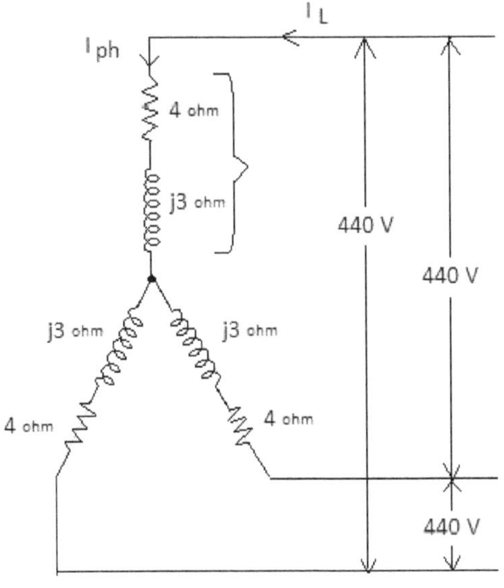

Figure 2.59

Line Voltage $V_L = 440\,\text{V}$

Per phase impedance $Z = R + jX = 4 + j3$

Hence, $R = 4\,\text{ohms}$ and $X = 3\,\text{ohms}$

(i) In star connection, Phase voltage

$$V_{Ph} = \frac{V_L}{\sqrt{3}} = \frac{440}{\sqrt{3}} = 254\,\text{V}$$

(ii) Phase impedance $Z = \sqrt{R^2 + X^2} = \sqrt{4^2 + 3^2} = 5\,\text{ohms}$

Phase current $I_{Ph} = \dfrac{V_{ph}}{Z_{ph}} = \dfrac{254}{5} = 50.8\,\text{A}$

(iii) Line current $I_L = I_{ph} = 50.8\,\text{A}$ (In star connection, $I_{Ph} = I_L$)

(iv) Power factor $Cos\phi = \dfrac{R}{Z} = \dfrac{4}{5} = 0.8$ lagging as the load is inductive

(v) Total Active Power $P = \sqrt{3} \times V_L \times I_L \times Cos\phi = \sqrt{3} \times 440 \times 50.8 \times 0.8 = 30.97\,\text{kW}$

(vi) Total Reactive Power $Q = \sqrt{3} \times V_L \times I_L \times Sin\phi = \sqrt{3} \times 440 \times 50.8 \times 0.6 = 23.22\,\text{kVAR}$

(vii) Total Apparent Power $S = \sqrt{3} \times V_L \times I_L = \sqrt{3} \times 440 \times 50.8 = 38.71\,\text{kVA}$

Exercise 2.3.2: A balanced Delta connected load of $2 + j3\,\text{ohms}$ per phase is connected to a balanced three phase 440 V supply. Find a) Phase voltage b) Phase current c) Line current d) Power factor e) Total Active Power f) Total Reactive Power and g) Total Apparent Power

Solution:

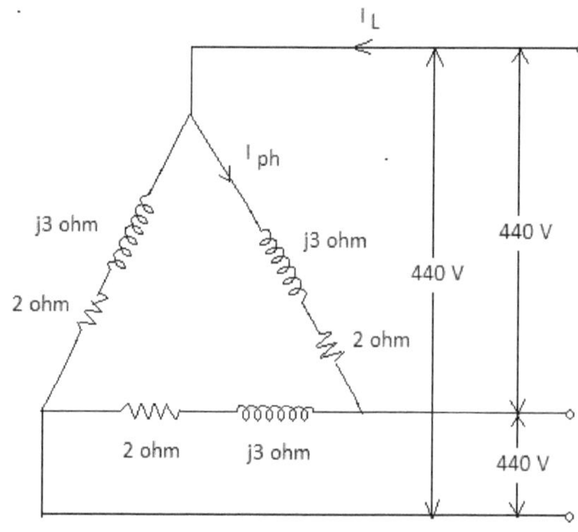

Figure 2.60

Balanced load has equal impedances in all three phases.

Line voltage $V_L = 440\,\text{V}$

Per phase impedance $Z = R + jX_L = 2 + j3$ ohms

Per Impedance magnitude $Z = \sqrt{R^2 + X^2} = \sqrt{2^2 + 3^2} = 3.6$ ohms

(a) Phase Voltage $V_{ph} = V_L = 440\,\text{V}$

(b) Phase current $I_{ph} = \dfrac{V_{ph}}{Z_{ph}} = \dfrac{440}{3.6} = 122.22\,\text{A}$

(c) Line current $I_L = \sqrt{3} \times I_{ph} = \sqrt{3} \times 122.22 = 211.6\,\text{A}$

(d) Power factor $Cos\phi = \dfrac{R}{Z} = \dfrac{2}{3.6} = 0.555$ lagging as the load is inductive

(e) Total Active Power $P = \sqrt{3} \times V_L \times I_L \times Cos\phi = \sqrt{3} \times 440 \times 211.6 \times 0.555 = 89.49\,\text{kW}$

(f) Total Reactive Power $Q = \sqrt{3} \times V_L \times I_L \times Sin\phi = \sqrt{3} \times 440 \times 211.6 \times 0.8317 = 134.12\,\text{kVAR}$

(g) Total Apparent Power $S = \sqrt{3} \times V_L \times I_L = \sqrt{3} \times 440 \times 211.6 = 161.25\,\text{kVA}$

Exercise 2.3.3: A three phase balanced delta connected load of $4 + j8$ per phase is connected across a 400 V, 3 phase supply. Determine the Phase current and Line currents. Also calculate the total power drawn from the source.

Solution:

Line Voltage $V_L = 400\,\text{V}$

Per phase Impedance $Z = R + jX_L = 4 + j8$ ohms

$$Z = \sqrt{R^2 + X^2} = \sqrt{4^2 + 8^2} = 8.94\,\text{ohms}$$

Phase Voltage $V_{Ph} = V_L = 400\,\text{V}$ (In delta connection)

(a) Phase Current $I_{ph} = \dfrac{V_{Ph}}{Z_{Ph}} = \dfrac{400}{8.94} = 44.74\,\text{A}$

(b) Line Current $I_L = \sqrt{3} \times I_{ph} = \sqrt{3} \times 44.74 = 77.48\,\text{A}$

(c) Power Factor $Cos\phi = \dfrac{R}{Z} = \dfrac{4}{8.94} = 0.447$ lagging as the load is inductive

(d) Power drawn $P = \sqrt{3} \times V_L \times I_L \times Cos\phi = \sqrt{3} \times 400 \times 77.48 \times 0.447 = 23.99\,\text{kW}$

Exercise 2.3.4: calculate the active and reactive current components in each phase of a star connected 5000 V, 3 phase alternator (AC source) supplying 1500 KW at 0.8 pf lagging. If the total current remains the same when the power factor of the load is increased to 0.95, find the new power.

Solution:

Line Voltage $V_L = 5000$ V

Active Power $P = 1500$ kW

Power Factor $Cos\phi = 0.8$ lagging (it means the load is inductive)

Total Active power $P = \sqrt{3} \times V_L \times I_L \times Cos\phi$

$$1500 \times 10^3 = \sqrt{3} \times 5000 \times I_L \times 0.8$$

$$I_L = \frac{1500 \times 10^3}{\sqrt{3} \times 5000 \times 0.8} = 216.5 \text{ A}$$

In star connection, $I_{ph} = I_L$. Hence, the phase current $I_{ph} = 216.5$ A

Active component of the phase current $= I_{ph} \times Cos\phi = 216.5 \times 0.8 = 173.2$ A

Reactive component of the phase current $= I_{ph} \times Sin\phi = 216.5 \times 0.6 = 129.9$ A

If the power factor is increased to 0.95 keeping the current constant,

$I_{ph} = 216.5$ A, $Cos\phi = 0.95$ and $V_L = 5000$ V (supply voltage does not change)

New Power $P_{new} = \sqrt{3} \times V_L \times I_L \times Cos\phi_{new} = \sqrt{3} \times 5000 \times 216.5 \times 0.95 = 1781.1$ kW

Exercise 2.3.5: A balanced star connected load of impedance $3 + j4$ ohms per phase is connected to a balanced 3 phase, 50 Hz supply with line voltage of 381 V. Write the expressions for instantaneous values of phase voltages and phase currents.

Solution:

Per phase Impedance $Z = R + jX = 3 + j4$ ohms

$$Z = \sqrt{R^2 + X^2} = \sqrt{3^2 + 4^2} = 5 \text{ ohms}$$

Power Factor $Cos\phi = \dfrac{R}{Z} = \dfrac{3}{5} = 0.6$ lagging

Then the power factor angle $\phi = Cos^{-1}(0.6) = 53.13^0$

The phase current lags its phase voltage by 53.13^0.

Line voltage $V_L = 381$ V

$$\text{Phase Voltage } V_{ph} = \frac{V_L}{\sqrt{3}} = \frac{381}{\sqrt{3}} = 219.97 \text{ V}$$

Frequency $f = 50$ Hz

Angular Frequency $\omega = 2\pi f = 2\pi \times 50 = 314$ rad/sec

$$\text{Phase Current } I_{ph} = \frac{V_{ph}}{Z_{ph}} = \frac{219.97}{5} = 44 \text{ A}$$

$$\text{Maximum value of Phase Voltage } V_m = \frac{219.97}{0.707} = 311.1 \text{ V}$$

$$\text{Maximum value of Phase Current } I_m = \frac{44}{0.707} = 62.2 \text{ A}$$

Expression for instantaneous values of phase voltages by taking phase 'a' as reference are:

$$v_a = V_m Sin\omega t = 311.1 Sin314t$$

$$v_b = V_m Sin\left(\omega t - 120^0\right) = 311.1 Sin\left(314t - 120^0\right)$$

$$v_c = V_m Sin\left(\omega t - 240^0\right) = 311.1 Sin\left(314t - 240^0\right)$$

Expression for instantaneous values of phase currents are:

$$i_a = I_m Sin\left(\omega t - \phi\right) = 62.2 Sin\left(314t - 53.13^0\right)$$

$$i_b = I_m Sin\left(\omega t - 120^0 - \phi\right) = 62.2 Sin\left(314t - 120^0 - 53.13^0\right)$$

$$= 62.2 Sin\left(314t - 173.13^0\right)$$

$$i_c = I_m Sin\left(\omega t - 240^0 - \phi\right) = 62.2 Sin\left(314t - 240^0 - 53.13^0\right)$$

$$= 62.2 Sin\left(314t - 293.13^0\right)$$

Exercise 2.3.6 : Three identical coils connected in star and takes a total power of 1.8 KW at 0.3 pf lagging from a 3 Phase, 500 V, 50 Hz supply. Calculate a) Line current b) VAR consumed c) Resistance and Inductance of the coil

Solution:

Line Voltage $V_L = 500$ V

Frequency $f = 50\,Hz$

Power factor $Cos\phi = 0.3$ lagging

Total power consumed $P = 18\,kW$

(a) Total Power input $P = \sqrt{3} \times V_L \times I_L \times Cos\phi$

$$1.8 \times 10^3 = \sqrt{3} \times 500 \times I_L \times 0.3$$

$$I_L = \frac{1.8 \times 10^3}{\sqrt{3} \times 500 \times 0.3} = 6.93\,A$$

(b) $Cos\phi = 0.3$ and hence, $Sin\phi = 0.954$

Total Reactive power consumed

$$Q = \sqrt{3} \times V_L \times I_L \times Sin\phi$$

$$= \sqrt{3} \times 500 \times 6.93 \times 0.954 = 5.725\,kVAR$$

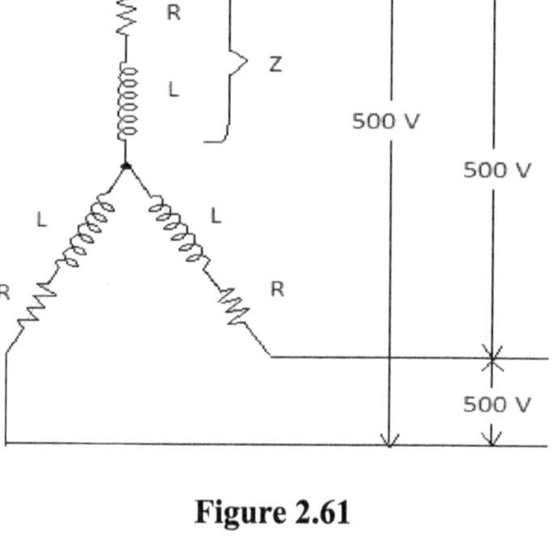

Figure 2.61

(c) Phase Voltage $V_{ph} = \dfrac{V_L}{\sqrt{3}} = \dfrac{500}{\sqrt{3}} = 288.68\,V$

Phase current $I_{Ph} = I_L = 6.93\,A$

Per Phase Impedance $Z = \dfrac{V_{Ph}}{I_{Ph}} = \dfrac{288.68}{6.93} = 41.6\,ohms$

The corresponding impedance triangle is

Resistance of the coil $R = Z \times Cos\phi = 41.6 \times 0.3 = 12.49\,ohms$

Inductive Reactance of the coil $X_L = Z \times Sin\phi = 41.6 \times 0.954 = 39.68$ ohms

$$X_L = 2\pi f L$$

$$39.68 = 2\pi f L$$

$$= 2\pi \times 50 \times L$$

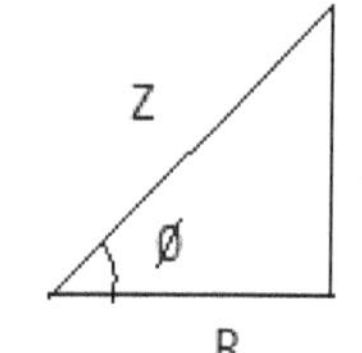

Figure 2.62

$$L = \frac{39.68}{2\pi \times 50} = 0.126\,H$$

Exercise 2.3.7 : Three identical coils are connected in star to a three phase 50 Hz supply. If the line current is 10 A, total power consumed is 12 KW and Volt-Ampere input is 15 KVA, find the line voltage V_L Phase voltage V_{ph}, VAR input, Resistance and Inductance of the coil

Solution:

In star connection, $I_L = I_{Ph} = 10\,A$

Active Power consumed $P = 12\,kW$

Apparent Power $S = 15\,kVA$

Total Power consumed $P = \sqrt{3} \times V_L \times I_L \times Cos\ \phi$

In this expression, $\sqrt{3} \times V_L \times I_L$ is called Apparent Power (S)

$$12000 = 15000 \times Cos\ \phi$$

$$Cos\ \phi = \frac{12000}{15000} = 0.8$$

And, Total Power consumed $P = \sqrt{3} \times V_L \times I_L \times Cos\ \phi$

$$12000 = \sqrt{3} \times V_L \times 10 \times 0.8$$

$$V_L = \frac{12000}{\sqrt{3} \times 10 \times 0.8} = 866\ \text{V}$$

Phase Voltage $V_{Ph} = \frac{V_L}{\sqrt{3}} = \frac{866}{\sqrt{3}} = 500\ \text{V}$ (star connection)

Reactive Power consumed $Q = \sqrt{3} \times V_L \times I_L \times Sin\ \phi$

$$= \sqrt{3} \times 866 \times 10 \times 0.6 = 8999$$

Impedance per phase $Z = \frac{V_{Ph}}{I_{Ph}} = \frac{500}{10} = 50$

Resistance of the coil $R = Z \times Cos\phi = 50 \times 0.8 = 40$

Inductive reactance of the coil $X_L = Z \times Sin\phi = 50 \times 0.6 = 30$

$$X_L = 2\pi f L$$

$$30 = 2\pi f L$$

$$= 2\pi \times 50 \times L$$

$$L = \frac{30}{2\pi \times 50} = 0.095\ \text{H}$$

Exercise 2.3.8: A balanced 3 phase star connected load of 100 KW takes a leading line current of 80 A with a line voltage of 1100 V, 50 Hz. Find the values of R and C per phase.

Solution:

Line Voltage $V_L = 1100\ \text{V}$

Frequency $f = 50\ \text{Hz}$

Line current $I_L = I_{Ph} = 80\ \text{A}$

Power consumed $P = 100\,\text{kW}$

Power consumed $P = \sqrt{3} \times V_L \times I_L \times Cos\ \phi$

$$100 \times 10^3 = \sqrt{3} \times 1100 \times 80 \times Cos\ \phi$$

$$Cos\ \phi = \frac{100 \times 10^3}{\sqrt{3} \times 1100 \times 80} = 0.656\ \text{leading}$$

Phase Voltage $V_{Ph} = \dfrac{V_L}{\sqrt{3}} = \dfrac{1100}{\sqrt{3}} = 635.1\,\text{A}$

Per phase impedance $Z_{ph} = \dfrac{V_{Ph}}{I_{Ph}} = \dfrac{635.1}{80} = 7.398\ \text{ohms}$

$$Z = R - jX_C$$

The impedance triangle is shown below

$$R = Z \times Cos\phi = 7.938 \times 0.656 = 5.2\ \text{ohms}$$

$$X_C = Z \times Sin\phi = 7.938 \times 0.754 = 6\ \text{ohms}$$

$$X_C = \frac{1}{2\pi fC}$$

$$6 = \frac{1}{2\pi \times 50 \times C}$$

$$C = \frac{1}{1884} = 0.53\,\text{mF}$$

Exercise 2.3.9 : A 440 V, 50 Hz, three phase supply is applied to a balanced delta connected load, each phase comprising of a resistor of 40 ohms and a capacitor of 40 μF in series. Calculate a) the line and phase currents b) total active power c) total reactive power d) total apparent power

Solution:

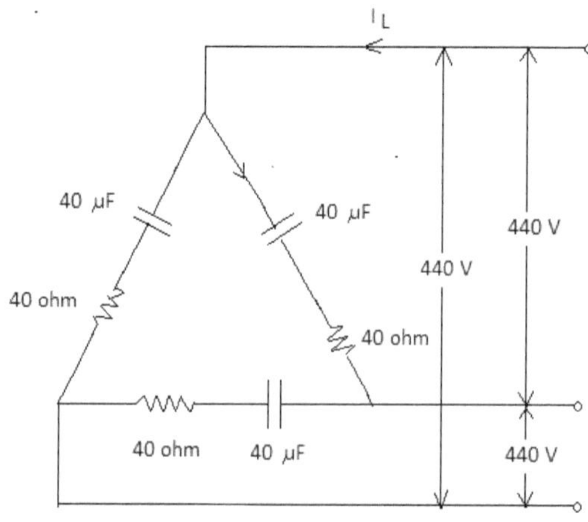

Figure 2.63

In delta connection, $V_{Ph} = V_L = 440$ V

Resistance of each phase $R = 40$ ohms

Capacitive Reactance of each phase $X_C = \dfrac{1}{2\pi fC} = \dfrac{1}{2\pi \times 50 \times 40 \times 10^{-6}} = 79.6$ ohms

(a) Per phase impedance $Z = R - jX_C$ ohms

Or $Z = \sqrt{R^2 + X_C^2} = \sqrt{40^2 + 79.6^2} = 89$ ohms

Phase current $I_{Ph} = \dfrac{V_{Ph}}{Z_{Ph}} = \dfrac{440}{89} = 4.94$ A

Line current $I_L = \sqrt{3} \times I_{Ph} = \sqrt{3} \times 4.94 = 8.56$ A

(b) Power factor $Cos\phi = \dfrac{R}{Z} = \dfrac{40}{89} = 0.449$ and hence, $Sin\phi = 0.893$

Total active power $P = \sqrt{3} \times V_L \times I_L \times Cos\ \phi = \sqrt{3} \times 440 \times 8.56 \times 0.449 = 2.932$ kW

(c) Total Reactive power $Q = \sqrt{3} \times V_L \times I_L \times Sin\ \phi = \sqrt{3} \times 440 \times 8.56 \times 0.893 = 5.828$ kVAR

(d) Total Apparent Power $S = \sqrt{3} \times V_L \times I_L = \sqrt{3} \times 440 \times 8.56 = 6.523$ kVA

Exercise 2.3.10: A delta connected balanced 3 phase load is supplied from a 3 phase 500 V supply. The line current is 25A and the power drawn by the load is 15 KW. Calculate the impedance of each branch of the load in complex form.

Solution:

Line voltage $V_L = 500$ V

Line Current $I_L = 25$ A

Power consumed $P = 15$ kW

Total active power $P = \sqrt{3} \times V_L \times I_L \times Cos\ \phi$

$$15 \times 10^3 = \sqrt{3} \times 500 \times 25 \times Cos\ \phi$$

$$Cos\ \phi = \dfrac{15 \times 10^3}{\sqrt{3} \times 500 \times 25} = 0.6928$$

And hence, $Sin\phi = 0.721$

$$I_{Ph} = \dfrac{I_L}{\sqrt{3}} = \dfrac{25}{\sqrt{3}} = 14.43 \text{ A}$$

Therefore $Z = \dfrac{V_{Ph}}{I_{Ph}} = \dfrac{500}{14.43} = 34.65$ ohms

Complex form of $Z = R + jX$

$$R = Z \times Cos\phi = 34.65 \times 0.6928 = 24 \text{ ohms}$$

$$X_L = Z \times Sin\phi = 34.65 \times 0.721 = 24.98 \text{ ohms}$$

Therefore the complex form of Z is $Z = 24 + j24.98$ ohms

Exercise 2.3.11: There are 3 identical coils each having a resistance of 30 ohms and inductance of 0.127H connected across 500V, 50 Hz, 3 phase supply. Find the current flowing through each coil and total power consumed when these coils are connected in a) Star and b) Delta. Make the conclusions.

Solution:

Resistance $R = 30$ ohms

Inductance $L = 0.127$ H

Inductive Reactance $X_L = 2\pi fL = 2\pi \times 50 \times 0.127 = 39.87$ ohms

(a) When the coils are placed in star

$$V_{Ph} = \frac{V_L}{\sqrt{3}} = \frac{500}{\sqrt{3}} = 288.68 \text{ V}$$

Impedance per phase

$$Z = \sqrt{R^2 + X_L^2} = \sqrt{30^2 + 39.87^2} = 49.88 \text{ ohms}$$

Current flowing through each coil is the phase current.

$$I_{Ph} = \frac{V_{Ph}}{Z_{Ph}} = \frac{288.68}{49.88} = 5.77 \text{ A}$$

Power factor $Cos\phi = \dfrac{R}{Z} = \dfrac{30}{49.88} = 0.601$

Total power consumed $P = \sqrt{3} \times V_L \times I_L \times Cos\ \phi = \sqrt{3} \times 500 \times 5.77 \times 0.607 = 3 \text{ kW}$

(b) When the coils are placed in Delta:

$$V_{Ph} = V_L = 500 \text{ V}$$

Current flowing through each coil $I_{Ph} = \dfrac{V_{Ph}}{Z_{Ph}} = \dfrac{500}{49.88} = 10 \text{ A}$

Line current $I_L = \sqrt{3} \times I_{Ph} = \sqrt{3} \times 10 = 17.32 \text{ A}$

Total power consumed $P = \sqrt{3} \times V_L \times I_L \times Cos\ \phi = \sqrt{3} \times 500 \times 17.32 = 9 \text{ kW}$

Conclusions:

- Power factor does not change from Star to Delta connection as the coil parameters R and L do not change with connection
- The coil currents increase by √3 times from Star to Delta connection
- The power consumption increases by 3 times as the connection changes from Star to Delta.

Review Questions

Short Answer (Two marks) Questions

1. What are the advantages of sinusoidal wave over other alternating waves?

2. Define Cycle and Time period of an alternating quantity

3. How do you represent the instantaneous value of an alternating quantity in terms of its maximum value and frequency?

4. Show the voltage and current phasors in a pure resistive , inductive and capacitive circuits

5. Find the phase angle between voltage and current whose instantaneous values are represented as $v = 250 Sin\left(314t - 15^0\right)$ and $i = 10 Sin\left(314t + 45^0\right)$ *(Ans: 60⁰)*

6. Define Form Factor and Peak Factor

7. Express the phasor $8 - j6$ in Polar and Exponential forms *(Ans: $10\angle -36.86^0$, $10e^{-j36.86^0}$)*

8. Does a Pure Inductor consume active power? Why?

9. Show the impedance triangle of series RC circuit

10. Differentiate between Star and Delta Connections

11. What is the phase difference between any i) two phase voltages and ii) two line voltages?

Essay (Six marks) Questions

1. The maximum value of a Sinusoidal alternating voltage whose frequency is 50 Hz is 300V. Find the voltage at the instants (i) t = 5 msec   (ii) t = 10 m sec and (iii) t = 12 msec.

   (Ans: 300V, 0V, -175.87V)

2. Derive the expressions for Average and RMS values of a Sinusoidal alternating quaintly

3. Discuss on the analysis an AC circuit having a Pure Resistance and Inductance in series

4. Is there any possibility for a series RLC circuit to operate at unity power factor? Explain

5. In a series R-L-C circuit, R = 6 ohms, L = 0.02 H and C = 300 µF is connected to 230V, 50Hz AC supply. Find (i) the supply current (ii) the voltage across each element and (iii) power factor angle Also draw the phasor diagram. *(Ans: 31.08A, 186.5V, 195.18V, 329.75V, 54.17⁰)*

6. Obtain the relation between Phase and Line values of voltage and currents in a balanced three delta connected system.

CHAPTER 3

Transformers and DC Machines

3.1 Transformers

Transformer is a static machine which changes the voltage levels from one to other without changing frequency. It transforms the electrical energy through magnetic field between two circuits which are at different voltages.

Uses of transformers:

1. It is used in Power systems for changing voltage levels.

2. Transformers are used to reduce the transmission losses by increasing the transmission voltages. As the transmission voltage increases, the current through the transmission lines decreases for a given power, the losses are thus reduced.

3. Used to couple different stages in cascaded circuits like amplifiers

4. Used to maintain the electrical isolation between the circuits

5. Used to allow the flow of AC and to block DC flow

3.1.1 Working Principle

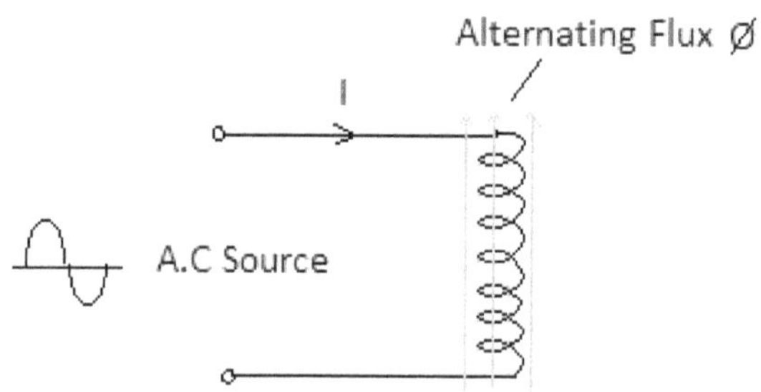

Figure 3.1 Alternating flux produced by AC

Consider a coil connected to an Alternating Current Source as shown in figure 3.1. As the current flowing through the coil changes with respect to time, it produces an alternating flux. The space around the coil is filled with alternating flux. If another coil having N_2 number of turns is placed nearer to it so that the axes of two windings coincide each other, it links with the flux produced by first coil. It is depicted in figure 3.2. As the flux linkages change, an emf E_2 is induced in it. This emf is proportional to the rate of change of flux linkages. This alternating flux also induces an emf of E_1 in the first coil.

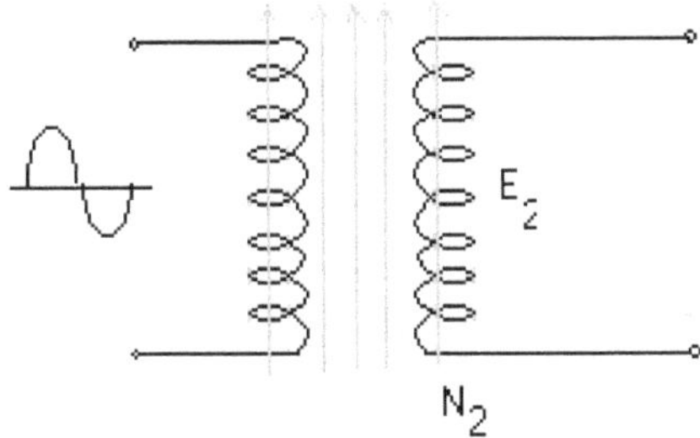

Figure 3.2 Transformer Principle

To see that maximum of flux produced by the first coil reaches the second winding or to arrest the leakage of magnetic flux, a good magnetic circuit is provided between these two windings. Hence, Iron core is placed between two windings as shown in figure 3.3 below. The iron core carries the flux Ø which is of alternating nature. Hence, a transformer works on principle of '*Electro-magnetic induction*'.

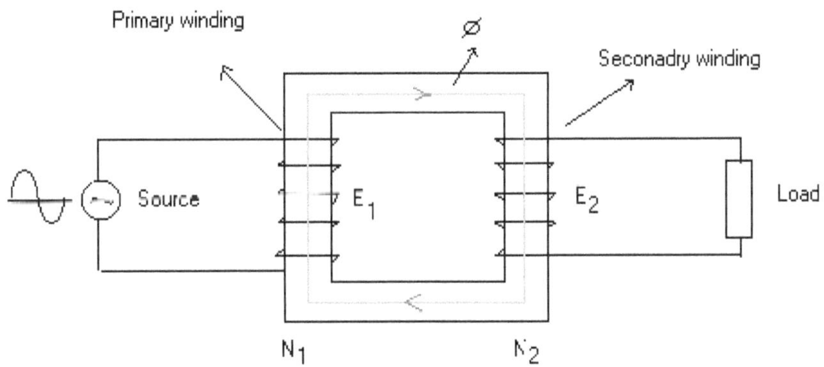

Figure 3.3 Transformer with an iron core

The winding which is connected to source is called '*Primary Winding*' and the winding which is connected to load is called '*Secondary Winding*'. E_1 and E_2 are the emfs induced in primary and Secondary windings respectively by the flux Φ. If $N_2 > N_1$, then, $E_2 > E_1$, then the transformer is called '*Step-Up transformer*'. Similarly, if $N_2 < N_1$, then, $E_2 < E_1$, then the transformer is called '*Step-Down transformer*'. If $N_2 = N_1$, that transformer is called '*Unit transformer*' which is exclusively used for electrical isolation. A Step-Up transformer can be used as a Step-down transformer or vice-versa.

Transformer does not work on DC supply as it does not produce alternating flux. The direct current maintains the magnitude and direction constant with respect to time.

3.1.2 Brief Construction of Transformers

The main parts of a transformer are 1. Primary Winding 2.Secondary windings and 3. Iron Core. The figure 3.4 gives an elementary idea of construction of a transformer. The primary and secondary windings which are made with copper carry the electric current continuously and produces some losses. These losses are called *Electrical Losses or Copper Losses*. Similarly, the Iron core carries magnetic flux and causes for some power loss in it. This loss is called *Magnetic Lossor Iron Loss*. The heat produced by these losses may deteriorate the insulation of transformer. Hence, the transformer construction must ensure the efficient removal of heat produced. The transformer oil is used to provide the necessary cooling. The transformer oil also acts as insulation between the windings and winding and iron core.

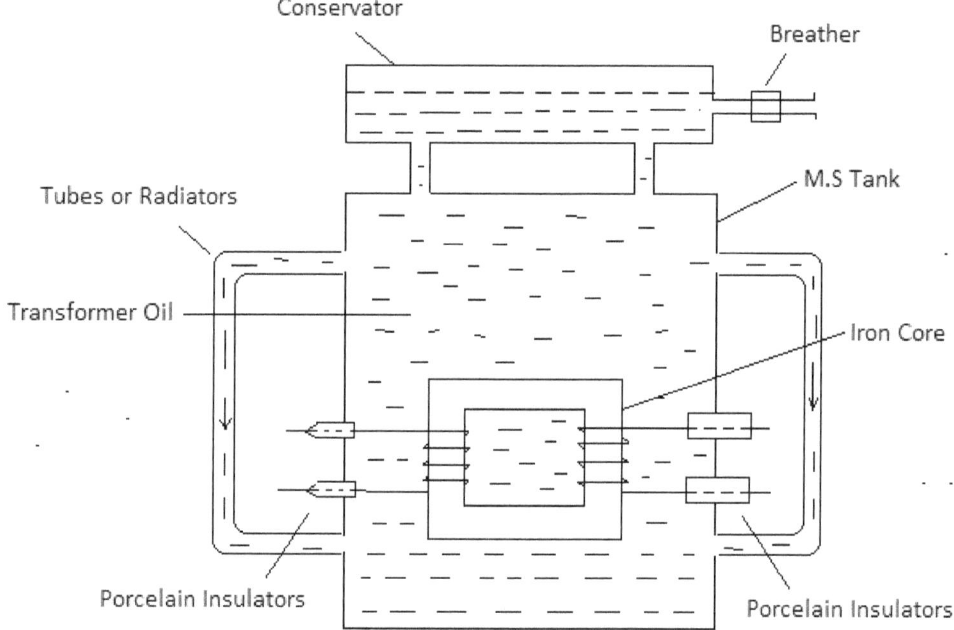

Figure 3.4 Basic Construction of a Transformer

The Iron core is made up of Silicon Steel laminations to reduce the Iron losses (mainly eddy current losses). The purpose of radiators is to cool the oil by exposing it open atmosphere. The container or tank is made up of Mild Steel. The Porcelain insulators called Bushings are used to bring out live terminals from primary and secondary windings.

3.1.3 Emf Equation

Consider a transformer having N_1 turns in primary winding and N_2 turns in secondary winding. Let the flux passing through the core be $\emptyset$ webers.

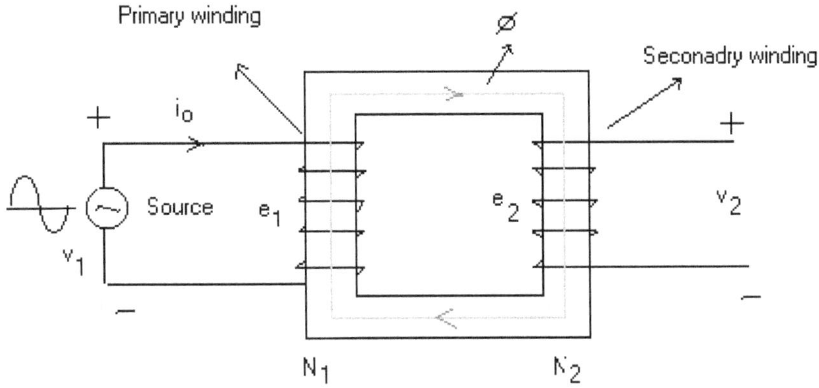

Figure 3.5 Transformer

Let Ψ_1 be the flux linkages with primary winding $= N_1\phi$

Let Ψ_2 be the flux linkages with secondary winding $= N_2\phi$

As the primary winding is excited by a sinusoidal alternating voltage source, the flux produced in the core is also sinusoidal. Therefore, the flux in the core at any time instant 't' is

$\phi = \phi_m Sin\omega t$

The instantaneous value of emf induced in primary winding $e_1 = \dfrac{d\psi_1}{dt} = N_1 \dfrac{d\phi}{dt}$

$$= N_1 \frac{d}{dt}\left(\phi_m Sin\omega t\right) = N_1\phi_m\omega\left[Cos\omega t\right]$$

$$= N_1\phi_m\omega Sin\left(\omega t - 90^0\right)$$

It is cleared from above equation that the maximum value of emf induced in primary winding is $E_{1_{max}} = N_1\phi_m\omega$

And the RMS value of emf induced in primary winding $E_1 = E_{1_{Max}} \times Form\ Factor$

Form Factor is 0.707 for Sine waves

Then, $E_1 = N_1\phi_m\omega \times 0.707$

$\qquad = 2\pi f \times N_1\phi_m \times 0.707$ (since $\omega = 2\pi f$)

Or $E_1 = 4.44\phi_m f N_1$

As the winding resistances are neglected, $E_1 = V_1$

$\qquad E_1 = V_1 = 4.44\phi_m f N_1$ $\hspace{4cm}$ .....(1)

Similarly, the RMS value of emf induced in Secondary winding is given by

$$V_2 = E_2 = 4.44\phi_m f N_2 \qquad \qquad .....(2)$$

By dividing equation 2 with 1, we get

$$\frac{V_2}{V_1} = \frac{E_2}{E_1} = \frac{4.44\phi_m f N_2}{4.44\phi_m f N_1} = \frac{N_2}{N_1} = k$$

Where 'k' is called *Transformation Ratio.*

And it is also cleared from above equation that

$$\frac{V_2}{V_1} = \frac{E_2}{E_1} = \frac{N_2}{N_1} \qquad \text{or} \qquad \frac{V_1}{N_1} = \frac{V_2}{N_2}$$

It is implied that the Volt per Turn on primary is equal to Volt per Turn on secondary. If primary winding having N_1 turns is given a voltage of V_1 volts, the volt per turn value will be fixed at $\dfrac{V_1}{N_1}$ and this value must be maintained same on secondary side also. The secondary voltage is given by

$$V_2 = N_2 = No.\ of\ Secondary\ turns \times Volt\ per\ turn$$

$$= N_2 \times \frac{V_1}{N_1}$$

$$= k \times V_1$$

This is the reason, a transformer having N_2 value greater than N_1 supplies more voltage on secondary side than primary and hence, it is called '*Step-Up transformer*' and similarly a transformer whose N_2 value is less than N_1 acts as '*Step-Down transformer*'.

As the transformer retains the power rating same during the transformation, the power or kVA value is maintained same on both primary and secondary sides.

KVA on Primary side = KVA on Secondary side

$$1000 \times V_1 \times I_1 = 1000 \times V_2 \times I_2$$

Or $\qquad V_1 \times I_1 = V_2 \times I_2$

$$\frac{I_1}{I_2} = \frac{V_2}{V_1} = k$$

Exercise 3.1.1: A 230/460 V transformer is excited by a 230 V, 50 Hz AC supply on primary side. The secondary winding is having 240 number of turns. Find the number of primary winding turns and the maximum value of operating flux in the core.

Solution :

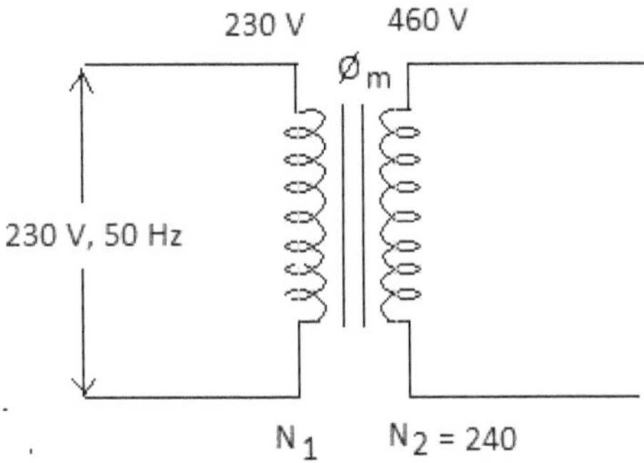

Figure 3.6

Transformation Ratio $k = \dfrac{V_2}{V_1} = \dfrac{460}{230} = 2$

The number of secondary turns $N_2 = 240$

The number of primary turns $N_1 = ?$

Since, $k = \dfrac{N_2}{N_1} = \dfrac{240}{N_1} = 2$ or $N_1 = 120$

Since, $V_1 = 4.44 \phi_m f N_1$

$$\phi_m = \dfrac{V_1}{4.44\, f N_1} = \dfrac{230}{4.44 \times 50 \times 120} = 0.0086 \text{ wb or } 8.6 \text{ mwb}$$

Exercise 3.1.2 : A 200 KVA, 11000/415 V, 50 Hz Single Phase Transformer has 2120 turns of primary winding. Calculate the number of secondary turns, full load secondary current, primary current and maximum value of core flux.

Solution:

Primary Voltage $V_1 = 11000$ V

Secondary Voltage $V_2 = 415$ V

Supply frequency $f = 50$ Hz

Number of primary turns $N_1 = 2120$

Transformation Ratio $k = \dfrac{N_2}{N_1} = \dfrac{V_2}{V_1} = \dfrac{415}{11000} = 0.03773$

Since, $\dfrac{N_2}{2120} = 0.03773$ or $N_2 = 0.03773 \times 2120 = 79.98$

As the number of turn should be an integer, we take the number of secondary turns as

$$N_2 = 80$$

The capacity of the transformer = 200 kVA (kilo Volt Amperes)

The full load primary current $I_1 = \dfrac{kVA \times 1000}{V_1} = \dfrac{200 \times 1000}{11000} = 18.18 \text{ A}$

The full load secondary current $I_2 = \dfrac{kVA \times 1000}{V_2} = \dfrac{200 \times 1000}{415} = 481.9 \text{ A}$

The primary voltage $\qquad V_1 = 4.44 \phi_m f N_1 \quad (\text{or} \quad V_2 = 4.44 \phi_m f N_2)$

$$11000 = 4.44 \times \phi_m \times 50 \times 2120$$

Or $\qquad \phi_m = \dfrac{11000}{4.44 \times 50 \times 2120} = 0.0233 \text{ wb}$

The ϕ_m value can also be obtained using secondary voltage

$$\phi_m = \dfrac{V_2}{4.44 \, f N_2} = \dfrac{415}{4.44 \times 50 \times 80} = 0.0233 \text{ wb}$$

Exercise 3.1.3 : In a 220/3000 Volt, 50 Hz Single Phase transformer, the maximum flux density is 2 wb/m². The emf per turn is 10 V. Determine a) Number of primary and secondary turns b) cross sectional area of the core.

Solution :

Primary Voltage $V_1 = 220 \text{ V}$

Secondary Voltage $V_2 = 3000 \text{ V}$

Frequency $f = 50 \text{ Hz}$

Volt per turn on Primary = Volt per turn on secondary

$$\dfrac{V_1}{N_1} = \dfrac{220}{N_1} = 10 \quad \text{or} \quad N_1 = \dfrac{220}{10} = 22$$

And also, $\qquad \dfrac{V_2}{N_2} = \dfrac{3000}{N_2} = 10 \quad \text{or} \quad N_2 = \dfrac{3000}{10} = 300$

a) $V_1 = 4.44 \phi_m f N_1$

$$220 = 4.44 \times \phi_m \times 50 \times 22 \quad \text{or} \quad \phi_m = \dfrac{220}{4.44 \times 50 \times 22} = 0.04504 \text{ wb}$$

Maximum Flux Density $B_m = 2 \text{ wb/m}^2$

Maximum flux Density $B_m = \dfrac{\phi_m}{A}$

$$2 = \dfrac{0.04504}{A}$$

or $\qquad A = \dfrac{0.04504}{2} = 0.02252 \text{ m}^2$

Exercise 3.1.4 : A 20 KVA Transformer has 600 turns on the primary and 60 turns on the secondary. The primary is connected to the 300 V, 50 Hz supply. Calculate a) the emf induced in the secondary winding b) the full load primary and secondary currents c) the maximum flux in the core.

Solution:

Capacity of the transformer $= 20\,\text{kVA}$ (kilo Volt Amperes)

Number of primary turns $N_1 = 600$

Number of secondary turns $N_2 = 60$

Primary Voltage $V_1 = 300\,\text{V}$

Frequency $f = 50\,\text{Hz}$

(a) Transformation ratio $k = \dfrac{N_2}{N_1} = \dfrac{60}{600} = 0.1$

Also, $k = \dfrac{V_2}{V_1} = \dfrac{V_2}{300} = 0.1$ or $V_2 = 300 \times 0.1 = 30\,\text{V}$

(b) The full load primary current $I_1 = \dfrac{kVA \times 1000}{V_1} = \dfrac{20 \times 1000}{300} = 66.67\,\text{A}$

The full load secondary current $I_2 = \dfrac{kVA \times 1000}{V_2} = \dfrac{20 \times 1000}{30} = 666.67\,\text{A}$

(c) $V_1 = 4.44\phi_m f N_1$

$$300 = 4.44 \times \phi_m \times 50 \times 600$$

$$\phi_m = \dfrac{300}{4.44 \times 50 \times 600} = 0.00225\,\text{wb}$$

3.1.4 Transformer on No-load

When transformer is operating under no load condition ie., its secondary is open circuited and not supplying any output, Primary winding also shall not take any current under ideal case. But, in practice, it absorbs a minimum current called *'No-load current(I_o)'* and it is shown in figure 3.7.

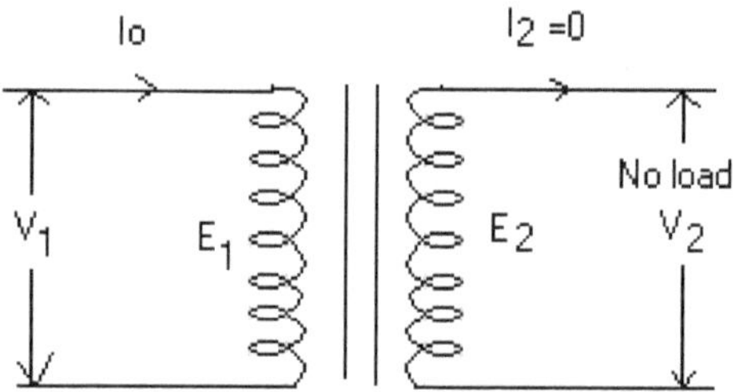

Figure 3.7 Transformer on No-Load

The purpose of this current is to establish the working flux Ø in the core. This flux producing current I_0 has to lag the supply voltage V_1 by 90^0. But in a practical transformer, there are some magnetic flux related losses called Iron losses present in the core. To meet these losses, the current I_0 lags the voltage V_1 by angle $Ø_0$ which is slightly less than 90^0 so that power consumed is not equal to zero.   (Otherwise, the active power consumed V_1 Io Cos $90°$ would have become zero).

The No-load Current I_o is resolved into two components. The component I_m is called *'Magnetizing Component'* and it produces the magnetic flux.

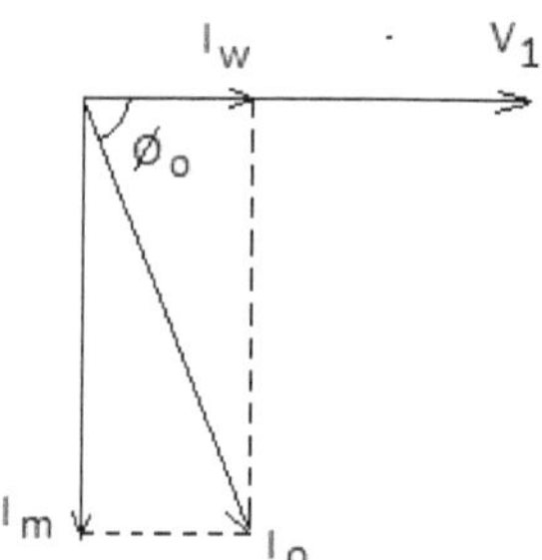

Figure 3.8 Phasor diagram under No-load

$$I_m = I_0 Sin\phi_0$$

And the component I_w is called *'Iron Loss component'* and it supplies the iron losses of core.

$$I_w = I_0 Cos\phi_0$$

These two components are shown in equivalent circuit in figure 3.9.

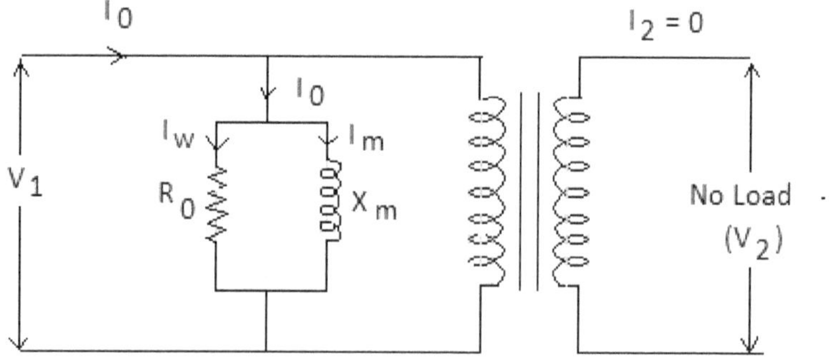

Figure 3.9 Transformer Equivalent Circuit

*Exercise 3.1.5:*A transformer takes a current of 2 A and consumes a power of 110W on no-load when it is excited from 220 V, 50 Hz supply. Find a) No-load power factor b) Magnetizing and Loss component of no-load current c) magnetizing reactance and iron loss equivalent resistance.

Solution:

The Primary voltage $V_1 = 220$ V

Frequency $f = 50$ Hz

No-load current $I_0 = 2$ A

No-load Power $W_0 = 110$ W

(a) No-load power input $W_0 = V_1 \times I_0 \times Cos\phi_0$

$$110 = 220 \times 2 \times Cos\phi_0$$

$$Cos\phi_0 = \frac{110}{220 \times 2} = 0.25 \text{ lagging}$$

(b) $Cos\phi_0 = 0.25$ and hence, $Sin\phi_0 = 0.968$

Magnetizing component $I_m = I_0 Sin\phi_0 = 2 \times 0.968 = 1.936$ A

Iron loss component $I_w = I_0 Cos\phi_0 = 2 \times 0.25 = 0.5$ A

(c) Magnetizing reactance $X_m = \dfrac{V_1}{I_m} = \dfrac{220}{1.936} = 113.63$ ohms

Iron loss equivalent resistance $R_0 = \dfrac{V_1}{I_w} = \dfrac{220}{0.5} = 440$ ohms

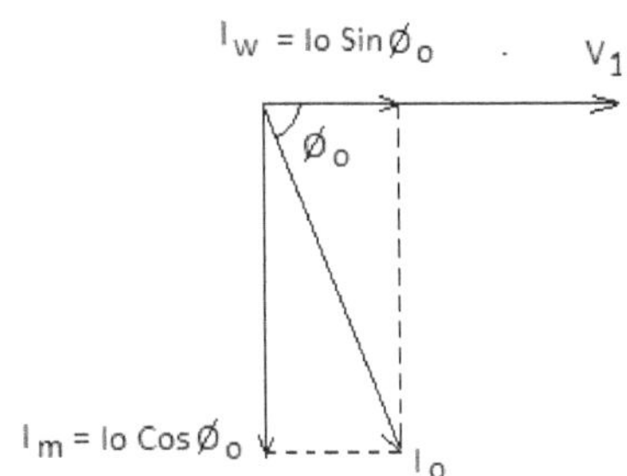

Figure 3.10

3.1.5   Ideal Transformer

The qualities of an Ideal Transformer are

(a) To have zero resistances on Primary and Secondary windings

(b) To have zero magnetic leakage flux between Primary and Secondary windings. Hence, the leakage reactances are assumed to be zero.

(c) To have zero reluctance for iron core

(d) To have zero copper losses and zero iron losses and hence to offer 100% efficiency

(e) To not produce any heat within the transformer

But, Ideal Transformers do not exist in reality.

3.1.6   Real Transformer or Practical Transformer

A Real or Practical Transformer is having

(a) Finite values of resistance in primary and secondary windings and these resistances are represented by R_1 and R_2.

(b) Finite value of flux leakage. The voltage that got reduced due to flux leakage is represented as voltage drops in leakage reactances X_1 and X_2.

(c) Finite value of reluctance in the iron core

(d) Finite values of copper losses and iron losses and hence does not offer 100% efficiency

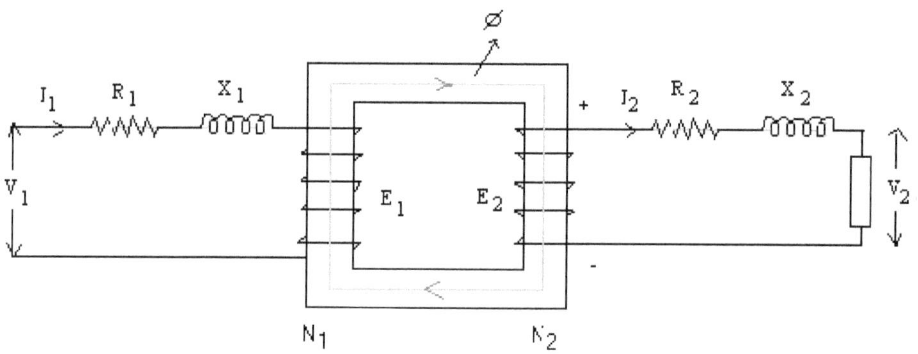

Figure 3.11 Practical Transformer

The equivalent circuit diagram of a real transformer is shown below in figure 3.12.

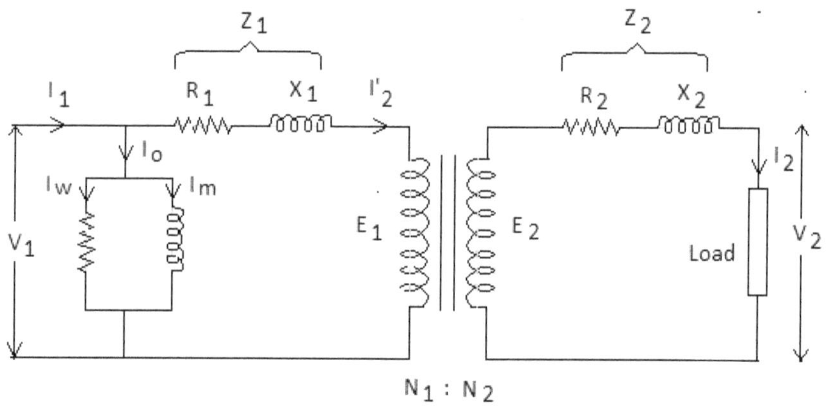

Figure 3.12 Complete equivalent circuit of a transformer

When a load is connected across the secondary winding, the load draws a current of I_2 from the secondary winding. This current must be supplied by the primary winding which is connected to the source. The current reflected from secondary to primary winding is I_2' and it is related to I_2 as

$$I_2' = k \times I_2$$

Where 'k' is the transformation ratio

The primary current I_1 is the phasor sum of this reflected current I₂' and the no load current I_o.

$$I_1 = I_0 + I_2'$$

If V_1 is the supply voltage on the primary side, the primary induced voltage is given by

$$E_1 = V_1 - I_1 Z_1$$

Where Z_1 is the primary leakage impedance $Z_1 = R_1 + jX_1$

The induced voltage in secondary winding $E_2 = k \times E_1$

And the secondary terminal voltage $V_2 = E_2 - I_2 Z_2$

Where Z_2 is the secondary leakage impedance $Z_2 = R_2 + jX_2$

3.1.7 Transformer Equivalent Circuit

The figure 3.13 shows the complete equivalent circuit diagram of transformer.

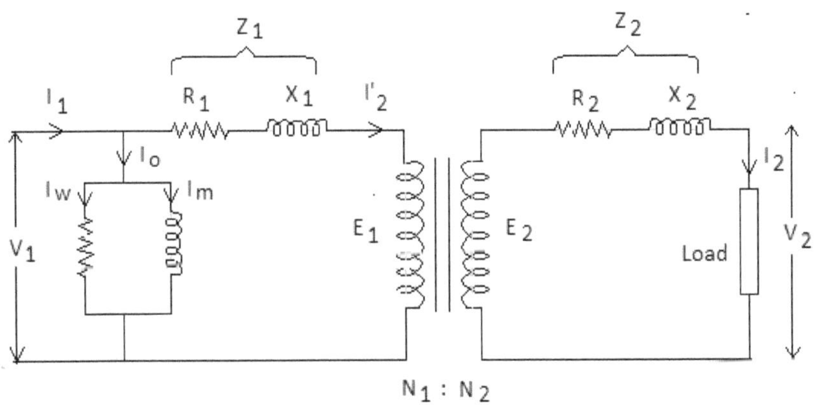

Figure 3.13 Complete equivalent circuit diagram

To make the analysis of the transformer simple, the primary circuit parameters are shifted to secondary side or secondary parameters are shifted to primary side. When the primary winding resistance R_1 and leakage reactance X_1 are shifted to secondary side, their equivalent values are represented by R_1' and X_1' respectively. Similarly, when the secondary parameters are moved to primary side, their equivalent values are given by R_2' and X_2'.

The primary resistance referred to secondary side is given by $R_1' = R_1 \times k^2$

The primary reactance referred to secondary side is given by $X_1' = X_1 \times k^2$

Similarly, the secondary resistance referred to primary side $R_2' = \dfrac{R_2}{k^2}$

And, the secondary reactance referred to primary side $X_2' = \dfrac{X_2}{k^2}$

The equivalent circuit of the transformer by referring the primary quantities to secondary side is obtained and shown in figure 3.14.

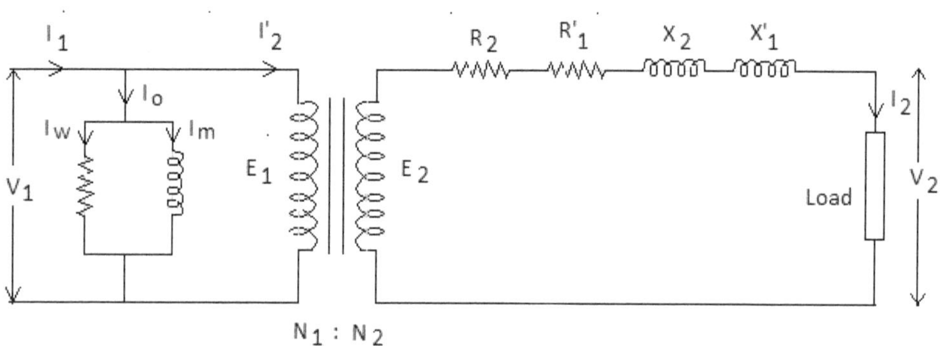

Figure 3.14 equivalent circuit referred to secondary side

Since, R_2 and R'_1 are in series, $R_{02} = R_2 + R_1'$

And X_2 and X'_1 are in series, $X_{02} = X_2 + X_1'$

Where R_{02} is the total resistance of the transformer referred to secondary side and X_{02} is the total reactance of the transformer referred to secondary side.

Then, the total impedance of the transformer referred to secondary side is given by

$$Z_{02} = R_{02} + jX_{02}$$

Hence, the equivalent circuit diagram is simplified and shown in figure 3.15.

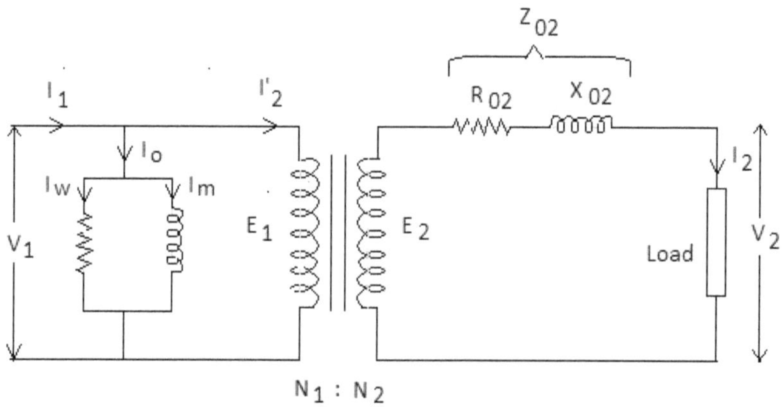

Figure 3.15 Simplified equivalent circuit

The No-load current I_0 when compared to I_1 can be neglected in above equivalent circuit, hence, the shunt branch can be omitted to make the transformer simpler. Then, the equivalent circuit is further simplified to a one which is shown in figure 3.16.

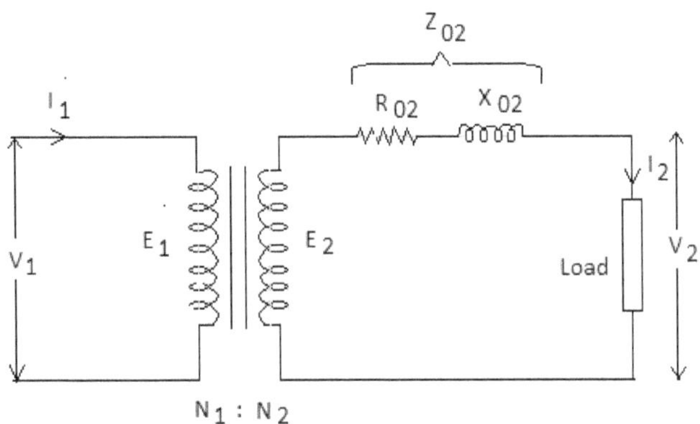

Figure 3.16 Final equivalent circuit with further simplification

Exercise 3.1.6: A 20 KVA, 50 Hz, 2000/200 V Transformer has leakage impedance of $0.42 + j5.2$ Ohm on H.V winding and $0.004 + j0.05$ Ohm on L.V winding. Find its equivalent leakage impedance referred to i) H.V side and ii) L.V side.

Solution :

Since, it is a 2000/200 V transformer, High Voltage (HV) side is primary side and Low Voltage (LV) side is secondary side.

Transformation Ratio of transformer $\quad k = \dfrac{V_2}{V_1} = \dfrac{200}{2000} = 0.1$

The leakage impedance on HV side or primary side $Z_1 = R_1 + jX_1 = 0.42 + j5.2$ ohms

The leakage impedance on LV side or secondary side $Z_2 = R_2 + jX_2 = 0.004 + j0.05$ ohms

Hence, $R_1 = 0.42$ ohms, $X_1 = 5.2$ ohms, $R_2 = 0.004$ ohms and $X_2 = 0.05$ ohms,

(a) To find total leakage impedance ref. to H.V side or primary side:

The total resistance of the transformer refer to primary $R_{01} = R_1 + R_2'$

$$= R_1 + \frac{R_2}{k^2} = 0.42 + \frac{0.004}{0.1^2} = 0.82 \text{ ohms}$$

The total leakage reactance of the transformer refer to primary $X_{01} = X_1 + X_2'$

$$= X_1 + \frac{X_2}{k^2} = 5.2 + \frac{0.05}{0.1^2} = 10.2 \text{ ohms}$$

Therefore, the total leakage impedance of the transformer refer to primary or H.V side

$$Z_{01} = R_{01} + jX_{01} = 0.82 + j10.2 \text{ ohms}$$

(b) To find total leakage impedance ref. to L.V side or secondary side:

The total resistance of the transformer refer to secondary $R_{02} = R_2 + R_1'$

$$= R_2 + R_1 k^2 = 0.004 + 0.42 \times 0.1^2 = 0.0082 \text{ ohms}$$

The total leakage reactance of the transformer refer to secondary $X_{02} = X_2 + X_1'$

$$= X_2 + X_1 k^2 = 0.05 + 5.2 \times 0.1^2 = 0.102 \text{ ohms}$$

Therefore, the total leakage impedance of the transformer refer to secondary or H.V side

$$Z_{02} = R_{02} + jX_{02} = 0.0082 + j0.102 \text{ ohms}$$

3.1.8 Voltage Regulation

The simplified equivalent circuit diagram of a transformer is shown in figure 3.17.

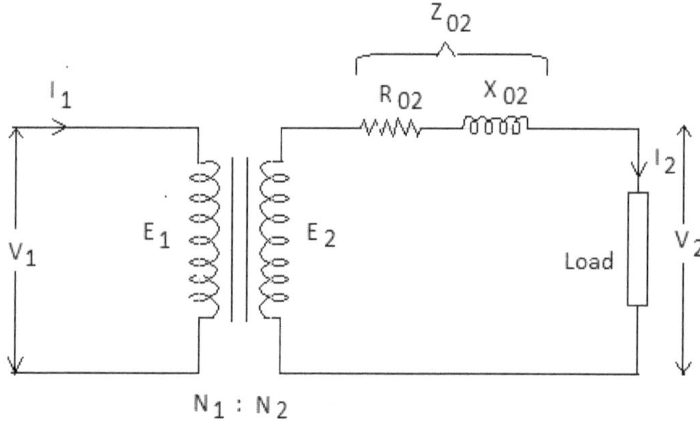

Figure 3.17 Simplified equivalent circuit

It is cleared from above circuit that under no-load conditions, the secondary current I_2 is zero. Also, the primary current I_1 is zero since the no-load current neglected. As there are no voltage drops under no-load, $V_1 = E_1$ and $E_2 = V_2$. The no-load terminal voltage is equal to secondary induced emf.

Under loaded conditions, the currents I_1 and I_2 flow through primary and secondary windings. The different voltage equations under these conditions are

$$E_1 = V_1 - I_1 Z_1$$

and $\qquad V_2 = E_2 - I_2 Z_2$

When the primary leakage impedance is shifted to secondary side, V_1 becomes equal to E_1. Then, Z_{02} represents the leakage impedance of the entire transformer in secondary side. The secondary terminal voltage is given by

$$V_2 = E_2 - I_2 Z_{02}$$

Where $E_2 = k \times E_1$ and $E_1 = V_1$

As the load increases, the total voltage drop in the transformer $I_2 Z_{02}$ increases, and hence V_2 changes. These changes in voltage V_2 with respect to load are accounted by Voltage Regulation.

The Voltage Regulation is defined as the change in secondary terminal voltage from no-load to full load expressed against the no-load voltage.

$$\% \text{ Voltage Regulation} = \frac{E_2 - V_2}{E_2} \times 100$$

The numerator $(E_2 - V_2)$ represents the total voltage drop in the transformer at a given current I_2 and power factor $Cos\phi$. This drop is predicted from the phasor diagram shown in figure 3.18.

The voltage V_2 is taken as reference. The current I_2 lags the V_2 by angle ϕ. The voltage drop in resistance $I_2 R_{02}$ is in phase with current

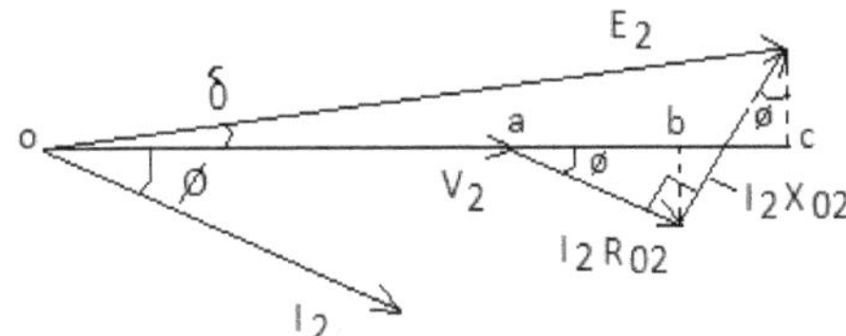

Figure 3.18 Transformer Phasor diagram

and the voltage drop in leakage reactance $I_2 X_{02}$ leads the current by 90^0 as shown. As the angle δ is very small, the magnitude of E_2 is taken equal to oc.

$$E_2 = oc = oa + ab + bc$$

$$= V_2 + I_2 R_{02} Cos\phi + I_2 X_{02} Sin\phi$$

Or $E_2 - V_2 = I_2 R_{02} Cos\phi + I_2 X_{02} Sin\phi$

Therefore the % *Voltage Regulation* $= \dfrac{E_2 - V_2}{E_2} \times 100$

$$= \frac{I_2 R_{02} Cos\phi + I_2 X_{02} Sin\phi}{E_2} \times 100$$

$$= \frac{I_2 [R_{02} Cos\phi + X_{02} Sin\phi]}{E_2} \times 100$$

The above expression for Voltage Regulation has been derived for a lagging power factor case. In general, the voltage regulation for any power factor is given by

$$\% \text{Voltage Regulation} = \frac{I_2 [R_{02} Cos\phi \pm X_{02} Sin\phi]}{E_2} \times 100$$

Where + for lagging power factors and − for leading power factors of load.

The Voltage Regulation can also be obtained in terms of primary quantities as follows.

$$\% \text{ Voltage Regulation} = \frac{I_1\left[R_{01}Cos\phi \pm X_{01}Sin\phi\right]}{V_1}\times100$$

Exercise 3.1.7: A 25 KVA, 6350/230 V, 50 Hz transformer is having 15 ohms resistance and 35 ohms leakage reactance on primary side. The corresponding value of secondary winding are 20 mΩ and 45 mΩ respectively. Find the percentage voltage regulation of the transformer when the transformer delivers a load current of a) 50 A at 0.8 pf lagging b) 75 A at 0.6 pf leading on secondary side.

Solution:

Capacity of the transformer $= 20\,\text{kVA}$

Primary Voltage $V_1 = 6350\,\text{V}$

Secondary Voltage $V_2 = 230\,\text{V}$

Supply frequency $f = 50\,\text{Hz}$

The primary resistance $R_1 = 15\,\text{ohm}$

The primary reactance $X_1 = 35\,\text{ohm}$

The secondary resistance $R_2 = 20\,\text{milli ohm}$

The secondary reactance $X_2 = 45\,\text{milli ohm}$

Transformation ratio $k = \dfrac{V_2}{V_1} = \dfrac{230}{6350} = 0.03622$

The total resistance of the transformer referred to secondary side

$$R_{02} = R_2 + R_1^{'} = R_2 + R_1 k^2 = 20\times10^{-3} + \left(15\times0.03622^2\right) = 0.0396\,\text{ohms}$$

The total reactance of the transformer referred to secondary side

$$X_{02} = X_2 + X_1^{'} = X_2 + X_1 k^2 = 45\times10^{-3} + \left(35\times0.03622^2\right) = 0.0909\,\text{ohms}$$

Total impedance of the transformer referred to secondary side

$$Z_{02} = R_{02} + jX_{02} = 0.0396 + j0.0909\,\text{ohms}$$

(a) When the load current is 50 A, pf is 0.8 lagging

$I_2 = 50\,\text{A}$ and $Cos\phi = 0.8$, $\phi = Cos^{-1}(0.8) = 36.86^0$ and $Sin\phi = Sin36.86^0 = 0.6$

$$\text{The Voltage Regulation} = \frac{I_2\left[R_{02}Cos\phi \pm X_{02}Sin\phi\right]}{E_2}\times100$$

Since, the load power factor is lagging, + ve sign is considered

Voltage Regulation

The Voltage Regulation $= \dfrac{I_2\left[R_{02}Cos\phi + X_{02}Sin\phi\right]}{E_2}\times 100$

$= \dfrac{50\left[0.0396\times 0.8 + 0.0909\times 0.6\right]}{230}\times 100$

$= 1.87\%$

(b) When the load current is 75 A, pf is 0.6 leading

$I_2 = 75$ A and $Cos\phi = 0.6$, $\phi = Cos^{-1}(0.6) = 53.13^0$ and $Sin\phi = Sin53.13^0 = 0.8$

The Voltage Regulation $= \dfrac{I_2\left[R_{02}Cos\phi \pm X_{02}Sin\phi\right]}{E_2}\times 100$

Since, the load power factor is leading, - ve sign is considered

Voltage Regulation $= \dfrac{I_2\left[R_{02}Cos\phi - X_{02}Sin\phi\right]}{E_2}\times 100$

$= \dfrac{75\left[0.0396\times 0.6 - 0.0909\times 0.8\right]}{230}\times 100$

$= -1.596\%$

-ve Sign represents that the secondary voltage increases with load and it may happen with leading power factors.

Alternative method:

The total resistance of the transformer referred to primary side

$$R_{01} = R_1 + R_2' = R_1 + \dfrac{R_2}{k^2} = 15 + \dfrac{20\times 10^{-3}}{0.03622^2} = 15 + 15.245 = 30.245 \text{ ohms}$$

The total reactance of the transformer referred to primary side

$$X_{01} = X_1 + X_2' = X_1 + \dfrac{X_2}{k^2} = 35 + \dfrac{45\times 10^{-3}}{0.03622^2} = 35 + 34.3 = 69.3 \text{ ohms}$$

Total impedance of the transformer referred to primary side

$$Z_{01} = R_{01} + jX_{01} = 30.245 + j69.3 \text{ ohms}$$

(a) When the load current is 50 A, pf is 0.8 lagging

$I_2 = 50$ A and $Cos\phi = 0.8$, $\phi = Cos^{-1}(0.8) = 36.86^0$ and $Sin\phi = Sin36.86^0 = 0.6$

The corresponding primary current $I_1 = k \times I_2 = 0.03622 \times 50 = 1.811\,A$

The Voltage Regulation $= \dfrac{I_1\left[R_{01}Cos\phi \pm X_{01}Sin\phi\right]}{V_1} \times 100$

Since, the load power factor is lagging, $+$ ve sign is considered

$$\text{Voltage Regulation} = \dfrac{I_1\left[R_{01}Cos\phi + X_{01}Sin\phi\right]}{V_1} \times 100$$

$$= \dfrac{1.811\left[30.245 \times 0.8 + 69.3 \times 0.6\right]}{6350} \times 100$$

$$= 1.87\%$$

(b) When the load current is 75 A, pf is 0.6 leading

$I_2 = 75\,A$ and $Cos\phi = 0.6$, $\phi = Cos^{-1}(0.6) = 53.13^0$ and $Sin\phi = Sin53.13^0 = 0.8$

The corresponding primary current $I_1 = k \times I_2 = 0.03622 \times 75 = 2.716\,A$

The Voltage Regulation $= \dfrac{I_1\left[R_{01}Cos\phi \pm X_{01}Sin\phi\right]}{V_1} \times 100$

Since , the load power factor is leading, -ve sign is considered

$$\text{Voltage Regulation} = \dfrac{I_1\left[R_{01}Cos\phi - X_{01}Sin\phi\right]}{V_1} \times 100$$

$$= \dfrac{2.716\left[30.245 \times 0.6 - 69.3 \times 0.8\right]}{6350} \times 100$$

$$= -1.595\%$$

Exercise 3.1.8: A 15 KVA, 200/400 V, 50 Hz transformer is having the leakage impedance of $0.2 + j0.6$ ohms on primary side and $0.8 + j2.6$ ohms on secondary side. Find the % voltage Regulation when transformer feeds half full load current to a resistive load on secondary side.

Solution:

Full Load Capacity of the transformer $= 15\,KVA$

Primary voltage $V_1 = 200\,V$

Secondary Voltage $V_2 = 400\,V$

Supply frequency $f = 50\,Hz$

Primary leakage impedance $Z_1 = R_1 + jX_1 = 0.2 + j0.6$ ohms

Secondary leakage impedance $Z_2 = R_2 + jX_2 = 0.8 + j2.6$ ohms

Transformation Ratio $k = \dfrac{V_2}{V_1} = \dfrac{400}{200} = 2$

The total leakage impedance of the transformer referred to secondary side

$$Z_{02} = R_{02} + jX_{02} = \left(R_2 + R_1'\right) + j\left(X_2 + X_1'\right) = \left(R_2 + R_1 k^2\right) + j\left(X_2 + X_1 k^2\right)$$

$$= \left(0.8 + 0.2 \times 2^2\right) + j\left(2.6 + j0.6 \times 2^2\right) = 1.6 + j5 \text{ ohms}$$

$$R_{02} = 1.6 \text{ ohms and } X_{02} = 5 \text{ ohms}$$

Full load current on secondary side $I_{2_{FL}} = \dfrac{kVA \times 10^3}{V_2} = \dfrac{15 \times 10^3}{400} = 37.5 \text{ A}$

Half Full load current $_2 = \dfrac{I_{2_{FL}}}{2} = \dfrac{37.5}{2} = 18.75 \text{ A}$

Since, the load is resistive, the power factor $Cos\phi = 1$, $\phi = Cos^{-1}(1) = 0^0$ and $Sin\phi = 0$

The Voltage Regulation $= \dfrac{I_2\left[R_{02}Cos\phi \pm X_{02}Sin\phi\right]}{E_2} \times 100$

$$= \dfrac{18.75\left[1.6 \times 1 \pm 5 \times 0\right]}{400} \times 100$$

$$= \dfrac{18.75 \times 1.6}{400} \times 100$$

$$= 7.5\%$$

3.1.9 Losses and Efficiency

A transformer transfers the electrical energy between two circuits through magnetic flux. It does not involve any mechanical energy or rotation. Hence, transformer does not consist of mechanical losses or friction losses. The different losses that occur in a transformer are:

1. *Copper Losses or Electrical Losses* (W_{Cu}) :

   (a) Primary Copper losses: These losses are due to the flow of current I_1 through primary winding having the resistance R_1 and are given by $I_1^2 R_1$

   (b) Secondary Copper losses: These losses are due to the flow of current I_2 through secondary winding having the resistance R_2 and are given by $I_2^2 R_2$

   Total Copper losses $W_{Cu} = I_1^2 R_1 + I_2^2 R_2 = I_1^2 R_{01} = I_2^2 R_{02}$

   These losses are current dependent and change as the load on the transformer changes. Hence, the copper losses are called '*Variable Losses*' with respect to load.

   If $W_{Cu_{FL}}$ are the copper losses at full load current, then the copper losses:

At Half full load are $W_{Cu_{HFL}} = \left[\dfrac{1}{2}\right]^2 \times W_{Cu_{FL}} = \dfrac{1}{4} \times W_{Cu_{FL}}$

At ¾ full load are $\left[\dfrac{3}{4}\right]^2 \times W_{Cu_{FL}} = \dfrac{9}{16} \times W_{Cu_{FL}}$

2. *Iron Losses or Magnetic Losses (W_i) :*

   (a) Hysteresis Losses (W_h) : These losses are due to magnetic reversals that take place in the iron core. The empirical formula for Hysteresis losses is given by

   $$W_h = K_1 B_m^{1.6} f$$

   (b) Eddy Current Losses (We) : The alternating flux passing though the iron core links with it and produces eddy currents. The eddy currents thus produced causes for eddy current losses. The core is laminated to reduce the eddy current losses. The empirical formula for eddy current losses is given by

   $$W_e = K_2 B_m^2 f^2$$

   Where K_1 and K_2 are constants. B_m is maximum flux density in iron core and 'f' is the supply frequency. The magnetic flux in iron core is maintained constant with respect to load. Hence, Iron losses are considered as *'Constant Losses'*.

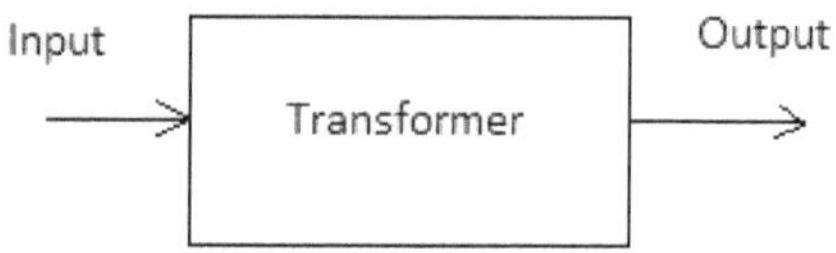

Figure 3.19 Transformer

The efficiency of a transformer is defined as the ratio of output to input.

$$\text{Efficiency } \eta = \frac{Output}{Input} \times 100$$

Since, the output is less than input, the efficiency is less than 100%.

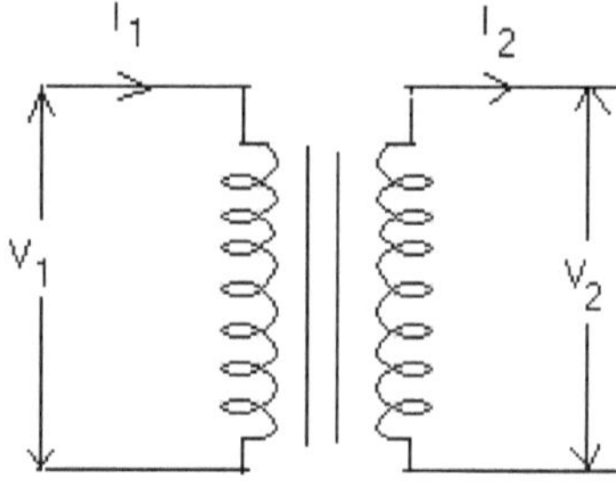

Figure 3.20 Input and outputs of transformer

The input $P_{in} = V_1 \times I_1 \times Cos\phi$

Total Losses $W_L = W_{Cu} + W_i = I_1^2 R_{01} + W_i$

The output P_{out} = Input – Output

$$P_{out} = V_1 \times I_1 \times Cos\phi - \left[I_1^2 R_{01} + W_i \right]$$

Efficiency $\eta = \dfrac{Output}{Input} \times 100$

$$= \frac{P_{out}}{P_{in}} = \frac{V_1 \times I_1 \times Cos\phi - \left[I_1^2 R_{01} + W_i \right]}{V_1 \times I_1 \times Cos\phi} \times 100$$

$$= \frac{V_1 \times I_1 \times Cos\phi}{V_1 \times I_1 \times Cos\phi} - \frac{I_1^2 R_{01}}{V_1 \times I_1 \times Cos\phi} - \frac{W_i}{V_1 \times I_1 \times Cos\phi}$$

$$= 1 - \frac{I_1 R_{01}}{V_1 \times Cos\phi} - \frac{W_i}{V_1 \times I_1 \times Cos\phi}$$

Condition for maximum efficiency:

$$\frac{d\eta}{dI_1} = 0$$

$$\frac{d}{dI_1}\left[1 - \frac{I_1 R_{01}}{V_1 \times Cos\phi} - \frac{W_i}{V_1 \times I_1 \times Cos\phi} \right] = 0$$

$$\left[0 - \frac{R_{01}}{V_1 \times Cos\phi} + \frac{W_i}{V_1 \times I_1^2 \times Cos\phi} \right] = 0$$

$$\frac{W_i}{V_1 \times I_1^2 \times Cos\phi} = \frac{R_{01}}{V_1 \times Cos\phi}$$

$$\frac{W_i}{I_1^2} = \frac{R_{01}}{1}$$

Or $W_i = I_1^2 R_{01}$

Constant Losses = Variable Losses

Exercise 3.1.9: A 10 KVA Single Phase transformer has an iron losses of 100 W and Full load copper losses of 200 W. The transformer feeds a load whose power factor is 0.8 lagging. Find the efficiency of the transformer at a) Full load b) ¾ th Full load and c) Half Full load

Solution:

Full load capacity of the transformer $= 10\,KVA$

Iron Losses $W_i = 100\,W$

Full load Copper losses $W_{Cu} = 200$ W

Load Power factor $= 0.8$ lagging

(a) At Full load:

Output $\quad P_{out} = 1000 \times kVA \times Cos\phi$

$$= 1000 \times 10 \times 0.8 = 8000 \text{ W}$$

Iron losses $\quad W_i = 100$ W

Full load Copper Losses $W_{Cu_{FL}} = 200$ W

Total Losses $W_L = W_{Cu_{FL}} + W_i = 200 + 100 = 300$ W

Input = Output + Total Losses $= P_{out} + W_L = 8000 + 300 = 8300$ W

Efficiency $\quad \eta = \dfrac{Output}{Input} \times 100$

$$= \dfrac{8000}{8300} \times 100 = 96.38\%$$

(b) At ¾ th Full load:

Output $\quad P_{out} = \left(\dfrac{3}{4}\right) \times 1000 \times kVA \times Cos\phi$

$$= \left(\dfrac{3}{4}\right) \times 1000 \times 10 \times 0.8 = 6000 \text{ W}$$

Iron losses $\quad W_i = 100$ W (remain constant with respect to load)

Copper Losses at ¾ th Full load $\quad = \left[\dfrac{3}{4}\right]^2 \times W_{Cu_{FL}} = \dfrac{9}{16} \times W_{Cu_{FL}} = \dfrac{9}{16} \times 200 = 112.5$ W

Total Losses $W_L = W_{Cu} + W_i = 112.5 + 100 = 212.5$ W

Input = Output + Total Losses $= P_{out} + W_L = 6000 + 212.5 = 6212.5$ W

Efficiency $\eta = \dfrac{Output}{Input} \times 100$

$$= \dfrac{6000}{6212.5} \times 100 = 96.57\%$$

(c) At Half Full load:

Output $P_{out} = \left(\dfrac{1}{2}\right) \times 1000 \times kVA \times Cos\phi$

$$= \left(\dfrac{1}{2}\right) \times 1000 \times 10 \times 0.8 = 4000$$

Iron losses $W_i = 100$ W (remain constant with respect to load)

Copper Losses at Half Full load $= \left[\dfrac{1}{2}\right]^2 \times W_{Cu_{FL}} = \dfrac{1}{4} \times W_{Cu_{FL}} = \dfrac{1}{4} \times 200 = 50$ W

Total Losses $W_L = W_{Cu} + W_i = 50 + 100 = 150$ W

Input = Output + Total Losses $= P_{out} + W_L = 4000 + 150 = 4150$ W

$$\text{Efficiency}\, \eta = \frac{Output}{Input} \times 100$$

$$= \frac{4000}{4150} \times 100 = 96.38\%$$

Exercise 3.1.10: A 25 KVA, 400/200 V, 50 Hz Transformer is having the resistance of 42 milli ohms on primary side and 10 milliohms on secondary side. The iron losses amount to 225 W. Find the efficiency of the transformer at ¾ th Full load, 0.707 pf lagging.

Solution:

Full Load Capacity $= 25$ KVA

Transformation Ratio $k = \dfrac{V_2}{V_1} = \dfrac{200}{400} = 0.5$

Primary Resistance $R_1 = 42$ milli ohms $= 0.042$ ohms

Secondary Resistance $R_2 = 10$ milli ohms $= 0.01$ ohms

Iron Losses $W_i = 225$ W

Total resistance of the transformer referred to secondary side $R_{02} = R_2 + R_1^{'}$

$$= 0.01 + 0.042 \times 0.5^2 = 0.0205 \text{ ohms}$$

At ¾ th Full load:

The secondary current $I_2 = \dfrac{3}{4} \times \dfrac{kVA \times 1000}{V_2} = \dfrac{3}{4} \times \dfrac{25 \times 1000}{200} = 93.75$ A

Output $P_{out} = \left(\dfrac{3}{4}\right) \times 1000 \times kVA \times Cos\phi = \left(\dfrac{3}{4}\right) \times 1000 \times 25 \times 0.707 = 13.256$ W

Copper losses $I_2^2 R_{02} = 93.75^2 \times 0.0205 = 180.17$ W

Total losses $W_L = W_{Cu} + W_i = 225 + 180.17 = 405.17$ W

Input $P_{in} = P_{out} + W_L = 13256 + 405.17 = 13661.17$ W

$$\text{Efficiency } \eta = \frac{Output}{Input} \times 100$$

$$= \frac{13256}{13661.17} \times 100 = 97.03\%$$

Exercise 3.1.11: A 15 KVA 230/460 V Transformer has the primary and secondary leakage impedances of $0.065 + j0.4$ and $0.25 + j1.0$ respectively. The transformer takes a power input of 225 watts on no-load. Find the secondary current when the transformer offers maximum efficiency and the corresponding KVA rating.

Solution:

Full load Capacity $:= 15\,KVA$

$$\text{Transformation Ratio } k = \frac{V_2}{V_1} = \frac{460}{230} = 2$$

Primary Leakage Impedance $Z_1 = R_1 + jX_1 = 0.065 + j0.4\,\text{ohms}$

Secondary Leakage Impedance $Z_2 = R_2 + jX_2 = 0.25 + j1.0\ \text{ohms}$

The no-load input $W_0 = 225\,\text{W}$

Under no-load conditions, the secondary current is zero and the primary current is negligible. Hence, copper losses are zero. The output power is also zero ($W_0 = V_2 \times I_2 \times Cos\phi = V_2 \times 0 \times Cos\phi = 0$). So, no-load input is equal to the losses.

$$W_0 = W_L = W_{Cu} + W_i = W_i \quad \text{(since, } W_{cu} \text{ are zero)}$$

The iron losses $W_i = 225\,\text{W}$

When transformer offers the maximum efficiency:

Constant losses = Variable losses (or)

$$\text{Iron Losses = Copper losses}$$

$$W_i = W_{Cu}$$

$$225 = I_2^2 R_{02}$$

Where R_{02} is the total resistance of the transformer referred to secondary side.

$$R_{02} = R_2 + R_1^{'} = R_2 + R_1 \times k^2$$

$$= 0.25 + 0.0625 \times 2^2 = 0.5$$

By substituting the value of R_{02} in above equation, we get

$$225 = I_2^2 R_{02}$$

$$225 = I_2^2 \times 0.5$$

$$I_2^2 = \frac{225}{0.5} = 450$$

$$I_2 = \sqrt{450} = 21.21\,\text{A}$$

The KVA corresponding to maximum efficiency $= \dfrac{V_2 \times I_2}{1000} = \dfrac{460 \times 21.21}{1000} = 9.756\,\text{kVA}$

Exercise 3.1.12: A 1000 VA, 300/150 V transformer having an iron losses of 25 W feeding a load whose power factor is 0.707. Find the maximum efficiency. Consider the total resistance of the transformer referred to secondary is 1 ohm.

Solution:

Full load Capacity $=1000\,\text{VA}$

Transformation Ratio $\quad k = \dfrac{V_2}{V_1} = \dfrac{150}{300} = 0.5$

Iron Losses $W_i = 25\,\text{W}$

Load power factor $Cos\phi = 0.707$

At maximum efficiency, iron losses = Copper losses

$$W_i = I_2^2 R_{02}$$

$$25 = I_2^2 \times 1 \quad \text{or} \quad I_2^2 = \frac{25}{1} = 25 \text{ or } I_2 = \sqrt{25} = 5\,\text{A}$$

KVA rating corresponding to maximum efficiency $= \dfrac{V_2 \times I_2}{1000} = \dfrac{150 \times 5}{1000} = 0.75\,\text{kVA}$

Output at maximum efficiency $= 0.75 \times Cos\phi = 0.75 \times 0.707 = 0.53025\,\text{W}$

Total losses at maximum efficiency $W_L = W_{Cu} + W_i = 2W_i = 2 \times 25 = 50\,\text{W}$

Input = output + losses

$$P_{in} = 530.25 + 50 = 580.25\,\text{W}$$

Maximum efficiency $\eta_{max} = \dfrac{Output}{Input} \times 100 = \dfrac{530.25}{580.25} \times 100 = 91.38\%$

Exercise 3.1.13 : A 15 KVA, 600/300 V, Transformer has an iron losses of 250 W and offers the maximum efficiency at half-full load. Find the total resistance of the transformer refer to a) primary side and b) secondary side

Solution:

Transformation Ratio $k = \dfrac{V_2}{V_1} = \dfrac{300}{600} = 0.5$

Maximum efficiency occurs at half-full load.

Secondary current at half-full load $= \dfrac{1}{2} \times \dfrac{15 \times 1000}{300} = 25\,\text{A}$

At maximum efficiency, $W_i = W_{Cu}$

$$250 = 25^2 \times R_{02} \quad \text{or} \quad R_{02} = \dfrac{250}{25^2} = 0.4\,\text{ohms}$$

Resistance of the transformer referred to primary side $R_{01} = \dfrac{R_{02}}{k^2} = \dfrac{0.4}{0.5^2} = 1.6\,\text{ohms}$

Exercise 3.1.14: The maximum efficiency of 98 % of a 20 KVA, 50 Hz, 2000/200 V Distribution Transformer occurs at full load, unity pf. Calculate the efficiency of the transformer at half full load, 0.8 pf lagging.

Solution:

Full load KVA $= 20\,\text{kVA}$

Maximum Efficiency $\eta_m = 98\%$

Power factor at maximum efficiency $= 1$

Output at maximum efficiency $= P_{out} = 1000 \times kVA \times Cos\phi = 1000 \times 20 \times 1 = 20000\,\text{W}$

Input at maximum efficiency $P_{in} = \dfrac{Output}{\eta} = \dfrac{P_{out}}{\eta} = \dfrac{20000}{0.98} = 20408\,\text{W}$

Total losses at maximum efficiency $W_L = P_{in} - P_{out} = 20408 - 20000 = 408\,\text{W}$

We know that Iron losses = Copper losses at maximum efficiency

Therefore, Iron losses (W$_i$) = full load copper losses ($W_{Cu_{FL}}$) $= \dfrac{W_L}{2} = \dfrac{408}{2} = 204\,\text{W}$

To find Efficiency at Half Full Load, 0.8 PF

Half full load output $P_{out} = \left(\dfrac{1}{2}\right) \times 1000 \times kVA \times Cos\phi$

$$= \left(\dfrac{1}{2}\right) \times 1000 \times 20 \times 0.8 = 8000\,\text{W}$$

Iron losses $W_i = 204\,\text{W}$ (maintained constant for different loads)

Copper losses at half full load $= \left[\dfrac{1}{2}\right]^2 \times W_{Cu_{FL}} = \dfrac{1}{4} \times W_{Cu_{FL}} = \dfrac{1}{4} \times 204 = 51$

Total Losses at half full load $W_L = W_{Cu} + W_i$

$$= 51 + 204 = 255 \text{ W}$$

Input at Half Full load $P_{in} = P_{out} + W_L = 8000 + 255 = 8255 \text{ W}$

$$\text{Efficiency } \eta = \dfrac{Output}{Input} \times 100 = \dfrac{8000}{8255} \times 100 = 96.91\%$$

Exercise 3.1.15: A 50 KVA transformer which feeds a unity power factor load gives a maximum efficiency of 98 % at ¾ full load. Find a) Iron losses b) Full load copper losses c) Full load efficiency.

Solution:

Full load KVA $= 50$

Maximum efficiency $\eta_m = 98\%$

The maximum efficiency occurs at ¾ th Full load

Output at ¾ th Full load $P_{out} = \left(\dfrac{3}{4}\right) \times 1000 \times KVA \times Cos\phi$

$$= \left(\dfrac{3}{4}\right) \times 1000 \times 50 \times 1 = 37500 \text{ W}$$

Input at ¾ th Full load $P_i = \dfrac{P_{out}}{\eta_m} = \dfrac{37500}{0.98} = 38260 \text{ W}$

Total losses at maximum efficiency $W_L = P_{in} - P_{out} = 38260 - 37500 = 765 \text{ W}$

(a) At maximum efficiency, iron losses = copper losses

Iron losses $W_i = \dfrac{W_L}{2} = \dfrac{765}{2} = 382.5 \text{ W}$

(b) As the maximum efficiency occurs at ¾ full load,

Copper losses at ¾ th Full load $= 382.5 \text{ W}$

Copper losses at ¾ full load $= \left[\dfrac{3}{4}\right]^2 \times W_{Cu_{FL}}$

$$382.5 = \dfrac{9}{16} \times W_{Cu_{FL}}$$

$$W_{Cu_{FL}} = 382.5 \times \dfrac{16}{9} = 680$$

(c) Full load output $P_{out} = 1000 \times kVA \times Cos\phi = 1000 \times 50 \times 1 = 50000 \text{ W}$

Total losses $W_L = W_{Cu} + W_i = 680 + 382.5 = 1062.5 \text{ W}$

Full load Input $P_i = FullloadOutput + totallosses$

$$P_{in} = P_{out} + W_L$$

$$= 50000 + 1062.5 = 51062.5$$

$$\eta = \frac{Output}{Input} \times 100 = \frac{50000}{51062.5} \times 100 = 97.9\%$$

3.2 DC Machines

3.2.0 Introduction

The Rotating Electrical Machines convert energy between mechanical and electrical forms. If a machine converts the mechanical energy into electrical energy or if a machine generates electrical energy from mechanical energy input, it is called 'Electric Generator'. Similarly, a machine which converts electrical energy into mechanical energy or if a machine supplies motion from electrical energy input is called 'Electric Motor'. The electrical machines are broadly classified into Direct Current (DC) Machines and Alternating Current (AC) Machines. DC machines are further classified into DC Generators and DC Motors.

3.2.1 DC Generators

DC Generators converts mechanical energy into DC electrical energy. A device which supplies mechanical energy input to generator is called 'Prime Mover'.

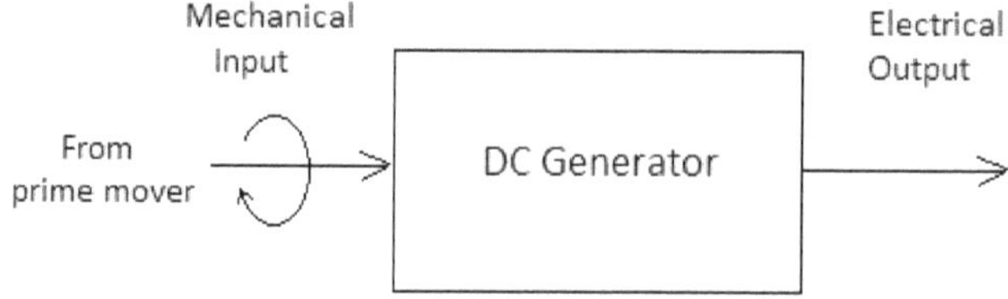

Figure 3.21 DC Generator

DC energy or power involves the voltages and currents of constant magnitudes and fixed directions.

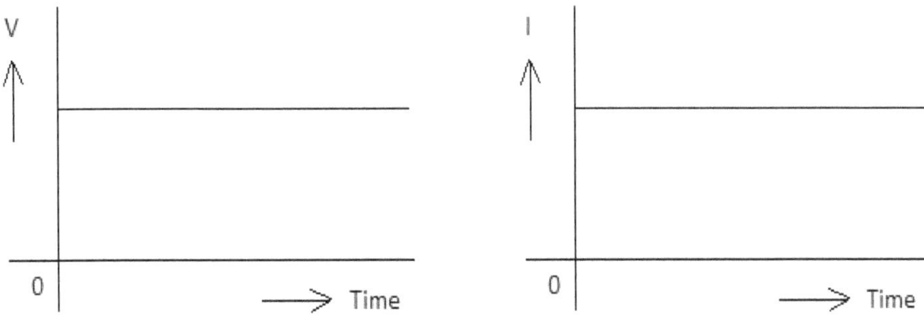

Figure 3.22 DC Voltages and Currents

3.2.2   DC Generator Principle

A DC Generator works based on Faraday's Laws of Electromagnetic Induction. According to the Faraday's laws of electromagnetic induction, whenever a conductor is rotated in a stationary magnetic field or a magnetic field is rotated around a stationary conductor, an Emf is generated in the conductor.

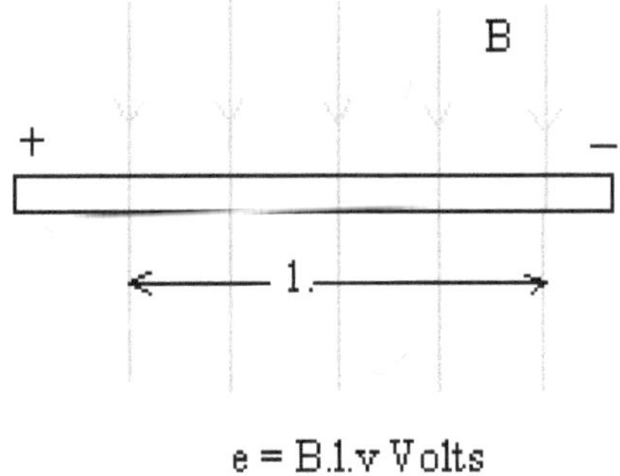

Figure 3.23 Moving conductor in a magnetic field

The magnitude of this emf is equal to the rate of change of flux linkages. The magnitude of this emf is given by

$$E = B \times l \times v \times Sin\theta$$

Where 'B' is flux density in wb/m^2, 'l' is the length of the conductor in meters, 'v' is the velocity of the conductor in m/sec and 'θ' is the angle between the flux direction and direction of motion of the conductor. This is illustrated in the following figure 3.24 where the cross-sectional view of conductor is shown.

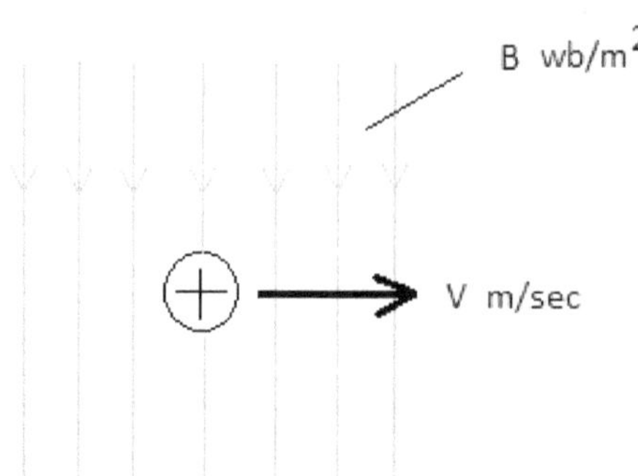

Figure 3.24 Obtaining the current direction using Fleming's Right Hand Rule

The $\oplus$ sign represents that the current flows away from the reader and the $\odot$ sign indicates that the current direction is towards the reader. The direction of current through a conductor for a given directions of flux and conductor motion is determined by Fleming's Right Hand Rule.

Fleming's Right Hand Rule: when the first three fingers of right hand are stretched so that they are mutually perpendicular, if thu**M**b represents the direction of **M**otion of the conductor and the**F**orefinger represents the **F**lux direction, then, them**I**ddle finger represents the direction of current (**I**).

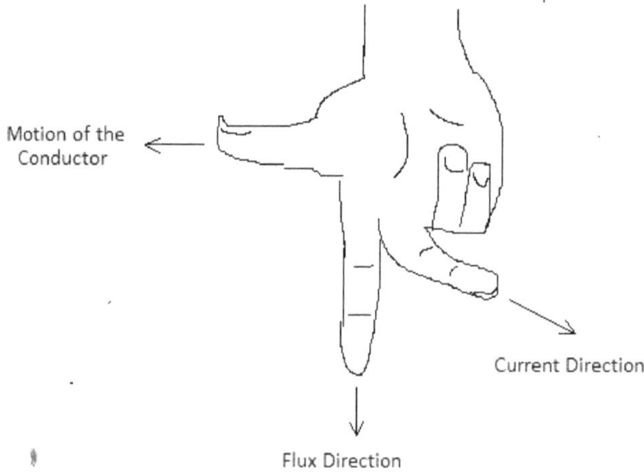

Figure 3.25 Fleming's Right Hand Rule

And also, it cleared from above expression that no emf is generated when the conductor is moved parallel to flux lines so that θ is zero. The emf induced becomes maximum when the conductor is moved perpendicular to flux lines ie., $\theta = 90^0$.

The operation of a DC generator is well understood with Simple Loop Generator.

3.2.3   Simple Loop Generator

Consider a single turn rectangular conductor loop ABCD rotating in a magnetic field of density 'B'wb/m^2 about its own axis. The two ends of coils are joined to two slip rings 'a' and 'b' which are insulated from each other. Two brushes are used to collect the current from rotating coil and this current is conveyed to an external load 'R' through two brushes placed on slip rings. The rotating coil is called as 'Armature winding' and magnets as 'Field Magnets'

Imagine the coil to be rotating in clockwise direction. As the coil rotates in the magnetic field, the flux linkages with the coil change. Hence, an emf is induced in the coil which is proportional to the rate of change of flux linkages($e = \dfrac{d\phi}{dt}$). In the position shown figure 3.26, as the coil sides AB and CD are moving parallel to flux lines, no emf is induced. Let us take this position as reference and angle of rotation of coil is measured from this position.

As the coil to continue to rotate, the rate of flux linkages increases till the position $90°$ is reached. At this position of 90^0, the coil sides are moving at right angle to flux lines and hence maximum emf is generated in them.

In the next quarter revolution, ie., from 90^0 to 180^0, the rate of change of flux linkages decreases, hence, emf induced gradually decreases and becomes zero at position of 180^0. So, we find that in first half revolution of coil, the emf induced in it increases from zero and becomes maximum at 90^0 and again decreases to zero. The direction of current according to Flemings left Hand Rule is ABMLCD. The direction of current through the load resistance is from 'M' to 'L'.

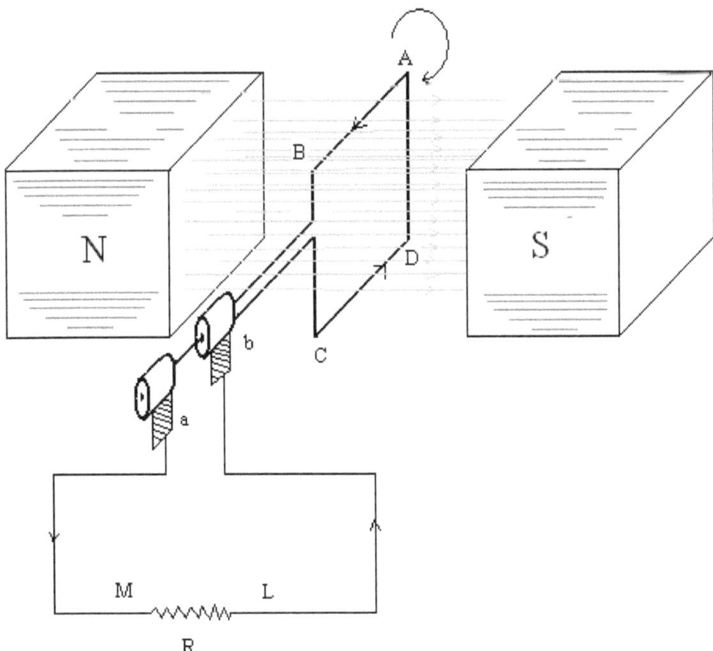

Figure 3.26 Simple Loop Generator

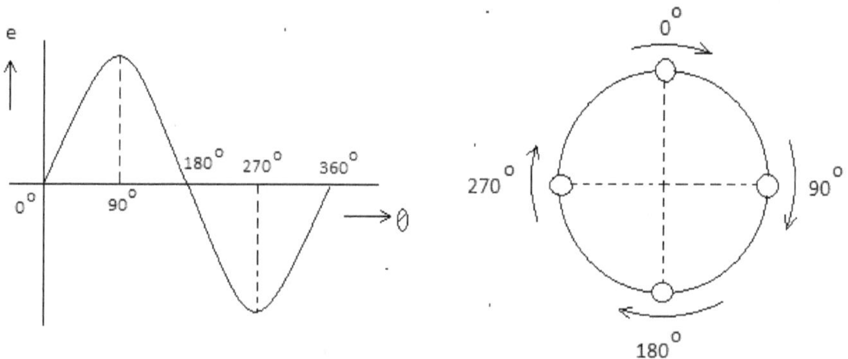

Figure 3.27 Production of alternating emf

In the next half revolution, ie., from 180 ^{0} to 360^0, the variations in induced emf are similar to those in first revolution, but in opposite direction ie., from L to M in external load resistance. So, it is cleared from above that the basic nature of emf induced in d.c generator is also alternating. To convert this alternating current into unidirectional, slip rings are replaced by *'Split Rings' or 'Commutator'*.

A two segment commutator is shown in figure 3.28. A commutator is made out of conducting cylinder which is cut into segments and these segments are insulated from each other by a thin sheet of Mica.

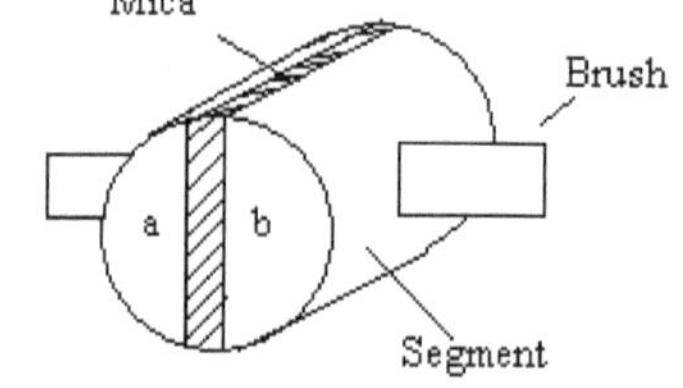

Figure 3.28 Commutator

The operation of above simple loop generator is explained here with Commutator. Even the segments 'a' and 'b' change their positions along with coil sides, the direction of current through external resistance will be maintained always from M to L as shown in figure 3.29 and it is found to be unidirectional.

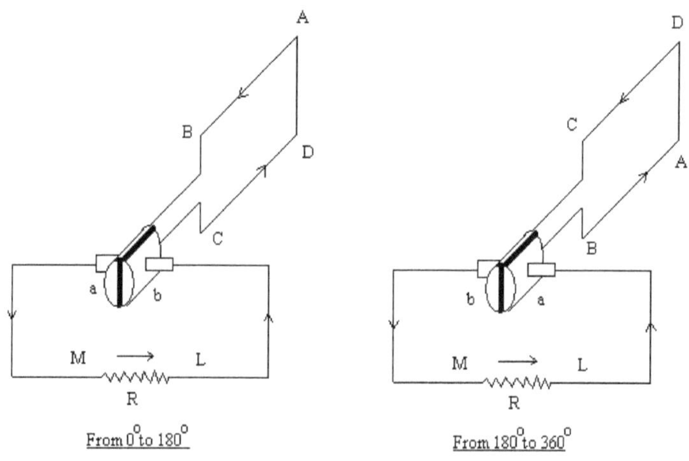

Figure 3.29 Action of Commutator

The direction of current is always positive irrespective of position of coil as shown in figure 3.30.

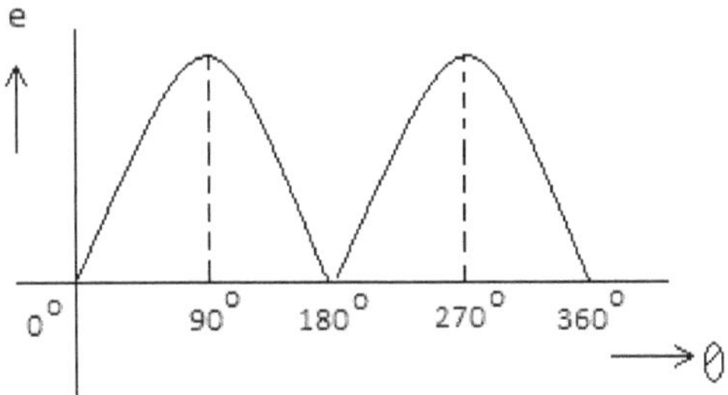

Figure 3.30 Unidirectional or Direct Current

3.2.4 Construction

The main requirements to establish a DC generator are:

1. Conductor(s)[a pool of conductors called 'Armature Winding']
2. Magnetic Field [produced by a winding called 'Field Winding']
3. Relative Motion between Conductor and Magnetic Field [Prime Mover]

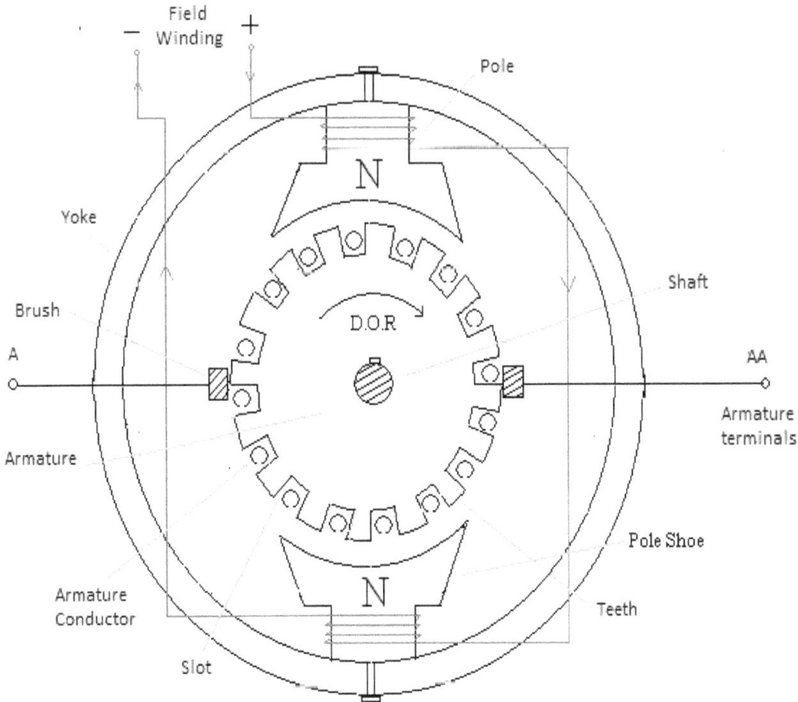

Figure 3.31 Basic Construction of a DC Machine

The figure 3.31 shows the brief construction of a DC generator (machine) which fulfils all above requirements. The DC machine is broadly classified into a) stator and b) rotor.

The main parts of stator are yoke, stator poles, pole shoes and field windings. The rotor includes an armature winding, armature core, commutator etc.

1. *Yoke:* The outer frame is called yoke which provides mechanical support to the poles and acts as protective cover for the whole machine. It also acts as return path for flux produced by poles. Hence, it must be made with a magnetic material ie., Cast Iron. But. for larger machines, Rolled Steel is employed.

2. *Stator Poles:* The pole cores and pole shoes are built of thin laminations of annealed Steel to reduce the eddy current loses. The thickness of lamination varies from 1mm to 0.25 mm.

3. *Pole Shoes:* Pole shoes are also made with steel laminations. The pole shoes serve two purposes. They spread out the flux in the air gap and they support the field coils.

4. *Field Windings:* These windings have got that name as they produce the required magnetic field. These windings are concentrated windings placed on stator poles and excited by direct current which is called excitation current. These windings are generally made up of Copper.

5. *Armature Core:* It houses armature conductors or coils. When the armature is rotated using a prime mover, the armature conductors cut the magnetic flux produced by poles and hence emf is generated in them. The armature core having low magnetic reluctance offers a good path for flux from 'N' pole to 'S' pole. This is made up of circular laminations of silicon steel. The thickness of laminations is maintained around 0.5mm. Generally, Varnish insulation is placed between the laminations. The armature is provided with air ducts to permit the axial flow of air though armature core for necessary cooling.

6. *Brushes:* Brushes are made up of Carbon or Graphite placed on commutator to collect the currents from armature winding.

7. *Armature Winding:* The emf induced in a single armature conductor is not sufficient for many applications. Hence, many armature conductors are used. These armature conductors are interconnected in series-parallel patterns to get the required magnitudes of voltages and currents. This pool of conductors is called *Armature Winding*. There are two types of armature windings. a) Lap Windings and b) Wave Windings. In Lap Windings, the number of parallel paths (A) in the winding is equal to the number of poles (P) whereas in Wave windings, it is always equal to 2. The following example explains the difference between lap and wave windings.

Let us consider a d.c machine having 100 conductors and 4 pole. When these conductors are rotated at rated speed, it is supposed that each conductor produces an emf of 2 V and carries a current of 5 A. These conductors can be connected in Lap or Wave pattern.

(a) When conductors are connected in Lap:

Since $A = P$ in Lap windings, there are four parallel paths. The number of conductors each

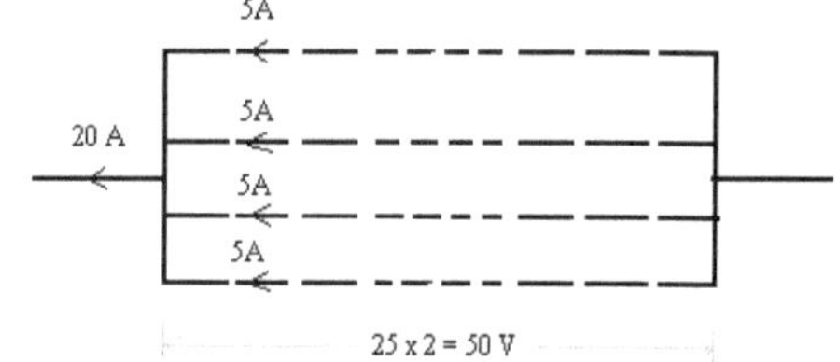

Figure 3.32 Lap Connection

parallel path is $\dfrac{100}{4}=25$. The emf generated out of any parallel path is $25\times2=50\,\text{V}$. Also, the total current supplied is equal to the current supplied by each parallel path multiplied by number of parallel paths and it is given by $5\times4=20\,\text{A}$.

(b) When conductors are connected in Wave:

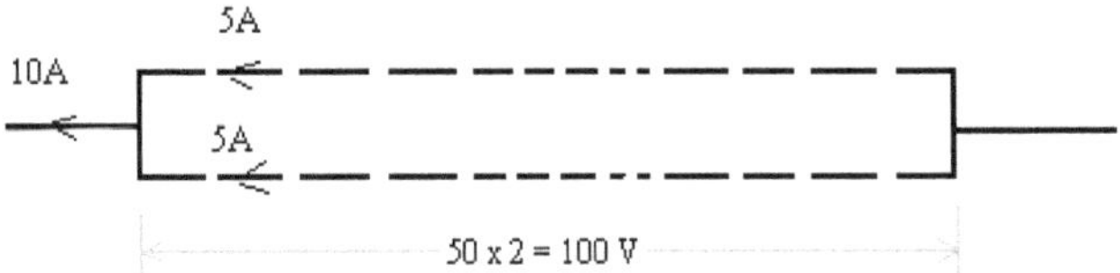

Figure 3.33 Wave Connection

Since $A=2$, the number conductors in each parallel path is $\dfrac{100}{2}=50$. The emf generated out of any parallel path is $50\times2=100\,\text{V}$. Also, the total current supplied is equal to the current supplied by each parallel path multiplied by number of parallel paths and it is given by $5\times2=10\,\text{A}$.

So, it concluded from above discussion that for low voltage-high current requirements, the Lap windings are used. And, wave windings are preferred for high voltage- low current applications.

Circuit Symbol of DC Generator:

We know that a DC machine is having two windings viz., the Field winding and Armature Winding. These windings are shown in circuit in figure 3.34. As the armature winding is rotating, it is represented as a circle on which two brushes are placed to collect the currents. The current supplied by armature winding is denoted by I_a.

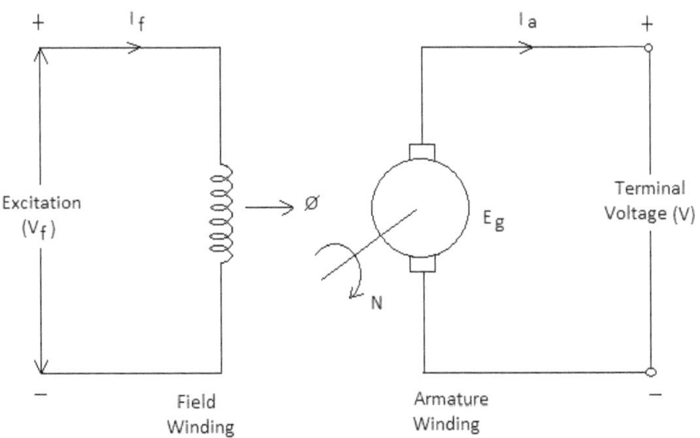

Figure 3.34 DC Generator Symbol

Let I_f be the current supplied to the Field Winding and the corresponding flux produced by each pole is Ø webers. The armature is rotated at speed of 'N' rpm using a prime mover and the emf induced in it is 'E_g' volts. The expression for the emf induced is derived as follows.

3.2.5   Emf Generated or EMF equation

Let ' ϕ ' be the flux per pole

'Z' be the total no. of armature conductors = No. of slots x no. of conductors per slot

'P' be the no. of poles

'A' be the no. of parallel paths

'N' be the speed of prime- mover

'e' be the average emf induced in a conductor

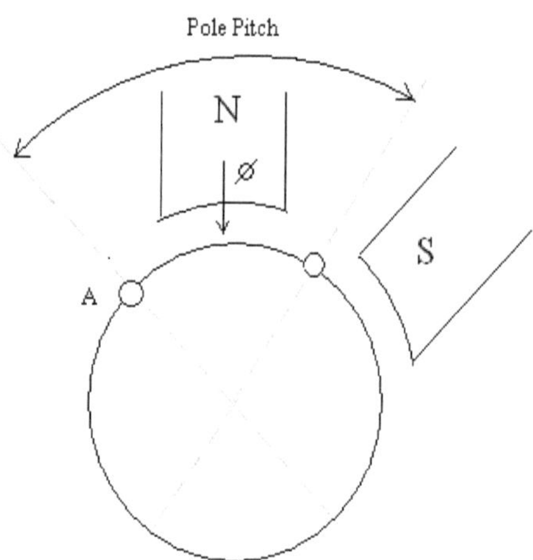

Figure 3.35 conductor passing through a Pole pitch

Therefore, ave. emf induced per conductor $e = \dfrac{d\phi}{dt}$

Let the conductor 'A' be moving through a distance of one pole pitch (one pole pitch is the distance covered by one pole). So, it links with flux of ϕ Weber. So, $d\phi = \phi$. Since, the conductor 'A' is moving at 'N' rpm, it takes one minute to complete 'N' revolutions.

For one revolution, it takes $\dfrac{60}{N}$ sec and it covers 'P' no. of pole pitches in one revolution.

Hence, the time taken for covering one pole pitch $dt = \dfrac{60}{NP}$ seconds.

Therefore, Emf generated / Conductor $e = \dfrac{d\phi}{dt} = \dfrac{\phi}{\left(\dfrac{60}{NP}\right)} = \dfrac{\phi NP}{60}$ Volts

Emf generated out of generator $= E =$ Emf induced per conductor $\times$No. of conductors per parallel path

$$E = \dfrac{\phi NP}{60} \times \dfrac{Z}{A}$$

Or $\qquad E = \dfrac{\phi ZN}{60} \times \dfrac{P}{A}$

Where $A = 2$ for Wave connected generator

$A = P$ for Lap connected generator

Exercise 3.2.1 *:* A Four Pole generator having wave connected armature has 51 slots, each slot containing 20 conductors. What will be the generated emf in the machine when the generator driven at 1500 rpm assuming the flux per pole as 7 mwb.

Solution :

No. of poles $P = 4$

Total no. of conductors $Z =$ No. of slots x conductors per slot

$\qquad = 51 \times 20 = 1020$

Flux per pole $\phi = 7 \times 10^{-3}$ wb

Speed of prime-mover or generator $N = 1500$ rpm

No. of parallel paths $A = 2$ (since wave connection)

Then, Emf generated $E_g = \dfrac{\phi ZN}{60} \times \dfrac{P}{A} = \dfrac{7 \times 10^{-3} \times 1020 \times 1500}{60} \times \dfrac{4}{2} = 178.5$ V

Exercise 3.2.2: A 6 pole DC generator is having 1200 lap connected armature conductors is producing an emf of 400 V in its armature winding. Find the speed of the generator if it is maintaining a flux of 10 mwb under each pole.

Solution:

No. of poles $P = 6$

No. of armature conductors $Z = 1200$

Type of armature winding is Lap. Hence, no. of parallel paths (A) = no. of poles (P)

Generated emf $E_g = 400\,\text{V}$

Flux per pole $\phi = 10 \times 10^{-3}$ wb

Speed of the generator $N = ?$

Emf generated $E_g = \dfrac{\phi ZN}{60} \times \dfrac{P}{A}$

or $\qquad N = \dfrac{E_g \times 60 \times A}{\phi \times Z \times P} = \dfrac{400 \times 60 \times 6}{10 \times 10^{-3} \times 1200 \times 6} = 2000 \text{ rpm}$

Exercise 3.2.3: A DC generator gives a voltage of 220 V at rated speed and flux. Find the generated voltage when the speed is increased by 60 % and the flux reduced to half.

Solution:

Let $\emptyset$ be the rated flux under each pole

N be the rated speed

Z be the total number of armature conductors

P be the number of poles

A be the number of parallel paths

$$\text{Emf generated } E_g = \frac{\phi ZN}{60} \times \frac{P}{A}$$

At rated speed and flux, $220 = \dfrac{\phi ZN}{60} \times \dfrac{P}{A}$ $\qquad\qquad$ .....(1)

When the speed is increased by 60 %, the revised speed $N_2 = 1.6 \times N$

When the flux is reduced to half, the revised value of flux $\phi_2 = 0.5 \times \phi$

The generated emf at these conditions, $E_{g_2} = \dfrac{\phi_2 ZN_2}{60} \times \dfrac{P}{A}$

$$E_{g_2} = \frac{0.5\phi \times Z \times 1.6N}{60} \times \frac{P}{A} \qquad\qquad ....(2)$$

Dividing equation (2) with (1),

$$\frac{E_{g_2}}{220} = \frac{0.5\phi \times Z \times 1.6N}{60} \times \frac{P}{A} \times \frac{60 \times A}{\phi ZN \times P}$$

$$\frac{E_{g_2}}{220} = 0.5 \times 1.6 = 0.8$$

Or $\qquad E_{g_2} = 220 \times 0.8 = 176 \text{ V}$

3.2.6   Classification of DC Generators

The DC Generators are classified into two types based on the interconnection between Field and Armature windings. These are a) Separately Excited DC Generators   and b) Self Excited DC Generators. The self- excited generators are further classified into i) Shunt ii) Series and iii) Compound generators. The connection diagrams and the relevant voltage and current equations are presented below for each type.

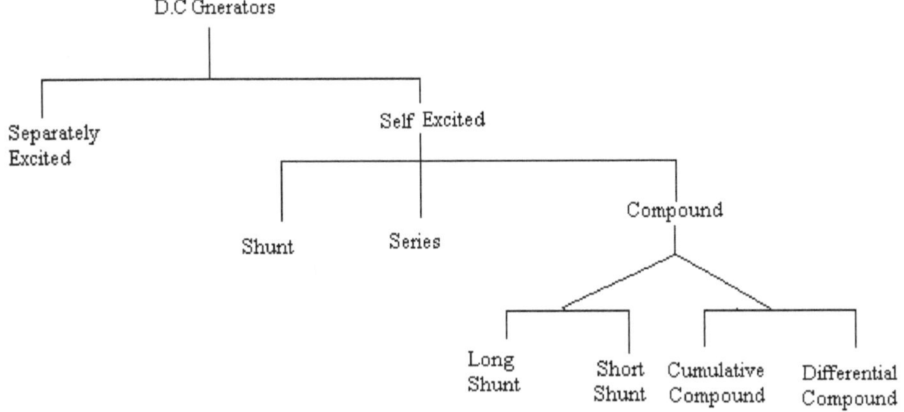

Figure 3.36 Classification of DC generators

1. **Separately excited generators:** These Separately excited generators are generators whose field windings are excited from a separate d.c source as shown in figure 3.37. There will be no electrical connection between armature and field windings.

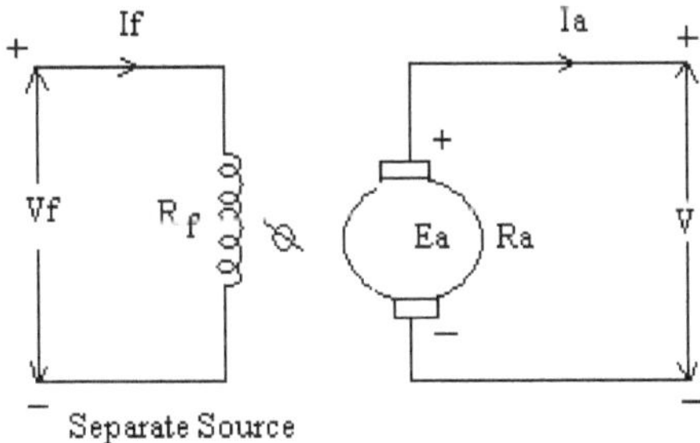

Figure 3.37 Separately Excited DC Generator

$$I_f = \frac{V_f}{R_f}$$

$$\phi \propto I_f$$

$$E_g \propto I_f N$$

$$V = E_g - I_a R_a - B.D$$

2. Self-excited generators: These generators are excited by the currents generated by generators themselves. Hence, some electrical connection exists between field and armature windings. There are three types of self excited generators named according to the manner in which their field coils are connected with armature windings

(a) Shunt Generator:

The field winding is connected in parallel or shunt to the armature winding. The figure 3.38 shows the connection diagram.

$$I_{sh} = \frac{V}{R_{sh}}$$

$$\phi \propto I_{sh}$$

$$I_a = I_L + I_{sh}$$

$$V = E_g - I_a R_a - B.D$$

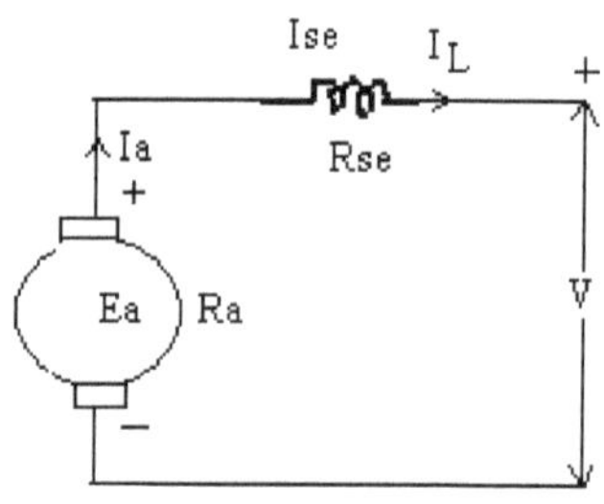

Figure 3.38 DC Shunt Generator

(b) Series Generator:

In this case, field winding is connected in series with the armature winding as shown in figure 3.39. As it carries full load current, it consists of relatively less number of turns of thick wire. Hence, the resistance of series field winding is less.

$$I_{se} = I_a = I_L$$

$$\phi \propto I_{se}$$

$$V = E_g - I_a R_a - I_{se} R_{se} - B.D$$

Figure 3.39 DC Series Generator

(c) Compound Generators:

These generators have both series and shunt field windings. These generators can be long shunt or short shunt compound generators.

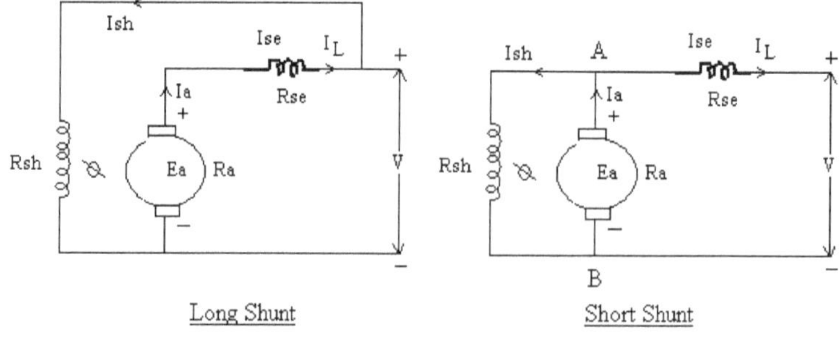

Figure 3.40 DC Compound Generator a) Long shunt b) Short Shunt

In long shunt compound generators,

$$I_{se} = I_a$$

$$I_a = I_L + I_{sh}$$

$$I_{sh} = \frac{V}{R_{sh}}$$

$$V = E_g - I_a R_a - I_{se} R_{se} - B.D$$

$$= E_g - I_a (R_a + R_{se}) - B.D$$

In Short shunt Compound generators,

$$I_L = I_{se}$$

$$I_a = I_{se} + I_{sh}$$

$$I_{sh} = \frac{V_{AB}}{R_{sh}} = \frac{E_g - I_a R_a}{R_{sh}} = \frac{V + I_{se} R_{se}}{R_{sh}}$$

$$V = E_g - I_a R_a - I_{se} R_{se} - B.D$$

$$= E_g - I_a (R_a + R_{se}) - B.D$$

3.3 DC Motors

So far, we have dealt with DC Generators where the given mechanical energy gets converted into electrical energy. For a given D.C generator, if prime mover is disconnected, and instead of taking DC electrical power from it, if DC electrical power is fed into the machine, it starts rotating and the machine gets converted into a *DC Motor. Thus, A DC Motor converts the DC electrical power or energy into mechanical power or energy.*

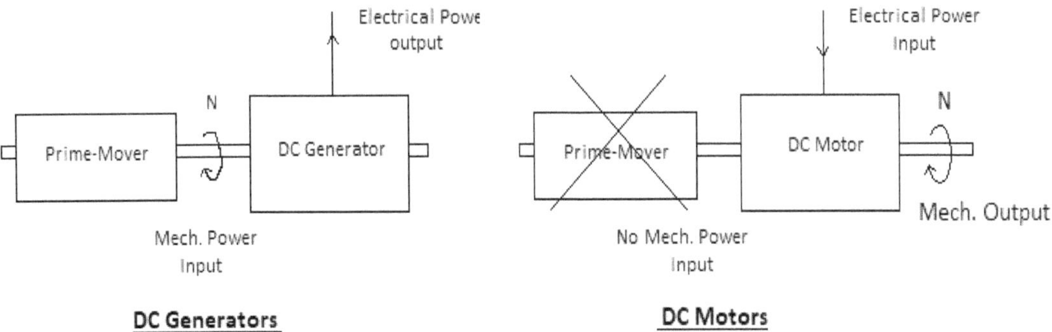

Figure 3.41 a) DC Generator b) DC generrator converted into a Motor

There is no constructional difference between a DC motor and DC generator. Same machine can be operated as a motor and also as a generator.

3.3.1 DC Motor Principle

The operation of a DC motor is based on the principle that when a current carrying conductor is placed in magnetic field, it experiences a force *(F)*.

Let *'I'* be the current passing through the conductor in Amperes. This current through a

conductor is represented as $\oplus$ or $\odot$. The $\oplus$ sign represents that current flows away from the reader

and the $\odot$ sign represents that the current direction is towards the reader.

Let *'B'* be the flux density in wb/m^2

Let *'l'* be the length of the conductor in meters

The magnitude of this force is given by $F = B \times I \times l$ Newton.

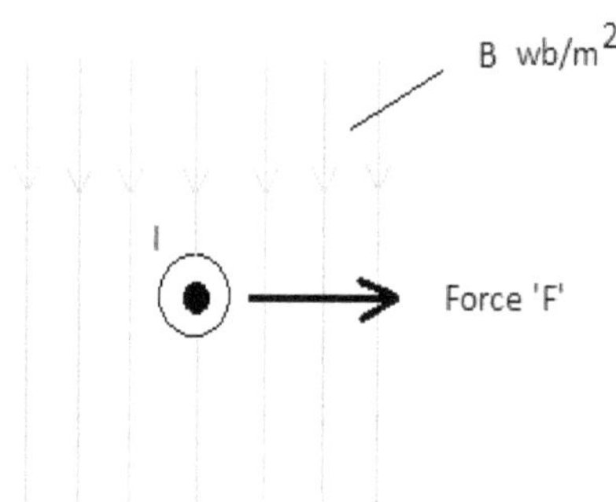

Figure 3.42 Fleming's Left Hand Rule

And the direction of force is given by Fleming's Left Hand Rule.

Fleming's left Hand Rule: when the first three fingers of left hand are stretched so that they are mutually perpendicular, if Forefinger represents the **F**lux direction and m**I**ddle finger represents the direction of current (**I**), then, the thu**M**b represents the direction of **M**otion of the conductor.

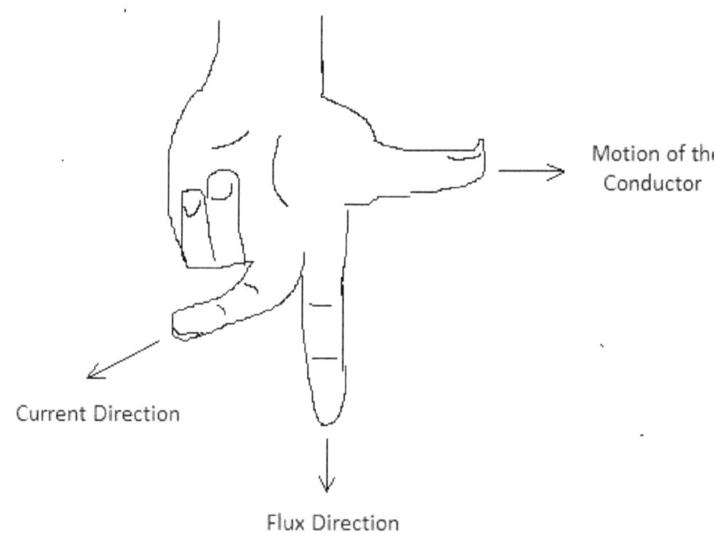

Figure 3.43 Fleming's Left Hand Rule

3.3.2 Back Emf

Consider a DC motor excited from a DC source of voltage 'V' and the corresponding current supplied to the armature winding is 'Iₐ'. The direction of currents supplied to armature conductors are assumed to be crosses($\oplus$) under North pole and dots (O) under South pole as shown in figure 3.44. By applying Fleming's Left Hand Rule, we came to know that the direction of rotation of the armature is anticlockwise. The motor continues to rotate in the same direction at a speed 'N' in the presence of flux 'Ø' as long as the armature is fed with the current I_a through Commutator and brushes.

As the armature conductors rotate at speed N in the flux Ø, it generates an emf E_g in the armature winding. This means that the generator action exists in a motor. The direction of this emf generated is determined using Fleming's Right Hand Rule and it is found to be in opposite direction with armature current direction. Since, this emf opposes the current supplied to the armature winding by source 'V', it is called 'Back Emf (E_b)'.

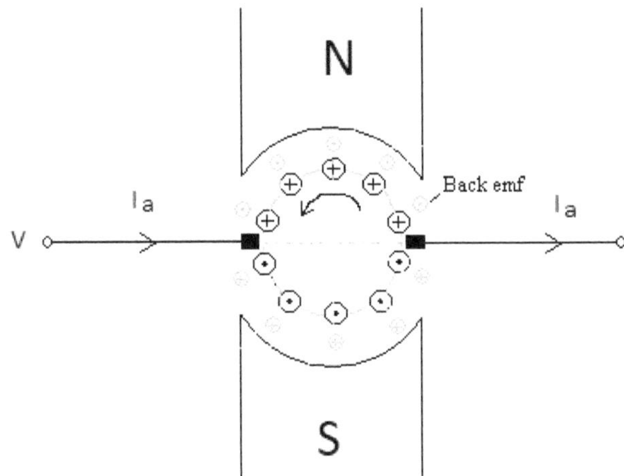

Figure 3.44 Back Emf

As the back emf is generated in a motor due to generator action, its expression is same as that of generated emf in DC generators.

$$\text{Back Emf} \quad E_b = \frac{\phi ZN}{60} \times \frac{P}{A}$$

DC Motor circuit model:

Out of the two windings in a DC motor, one winding (Field winding) is given a field current I_f and thus it produces a magnetic field of Ø in the air gap. The other winding (Armature winding) is provided with a supply voltage of V. A torque in produced in the armature due to the interaction between fluxes produced by field and armature windings. The rotor carrying armature winding rotates at a speed N.

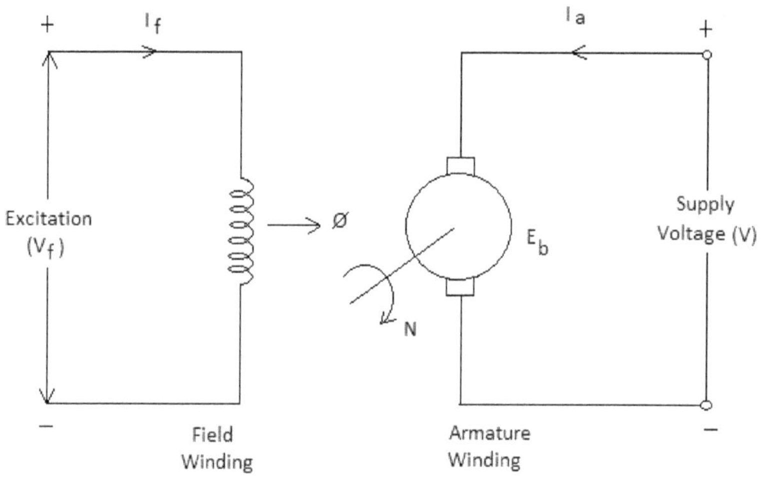

Figure 3.45 DC Motor Symbol

The voltage equation of DC motor is given by $V = E_b + I_a R_a$ (In Case of DC generators, it was $V = E_g - I_a R_a$)

Exercise 3.3.1: A 4 pole DC motor having 960 lap connected armature conductors rotates at a speed of 1400 rpm. The flux under each pole is 9 mwb. Find the Back emf generated in the armature winding.

Solution:

Number of poles $P = 4$

Total number of armature conductors $Z = 960$

Type of winding = Lap (hence, number of armature winding parallel paths A = P = 4)

Speed of the motor $N = 1400$ rpm

Flux under each pole $\phi = 9 mwb = 9 \times 10^{-3}$ wb

$$\text{Back Emf } E_b = \frac{\phi ZN}{60} \times \frac{P}{A} = \frac{9 \times 10^{-3} \times 960 \times 1400}{60} \times \frac{4}{4} = 201.6 \text{ V}$$

Exercise 3.3.2 : A 4 pole wave connected separately excited DC motor draws a current of 25 A from 220 V source and runs at 1500 rpm. The armature resistance is 0.8 ohms. Find a) the back emf b) number of armature conductors if the flux under reach pole is 20mwb.

Solution:

Number of Poles $P = 4$

Supply Voltage $V = 220$ V

Armature current $I_a = 25$ A

Type of the winding $=$ Wave (hence, $A=2$)

Speed of the motor $N=1500\,\mathrm{rpm}$

Armature resistance $R_a =0.8\,\mathrm{ohm}$

Flux per pole $\phi = 20.mwb = 20\times10^{-3}\,\mathrm{wb}$

(a) Back emf $E_b =V -I_aR_a =220-25\times0.8=220-20=200$ V

(b) Back emf $E_b =\dfrac{\phi ZN}{60}\times\dfrac{P}{A}$

$$200=\dfrac{20\times10^{-3}\times Z\times1500}{60}\times\dfrac{4}{2}$$

Or $\quad Z =\dfrac{200\times60\times2}{0.02\times1500}=800$

Exercise 3.3.3: A 6 pole Lap connected DC motor having 1020 armature conductors absorbs a current of 50 A from a DC source. The flux per pole is given as 18 mwb. Find the supply voltage to run the motor at 1200 rpm speed. The armature resistance 0.62 ohms and also consider the brush drop as 2V.

Solution:

Number of poles $P=6$

Type of winding $=$ Lap (hence, number of armature winding parallel paths $A=P=6$)

Number of armature conductors $Z=1020$

Armature current $I_a =50$ A

Flux per pole $\phi =18.mwb =18\times10^{-3}$ wb

Speed of the motor $N=1200\,\mathrm{rpm}$

Armature resistance $R_a =0.62\,\mathrm{ohm}$

Brush drop $=2$ V

Back Emf $\quad E_b =\dfrac{\phi ZN}{60}\times\dfrac{P}{A}=\dfrac{18\times10^{-3}\times1020\times1200}{60}\times\dfrac{6}{6}=367.2$ V

The supply voltage $V = E_b +I_aR_a +B.D=367.2+(50\times0.62)+2=400$ V

3.3.3   Classification of DC Motors

Like in DC Generators, the DC motors are classified based on the interconnection between Field and Armature Windings. These are a) Separately Excited DC Motors and b) Self Excited DC Motors. The Self- Excited Motors are further classified into i) Shunt ii) Series and

iii) Compound Motors. The connection diagrams and the relevant voltage and current equations are presented below for each type.

1. Separately excited D.C Motor :

In this type, the field winding is connected to a separate source V_f. There will be no electrical connection between field winding and armature windings. The voltage and current equations for this motor are presented below. As the back emf opposes the supply voltage 'V', it is also considered as voltage drop along with armature resistance drop I_aR_a and brush drop B.D (voltage drop at brushes due to brush contact resistance)

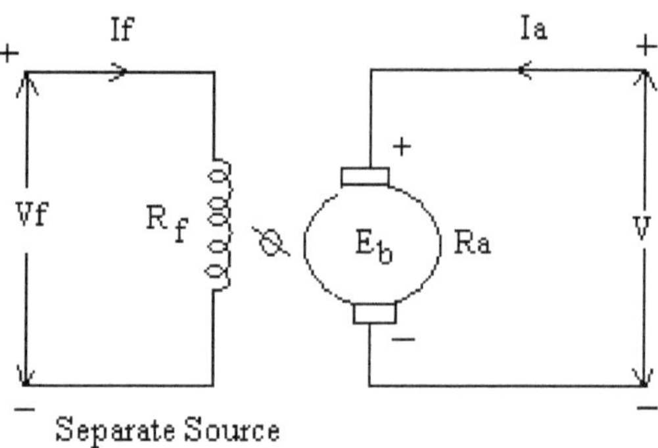

Figure 3.46 Separately excited DC motor

$$V = E_b + I_aR_a + B.D$$
$$I_a = I_L$$
$$I_f = \frac{V}{R_f}$$

The voltage and current equations for other types of DC motors also have been provided for the reference.

2. Self excited D.C Motors :

a. Shunt Motor:

$$V = E_b + I_aR_a + B.D$$
$$I_a = I_L - I_{sh}$$
$$I_{sh} = \frac{V}{R_{sh}}$$

b) Series Motor :

$$V = E_b + I_aR_a + I_{se}R_{se} + B.D$$
$$I_a = I_{se} = I_L$$

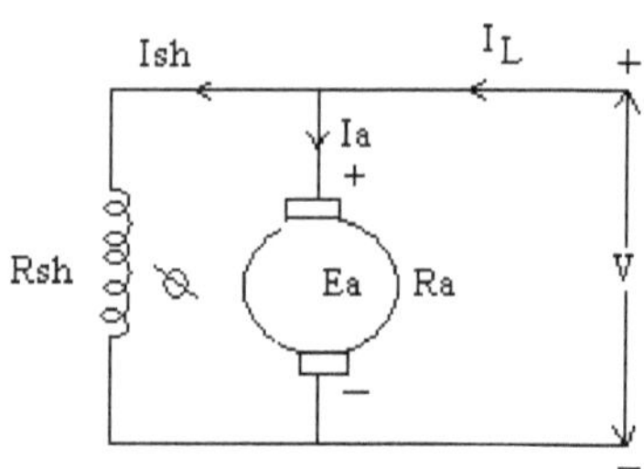

Figure 3.47 Shunt Motor

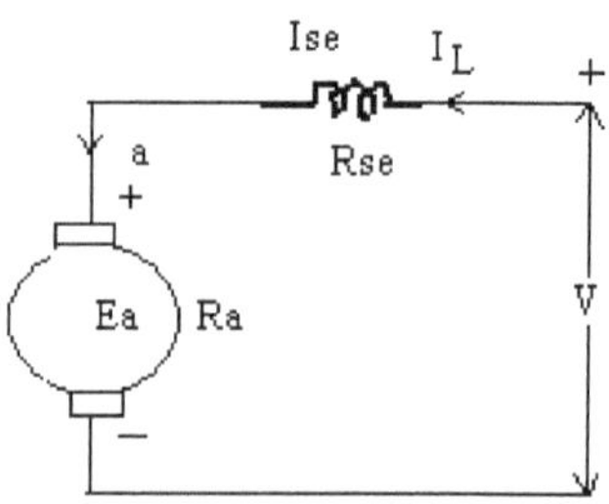

Figure 3.48 DC Series Motor

(c) Compound Motors :

(i) Long-Shunt Motor :

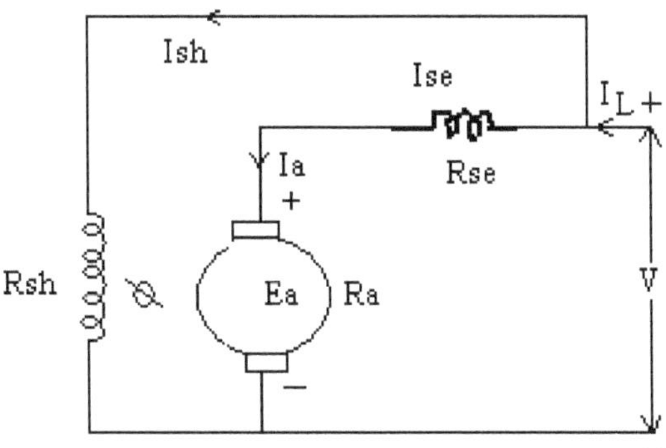

Figure 3.49 Long Shunt Compound Motor

$$V = E_b + I_a R_a + I_{se} R_{se} + B.D$$
$$I_a = I_L - I_{sh}$$
$$I_{sh} = \frac{V}{R_{sh}}$$

(ii) Short-shunt Motor:

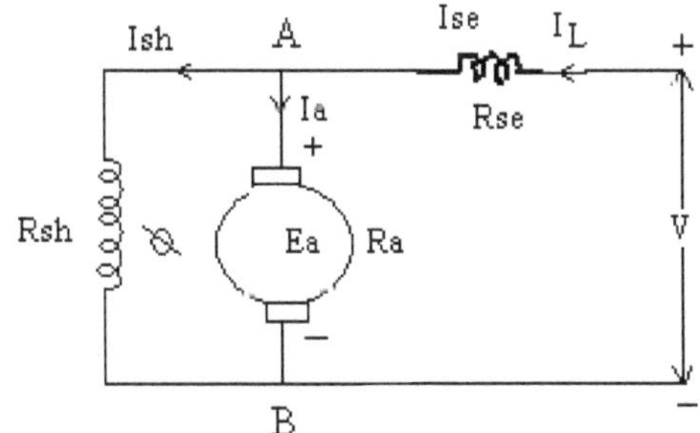

Figure 3.50 Short Shunt Compound Motor

$$V = E_b + I_a R_a + I_{se} R_{se} + B.D$$
$$I_a = I_L - I_{sh}$$
$$I_{sh} = \frac{V_{AB}}{R_{sh}} = \frac{V - I_{se} R_{se}}{R_{sh}} = \frac{E_b + I_a R_a}{R_{sh}}$$

3.3.4 Load Characteristics of a Separately Excited DC Motor

The operating performance of a DC motor say, Separately Excited DC motor is assessed by using three characteristics called *Load characteristics.*

a. Torque vs Armature Current Characteristics

b. Speed vs Armature Current Characteristics

c. Speed vs Torque Characteristics

1. *Torque vs Armature Current Characteristics:*

The connection diagram of a DC separately excited DC motor is shown in figure 3.51.

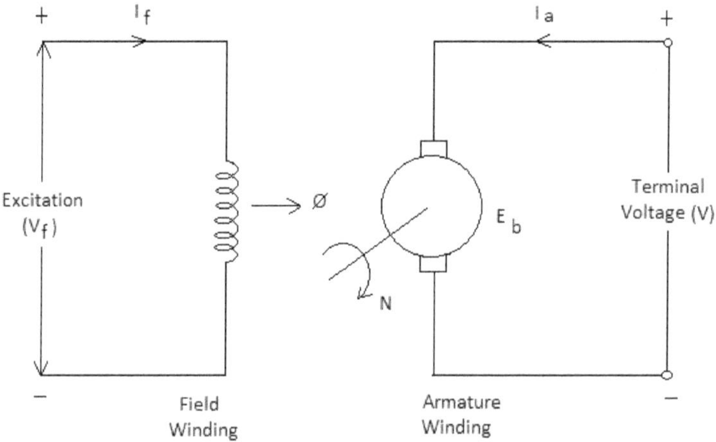

Figure 3.51 Separately Excited DC Motor

The supply voltage V is given by $V = E_b + I_a R_a + B.D$

Generally the brush drop is included in armature resistance drop $I_a R_a$.

Then, $\quad V = E_b + I_a R_a$

The electrical input to the armature winding $P_{in} = V \times I_a$

Armature copper losses $= I_a^2 R_a$

The net power $P_{in} - I_a^2 R_a = VI_a - I_a^2 R_a = I_a[V - I_a R_a] = I_a \times E_b = E_b \times I_a$

This power $E_b \times I_a$ is converted into mechanical power P_m. Hence, $P_m = E_b \times I_a \qquad .....(1)$

The mechanical power is expressed in mechanical quantities, Speed 'N' and Torque 'T'.

$$P_m = \frac{2\pi NT}{60} \qquad .....(2)$$

By equating the equations (1) and (2),

$$E_b \times I_a = \frac{2\pi NT}{60}$$

$$\frac{\phi ZN}{60} \times \frac{P}{A} \times I_a = \frac{2\pi NT}{60}$$

Or $\qquad T = \frac{1}{2\pi} \times \phi I_a \times \frac{ZP}{A}$

Or $\qquad T \propto \phi I_a$

As the field current I_f is maintained constant, the flux Ø is also maintained constant. Hence, the torque T is proportional to armature current. $T \propto I_a$

The Toque vs Armature current characteristics is shown below.

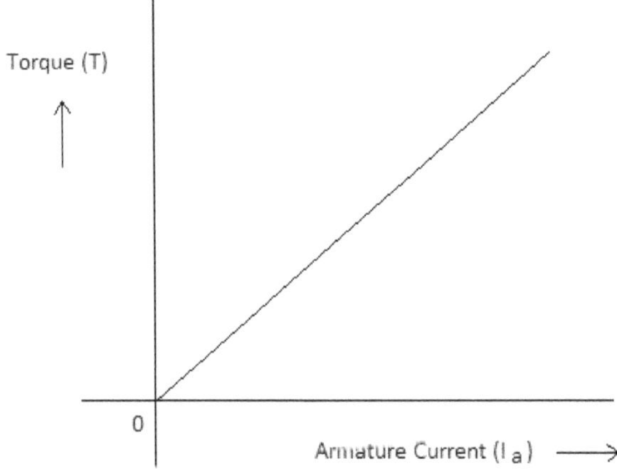

Figure 3.52 Torque vs Armature Current Characteristics

2. *Speed vs Armature Current Characteristics*

The voltage equation of a Separately Excited Motor is $V = E_b + I_a R_a$

Or $\qquad E_b = V - I_a R_a$ where $E_b = \frac{\phi ZN}{60} \times \frac{P}{A}$ or $E_b \propto \phi N$ or $E_b = k\phi N$

Where K is a constant.

Then, $\qquad k\phi N = V - I_a R_a$

Or $\qquad N = \frac{1}{k} \times \frac{V - I_a R_a}{\phi}$

Where V, Ø, R_a and k are constants.

When I_a is zero initially, the speed N is equal to $\dfrac{V}{k\phi}$. As then armature current increases, the drop I_aR_a increases and the speed slight decreases as shown in figure 3.53.

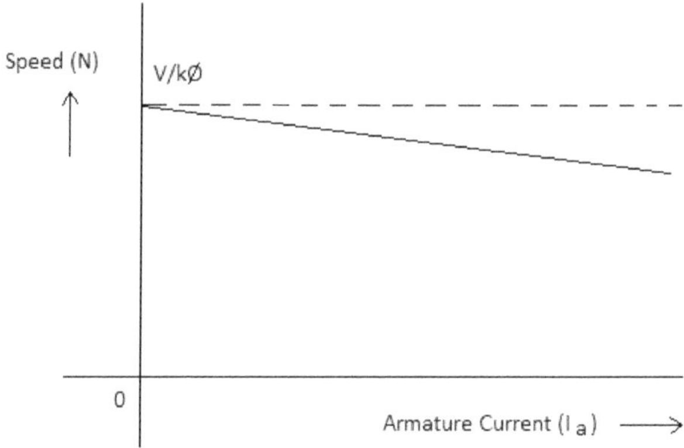

Figure 3.53 Speed vs Armature Current Characteristics

It is cleared from above characteristics that the speed decreases with load (or armature current I_a). These changes in speed with load variations are not desirable. There should be a regulation over the change in speed to be not beyond a tolerable limit called *Speed Regulation.*

Hence, *Speed Regulation is defined as the Change in speed from no-load to full load expressed against as percentage of no-load speed.*

$$\% \text{ Speed Regulation} = \frac{N_{NL} - N_{FL}}{N_{FL}} \times 100$$

For an ideal motor, the Speed Regulation is zero.

3. *Speed vs Torque Characteristics:*

Since, the Torque is proportional to armature current, the Speed vs Torque characteristics is similar to the Speed vs Armature current characteristics and is shown below.

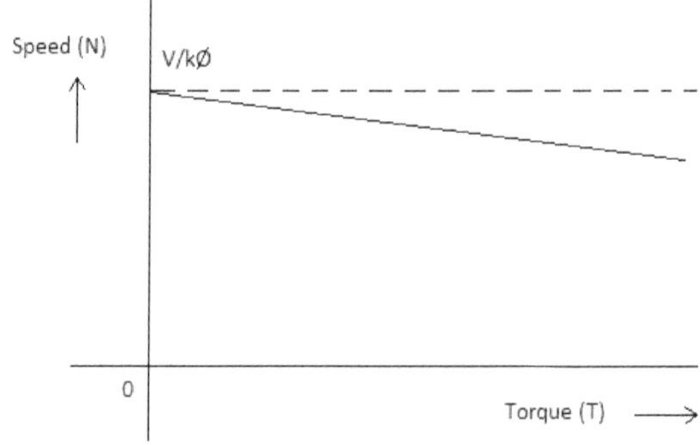

Figure 3.54 Speed vs Torque Characteristics

3.3.5 Speed Control of Separately Excited DC Motor

The speed of a DC motor needs to be controlled to control the process in industries. Consider a Separately excited DC motor to understand the principles of speed control.

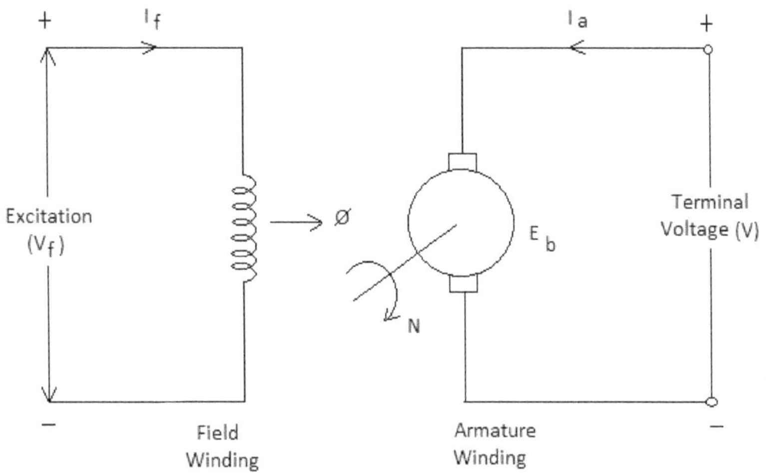

Figure 3.55 Separately Excited DC motor

The voltage equation of the armature circuit is given by is $V = E_b + I_a R_a$

Or $E_b = V - I_a R_a$ where $E_b = \dfrac{\phi ZN}{60} \times \dfrac{P}{A}$ or $E_b \propto \phi N$ or $E_b = k\phi N$

Where K is a constant.

Then, $k\phi N = V - I_a R_a$

Or $N = \dfrac{1}{k} \times \dfrac{V - I_a R_a}{\phi}$

Or $N \propto \dfrac{V - I_a R_a}{\phi}$

In above expression, the Supply Voltage *(V)* and Armature Resistance *(R_a)* are constant and the armature current *(I_a)* and flux *(Ø)* are variable quantities.   Hence, the speed of a DC motor can be controlled by controlling the magnetic field which is called *Field Control method* and by controlling the armature current (I_a) which is called *Armature Control method*. These two methods are briefly discussed below.

A. *Field Control Method:*

In this method, the speed is controlled by controlling the magnetic field *Ø* being armature voltage or supply voltage maintained constant.

$$N = \dfrac{1}{k} \times \dfrac{V - I_a R_a}{\phi}$$

Or $\qquad N \propto \dfrac{1}{\phi}$

The speed is inversely proportional to the magnetic flux. The speed increases as the flux value is decreased from its rated value. The field current I_f is controlled to control the magnetic field. Hence, a rheostat is connected in series with the field winding as shown.

Initially, the speed control resistance R_c is kept in minimum resistance position so that rated current flows through the field winding. The speed corresponding to this rated field current is the rated speed N_{rated}. As the resistance R_C is increased, the field current decreases, and the speed increases above its rated value.

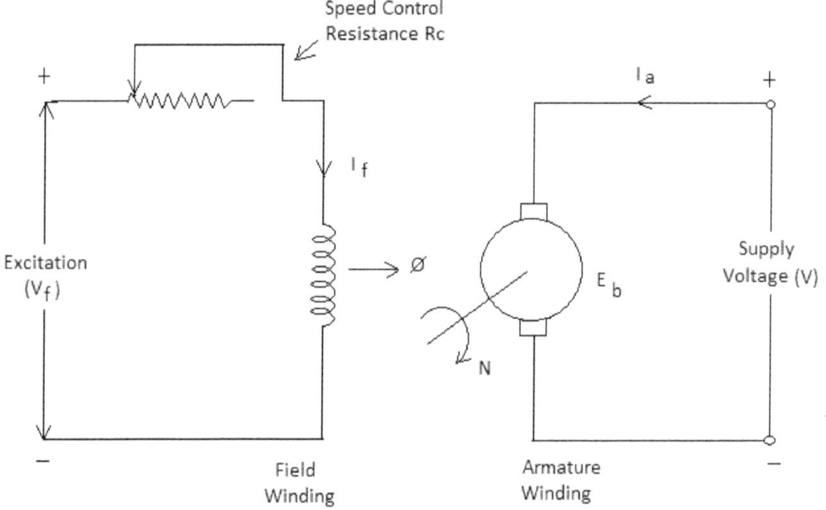

Figure 3.56 Field Control Method

The variations in speed (N) with change of field current (I_f) are shown in figure 3.57.

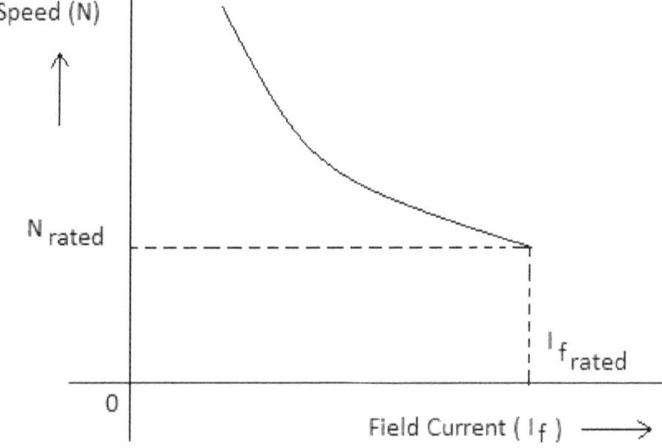

Figure 3.57 Field Current vs Speed Characteristics

The field control method is used to obtain the speeds above the rated value. If the field winding gets opened, the field current I_f becomes zero, the flux Ø become zero, and the speed become infinite or dangerously high which may damage the motor. Hence, the field winding should not get open circuited under any circumstances.

B. Armature Control Method:

In this method, the speed N is controlled by controlling I_a or indirectly controlling the voltage across the armature winding. The field current I_f is maintained constant in this method. A speed control resistance R_C is placed in armature circuit to drop some of the supply voltage and it is shown in figure.

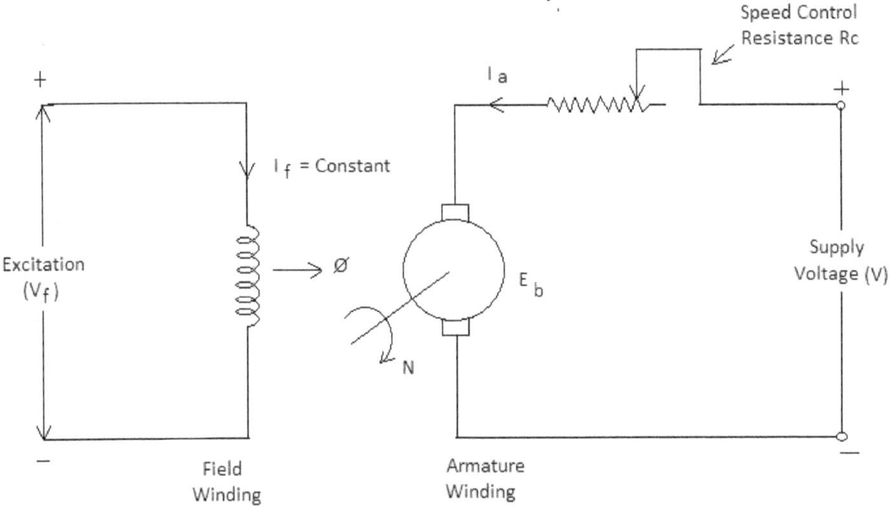

Figure 3.58 Armature Control Method

$$N = \frac{1}{k} \times \frac{V - I_a R_a}{\phi}$$

Since $I_a R_a$ drop is negligible when compared to the supply voltage V and Ø is also maintained constant by maintaining I_f constant,

$$N \propto V$$

To change the voltage across armature winding, a rheostat R_C is used. Initially, the rheostat is kept in minimum resistance position so that the voltage across armature winding *(E_b)* is its rated value and the motor runs at its rated speed. As the R_C value increases, the voltage across armature winding decreases, the speed decreases and it becomes zero as the voltage across armature winding (V) reduces to zero. The variation in speeds with respect to the changes in armature voltage are shown below.

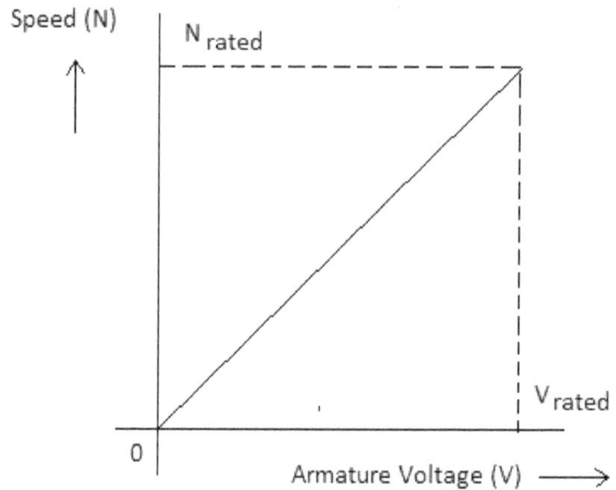

Figure 3.59 Armature Voltage vs Speed Characteristics

It is cleared from above characteristics that the armature control method is used to control the speed below its rated value.

Review Questions

Short Answer (Two marks) Questions

1. What is a Transformer?
2. List the uses of transformers
3. Write down the qualities of an Ideal Transformer
4. Differentiate between an Ideal Transformer and a practical transformer
5. Show the equivalent circuit of a transformer
6. Define Voltage Regulation
7. Specify the condition for maximum efficiency in a transformer
8. The Full load copper losses of a transformer are 200W. Find the copper losses at half full load.
9. Differentiate between a DC generator and a DC motor
10. State Fleming's Right Hand Rule
11. What are the major differences between Lap and Wave windings
12. Show the circuit symbol of a DC generator
13. Classify DC generators
14. Why Back Emf is called so?
15. State Fleming's Left Hand Rule
16. Show the torque vs armature current characteristic of DC separately excited motor

Essay (Six marks) Questions

1. Explain the working principle of a transformer

2. What are the main parts of a transformer? Explain briefly about its construction

3. Derive the emf equation of transformer

4. In a 110/440 Volt, 50 Hz Single Phase transformer, the maximum flux density is 1.2 wb/m^2. The emf per turn is 1.8 V. Determine a) Number of primary and secondary turns b) operating flux c) cross sectional area of the core. (Ans: 61,245, 8.12 mwb and 0.00676 m^2)

5. A 25 KVA, 220/440 V, 50 Hz transformer is having the leakage impedance of 0.2+j0.55 ohms on primary side and 0.8+j2.4 ohms on secondary side. Find the % voltage Regulation when transformer feeds ¾ full load current to a resistive load on secondary side. (Ans: 15.49%)

6. A 20 KVA, 400/200 V, 50 Hz Transformer is having the resistance of 48 milli ohms on primary side and 18 milliohms on secondary side. The iron losses amount to 200 W. Find the efficiency of the transformer at 1/2 Full load, 0.8 pf lagging. (Ans: 96.67%)

7. With the help of a neat sketch, explain the operation of a DC generator (or Simple Loop Generator)

8. Briefly explain the constructional aspects of a DC machine

9. Derive the expression for emf generated in DC generator

10. A 4 pole Lap connected DC motor having 1000 armature conductors absorbs a current of 25 A from a DC source. The flux per pole is given as 24 mwb. Find the supply voltage to run the motor at 1500 rpm speed. The armature resistance is 0.8 ohms and also consider the brush drop as 2V. (Ans: 622 V)

11. Explain the operating characteristics of Separately Excited DC motor

12. What are the methods available to control the speed of a Separately Excited dc motor? Explain in brief.

CHAPTER 4

Alternating Current Machines

4.1 Three Phase Induction Motors

4.1.1 Introduction

Induction Machines use Alternating Current (AC) supply for their operation. These machines find wide applicability as a motor in Industry. More than 85 % of industrial motors in use today are induction motors and hence it is generally referred as *work horse*. The reasons for Induction Motors to gain much popularity are simple and rugged construction, higher efficiency and practically constant speed.

4.1.2 Construction

The main parts of an Induction motor are i) Stator and ii) Rotor

(a) Stator: The stator has got the slots on its inner periphery as shown in figure 4.1. The stator houses a three phase distributed winding.

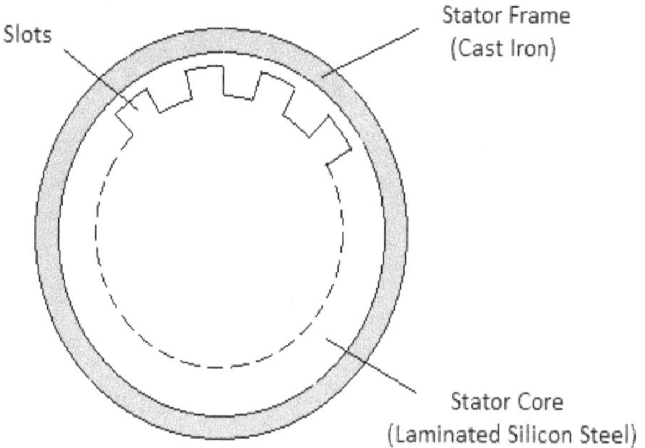

Figure 4.1 Cross sectional view of stator

(b) Rotor:

There are two types of rotors constructions generally employed.

1. Phase wound rotor   and   2. Squirrel Cage rotor.

In these two types, the rotor core is laminated and punched slots for accommodating rotor winding or rotor bars.

1. Phase wound rotor

This rotor is provided with slots on its outer periphery as shown in figure 4.2.A 3- phase distributed winding is placed in rotor slots. The rotor winding must be also designed for same number of poles as that of Stator winding.

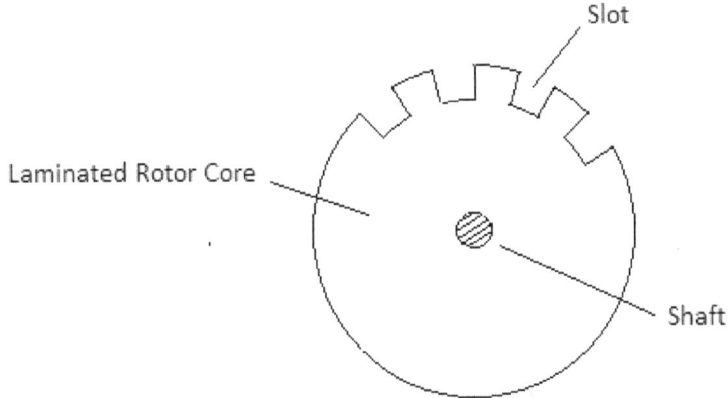

Figure 4.2 Cross sectional view of Phase Wound Rotor

The 3phase winding is connected in star or delta and the three terminals are brought out of the machine through three slip rings placed on shaft and brushes. An external resistance is added across rotor terminals in starting period to reduce the starting currents simultaneously to improve the starting torque. It is depicted from the figure 4.3.

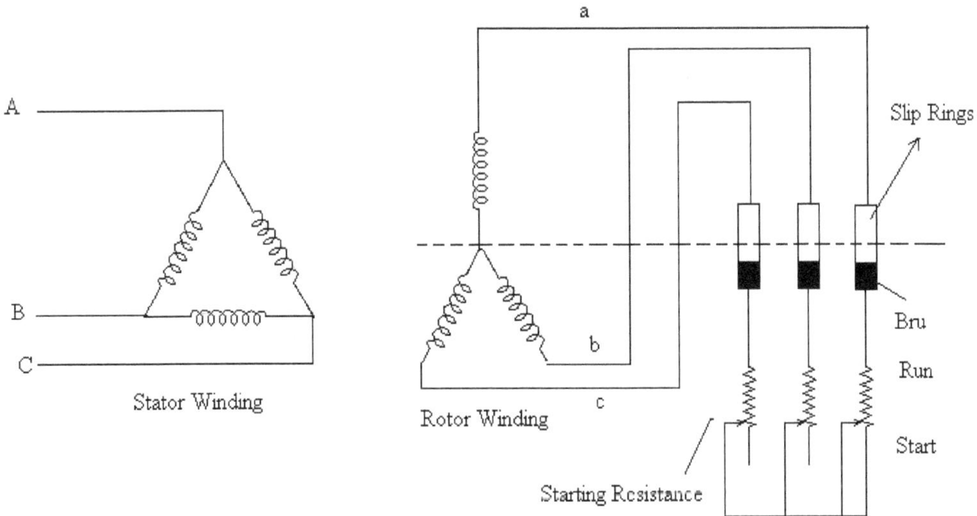

Figure 4.3 Slip Ring Induction Motor

An induction motor which makes use a wound rotor is called 'Slip Ring Induction Motor (SRIM)' and these rotors are used for large size machines where high starting torque is required.

2. Squirrel Cage rotor

The squirrel cage rotor has solid bars of conducting material placed in rotor slots and shorted through end rings on each side as shown in figure 4.4. *The rotor resembles squirrel cage, hence the name' Squirrel Cage Rotor'.*

Since the rotor is permanently short circuited, its resistance can't be changed. But this type of rotor adjusts itself for any no. of stator poles. This type of rotors cannot produce high staring torques as compared to slip ring induction motors, but their running performance is excellent.

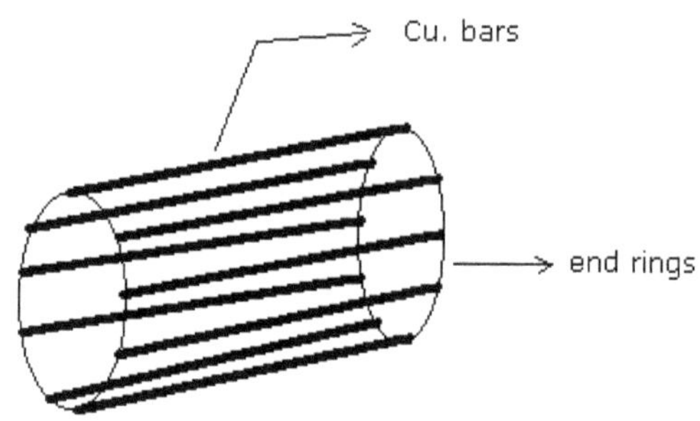

Figure 4.4 Squirrel Cage Rotor

An induction motor which makes use this type of rotors is called 'Squirrel Cage Induction Motor (SQIM)'.

4.1.3 Principle

A Three Phase Induction Motor is a Singly Excited Motor which excited only from its stator winding end. It works on the principle of induction. Hence, it has got the name *'Induction Motor'*. When the three phase winding placed on stator is excited by a three phase supply, it produces a magnetic field which is rotating in nature. This magnetic field is called *'Rotating Magnetic Field (RMF)'.The speed of this rotating magnetic field is called 'Synchronous Speed N_s'.* The expression for synchronous speed is given by

$$N_S = \frac{120 \times f}{P}$$

Where *'f'* is the supply frequency and *'P'* is total number of poles for which the stator winding is made.

The production of Rotating Magnetic Field (RMF) by a three phase winding has been briefly explained below. Consider a stator having three phase winding (three windings mutually displaced by 120^0 in space) excited by a three-phase supply.

At the instant 't_1', the current in phase 'a' is maximum and the currents passing through phases 'b' and 'c' are equal and negative. Hence, the flux concentrates more at phase 'a' winding and returns through phases 'b' and 'c' as shown in figure 4.6 (a).

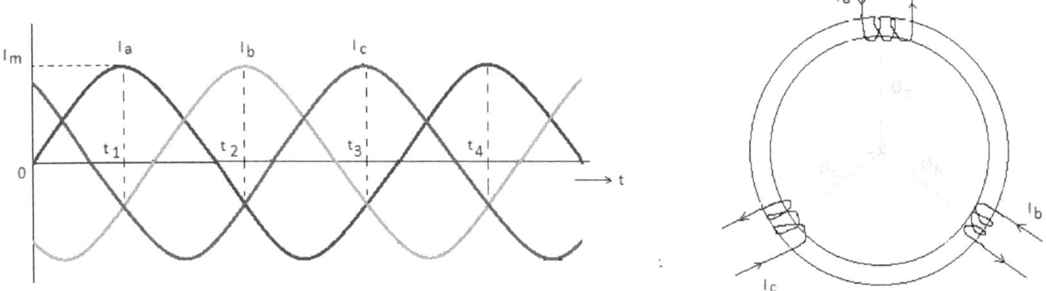

Figure 4.5 a) Three Phase Supply   b) Stator

Similarly, the flux focusses more at phases 'b' and phase 'c' at the instants t_2 and t_3 since their current magnitudes are maximum at those instants. These are depicted in the figures 4.6(b) and (c) respectively. The remaining phases act as return paths at each instant. As it happens continuously with respect to time, the flux is found to be rotating in clockwise direction at a speed called 'Synchronous Speed (N_S)'.

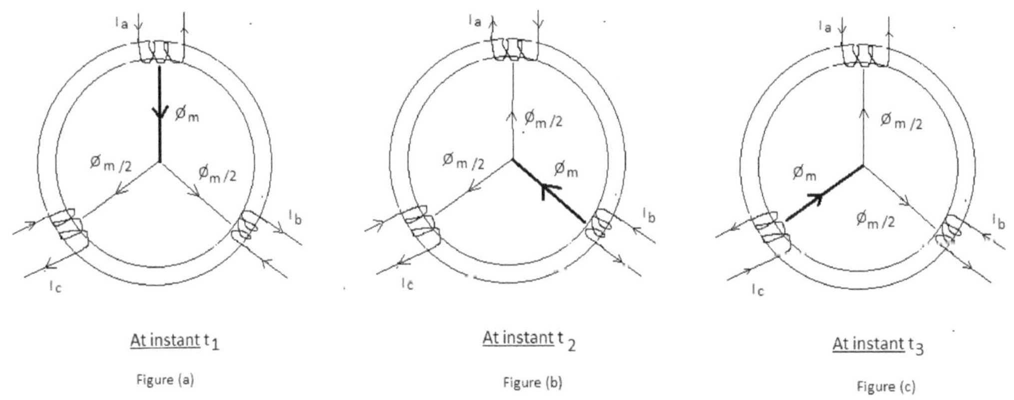

Figure 4.6 Production of Rotating Magnetic Field

The Rotating Magnetic Field produced by stator produces an emf E_2 in the rotor winding which is initially maintained at rest or zero speed. This emf gets converted into the current I_2 as the rotor winding is short circuited. According to Len'z law, this current opposes the cause of its generation. The cause is the relative speed between stator field and rotor and it is given by $N_S - 0$. To reduce the relative speed, either N_S should decrease or the rotor speed should increase. Since, the synchronous speed N_S is fixed as long as the supply frequency (f) and number of poles(P) are maintained constant, the rotor speed increases from zero.

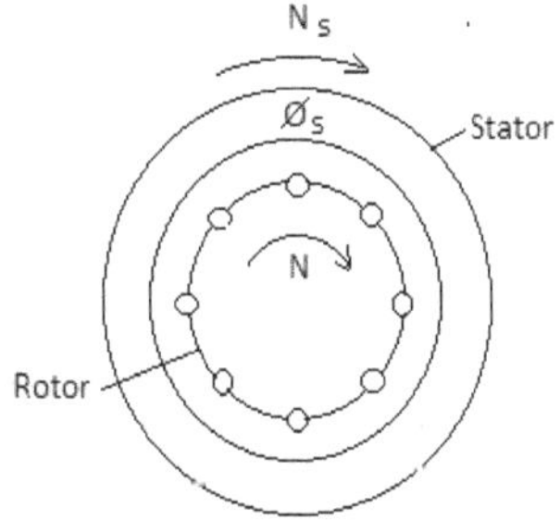

Figure 4.7 Stator and Rotor Fluxes

If rotor attains the speed of N_S to make the relative speed completely zero between stator flux and rotor, no emf and hence no current is produced in the rotor winding. Hence, rotor need not obey the Len's law. So, the rotor starts reducing its speed below N_S. Likewise, the rotor does not reach the synchronous speed N_S or does not come down to zero speed. It settles at a speed 'N' which is less than N_S. *An Induction Motor does not rotate at Synchronous Speed. Hence, it is called 'Asynchronous Motor'.*

The difference between Synchronous speed (N_s) and actual rotor speed (N) is called 'Slip speed'.

Slip speed $= N_S - N$

$$\% \text{ Slip } s = \frac{N_S - N}{N_S} \times 100$$

At stand-still (ie., when rotor speed $N = 0$), the value of slip

$$s = \frac{N_S - 0}{N_S} \times 100 = 1 \text{ or } 100\%$$

If the rotor reaches synchronous speed N_S ,ie., $N = N_S$, the value of slip

$$s = \frac{N_S - N}{N_S} \times 100 = \frac{N_S - N_S}{N_S} = 0$$

As the rotor speed changes, slip also changes. But, the synchronous speed N_S is maintained constant as long as frequency and number of poles are maintained constant. Since rotor emf E_2 produced is proportional to the slip speed, the frequency of rotor emf is given by

$$f_r = \frac{P \times Slip\ Speed}{120} = \frac{P \times (N_S - N)}{120}$$

Dividing and multiplying the above equation with N_S, we get

$$f_r = \frac{P \times (N_S - N)}{120} \times \frac{N_S}{N_S}$$

$$= \frac{P N_S}{120} \times \frac{N_S - N}{N_S}$$

$$= s \times f$$

$$\boxed{\textbf{\textit{Rotor frequency } (f_r) = \textit{stator frequency } (f) \times \textit{slip } (s)}}$$

The induction motor maintains much analogy with a transformer. Hence, it is called *Generalized Transformer.* The primary and secondary windings of transformer are analogous to stator and rotor windings of induction motor. *An Induction motor is a Rotating Transformer.* Hence, the same emf equation of transformer holds good for induction motor also.

The emf induced in rotor per phase at stand-still (when slip $s = 1$),

$$E_2 = 4.44 \times \phi_m \times f_r \times N_2 = 4.44 \times \phi_m \times s \times f \times N_2 = 4.44 \times \phi_m \times f \times N_2$$

Where N_2 is number of turns on rotor winding per phase and slip $s = 1$ in starting or at stand-still.

Similarly, the rotor emf under running conditions

$$= 4.44 \times \phi_m \times f_r \times N_2 = 4.44 \times \phi_m \times s \times f \times N_2 = s \times 4.44 \times \phi_m \times f \times N_2 = s \times E_2$$

Likewise, all the rotor quantities which are concerned to frequency change from starting to running conditions. The rotor leakage reactance at starting,

$$X_2 = 2\pi f L$$

The rotor reactance under running conditions,

$$= 2\pi f_r L = 2\pi \times sf \times L = s \times 2\pi f L = sX_2$$

Similarly, the leakage impedance at stand-still is $\sqrt{R_2^2 + X_2^2}$

Under running conditions, the leakage impedance is $\sqrt{R_2^2 + (sX_2)^2}$

The following table 4.1 shows how the different rotor quantities change from starting to running conditions.

Table 4.1 Rotor quantities during starting and running

Rotor parameter	At stand-still	Under running
Slip	1	0
Frequency	f	sf
Resistance	R_2	R_2
Leakage reactance	X_2	sX_2
Leakage impedance	$R_2 + jX_2$	$R_2 + jsX_2$
Rotor power factor	$\dfrac{R_2}{\sqrt{R_2^2 + X_2^2}}$	$\dfrac{R_2}{\sqrt{R_2^2 + (sX_2)^2}}$
Emf	E_2	sE_2
Current	$\dfrac{E_2}{\sqrt{R_2^2 + X_2^2}}$	$\dfrac{sE_2}{\sqrt{R_2^2 + (sX_2)^2}}$

Exercise 4.1.1: A 3-ph, 4-pole, 50 Hz induction motor has a full load speed of 1440 rpm. Calculate a) Synchronous Speed b) % slip and c) Rotor frequency at full load

Solution:

Number of Poles $P = 4$

Supply frequency or stator frequency $f = 50 \, \text{Hz}$

Speed of the motor at full load $N = 1440 \, \text{rpm}$

(a) Synchronous Speed $N_S = \dfrac{120 \times f}{P} = \dfrac{120 \times 50}{1500} = 1500 \, \text{rpm}$

(b) % Slip at Full load $s = \dfrac{N_S - N}{N_S} \times 100 = \dfrac{1500 - 1440}{1500} \times 100 = 4\%$

(c) Rotor frequency $f_r = s \times f = 0.04 \times 50 = 2 \, \text{Hz}$

Exercise 4.1.2: A 6 Pole, 3 phase Induction Motor rotates at 970 rpm and maintains a slip of 3%. Find a) Synchronous Speed b) Supply frequency and c) Rotor frequency

Solution:

Number of poles $P = 6$

Speed of the motor $N = 970 \, \text{rpm}$

Slip $s = 3\% = 0.03$

(a) Slip $s = \dfrac{N_S - N}{N_S}$

$0.03 = \dfrac{N_S - 970}{N_S}$

Or $970 = (1 - 0.03) N_S$

$N_S = \dfrac{970}{1 - 0.03} = 1000 \, \text{rpm}$

(b) Since $N_S = \dfrac{120 \times f}{P}$

$1000 = \dfrac{120 \times f}{6}$

Or supply frequency $f = \dfrac{1000 \times 6}{120} = 50 \, \text{Hz}$

(c) Rotor frequency $f_r = sf = 0.03 \times 50 = 1.5 \, \text{Hz}$

Exercise 4.1.3: A Center Zero Galvanometer is connected in rotor circuit of a $3 - \text{phase}$, 6 pole, 50 Hz slip ring induction motor. During its running, if galvanometer makes 90 oscillations per minute, calculate the rotor speed.

Solution :

Number of poles $P = 6$

Supply frequency $f = 50 \, \text{Hz}$

Synchronous Speed $N_S = \dfrac{120 \times f}{P} = \dfrac{120 \times 50}{6} = 1000\,\text{rpm}$

As rotor completes 90 oscillations per minute, the number of oscillations per sec is given

by $\dfrac{90}{60} = 1.5$

Therefore, the rotor frequency f_r = number of rotor cycles / sec $= 1.5\,\text{Hz}$.

$$\text{Slip s}\; s = \frac{f_r}{f} = \frac{1.5}{50} = 0.03\,\text{or}\,3\%$$

$$\text{And also slip}\; s = \frac{N_S - N}{N_S}$$

$$0.03 = \frac{1000 - N}{1000}$$

Or $\qquad\qquad N = 970\,\text{rpm}$

Exercise 4.1.4: A 3 Phase, 4 –pole, 50 Hz induction motor has the rotor leakage impedance of 2+j10 ohms per phase and rotor emf of 150 V per phase at stand-still. Calculate a) slip b) rotor frequency c) rotor power factor d) rotor current at i) stand-still ii) rotor speed of 1440 rpm.

Solution:

Number of poles $P = 4$

Supply frequency $f = 50\,\text{Hz}$

Rotor leakage impedance at stand-still $Z_2 = R_2 + jX_2 = 2 + j10\,\text{ohms}$

Rotor emf/phase at stand-still $E_2 = 150\,\text{V}$

Synchronous Speed $N_S = \dfrac{120 \times f}{P} = \dfrac{120 \times 50}{1500} = 1500\,\text{rpm}$

(i) At stand-still condition ($N = 0$ and slip $s = 1$)

(a) Slip $s = \dfrac{N_S - N}{N_S} = \dfrac{1500 - 0}{1500} = 1\,\text{or}\,100\%$

(b) Rotor frequency $f_r = s \times f = 1 \times 50 = 50\,\text{Hz}$

(c) Impedance $Z_2 = \sqrt{R_2^2 + (sX_2)^2} = \sqrt{2^2 + (10)^2} = 10.19\,\text{ohms}$

Rotor Power factor $= \dfrac{R_2}{Z_2} = \dfrac{2}{10.19} = 0.196$

(d) Rotor Emf $E_2 = 150\,\text{V}$

$$\text{Rotor current } I_2 = \frac{E_2}{Z_2} = \frac{150}{10.19} = 14.72 \text{ A}$$

(ii) At the rotor speed of 1440 rpm ($N = 1440 \text{ rpm}$)

(a) Slip $s = \dfrac{N_S - N}{N_S} = \dfrac{1500 - 14400}{1500} = 0.04 \text{ or } 4\%$

(b) Rotor frequency $f_r = s \times f = 0.04 \times 50 = 2 \text{ Hz}$

(c) Impedance $Z_2 = \sqrt{R_2^2 + (sX_2)^2} = \sqrt{2^2 + (0.04 \times 10)^2} = 2.04 \text{ ohms}$

$$\text{Rotor Power factor} = \frac{R_2}{Z_2} = \frac{2}{2.04} = 0.9803$$

(d) Rotor current $I_2 = \dfrac{sE_2}{Z_2} = \dfrac{0.04 \times 150}{2.04} = 2.94 \text{ A}$

Exercise 4.1.5: A 3 Phase, 4 pole, 50 Hz induction motor has star connected rotor winding having a resistance of 0.2 ohm /ph and a stand-still leakage reactance of 1 ohm/ph. When the stator is energized at rated voltage and frequency, the rotor induced emf at stand-still is found to be 208 V between any two slip rings. Also a star connected external resistance of 1 ohm/ph is connected in rotor circuit during starting period. Calculate rotor current, rotor power factor at i) stand-still ii) rotor speed of 1420 rpm.

Solution :

Number of poles $P = 4$

Supply frequency $f = 50 \text{ Hz}$

Synchronous Speed $N_S = \dfrac{120 \times f}{P} = \dfrac{120 \times 50}{1500} = 1500 \text{ rpm}$

Rotor resistance /phase $R_2 = 0.2 \text{ ohm}$

Stand – Still Rotor leakage reactance /phase $X_2 = 1 \text{ ohm}$

Stand-Still rotor emf /ph $E_2 = \dfrac{208}{\sqrt{3}} = 120 \text{ V(since rotor is star connected)}$

(i) At Starting or Stand-Still (ie., $N = 0$, $s = 1$)

Rotor emf /phase $= 120 \text{ V}$,

Rotor leakage impedance /phase $Z_2 = \sqrt{(R_2 + R_{ex})^2 + X_2^2} = \sqrt{(0.2 + 1)^2 + 1^2} = 1.562 \text{ ohms}$

Rotor current /ph $I_2 = \dfrac{E_2}{Z_2} = \dfrac{120}{1.562} = 76.9 \text{ A}$

Rotor Power factor $= \dfrac{R_2 + R_{ex}}{Z_2} = \dfrac{1.2}{1.562} = 0.768$

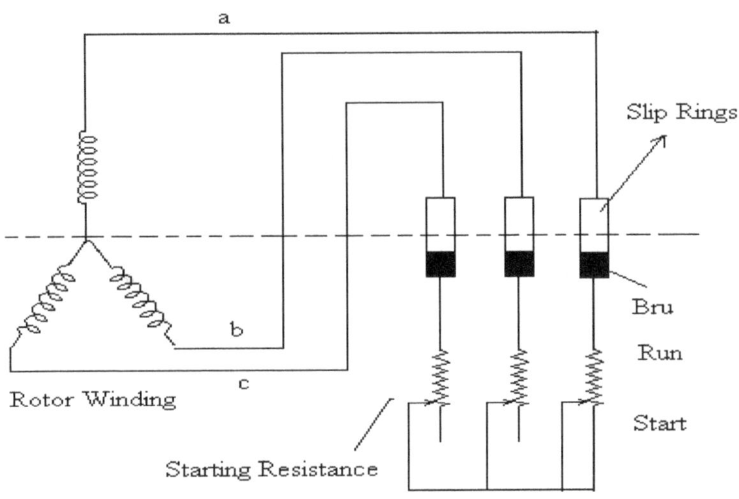

(ii) At rotor speed of 1440 rpm

$$s = \frac{N_S - N}{N_S} = \frac{1500 - 14400}{1500} = 0.04$$

Rotor emf /phase $sE_2 = 0.04 \times 120 = 4.8 \text{ V}$

Rotor leakage impedance /phase $R_2 + jsX_2 = 0.2 + j(0.04 \times 1) = 0.2 + j0.04 \text{ ohms}$

The magnitude of Z_2 is $Z_2 = \sqrt{0.2^2 + 0.04^2} = 0.2055 \text{ ohms}$

Rotor current /phase $= \dfrac{sE_2}{Z_2} = \dfrac{4.8}{0.2055} = 23.35 \text{ A}$

Rotor Power factor $= \dfrac{R_2}{Z_2} = \dfrac{0.2}{0.2055} = 0.973$

4.1.4   Torque – Slip Characteristics

The torque is produced in an induction motor due to the interaction between stator and rotor fluxes. Hence, the torque is always proportional to the product of stator and rotor fluxes and also the rotor power factor.

$$\text{Torque } T \propto \phi_S \times \phi_R \times Cos\theta_2$$

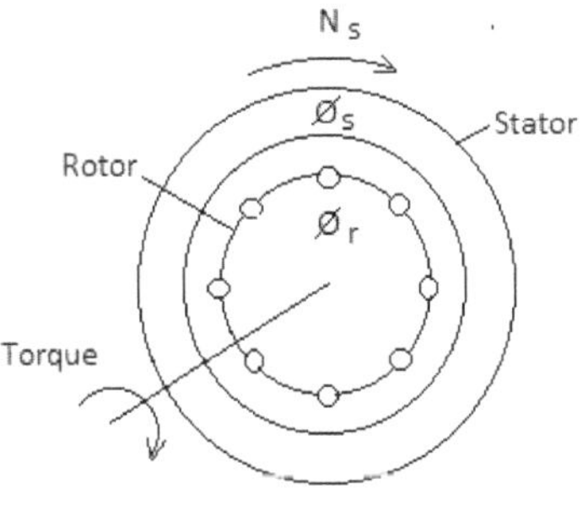

Figure 4.8 Stator and Rotor Fluxes

Where $Ø_S$ is the stator flux which is proportional to the supply voltage V, $Ø_R$ is the rotor flux which depends up on the rotor current. The rotor current at any instant is given by $I_2 = \dfrac{sE_2}{Z_2}$.

But, the stand-still rotor emf E_2 is proportional to stator flux which in turn proportional to supply voltage V. And the rotor power factor $Cos\ \Theta_2$ is given by $\dfrac{R_2}{Z_2}$.

Then, the torque produced in induction motor $T \propto V \times \dfrac{sE_2}{Z_2} \times \dfrac{R_2}{Z_2}$

Or $\qquad T \propto V \times \dfrac{sV}{Z_2} \times \dfrac{R_2}{Z_2}$ (since E_2 is proportional to V)

Or $\qquad T = K \times V^2 \times s \times \dfrac{R_2}{Z_2^2}$

$$T = \dfrac{KsV^2 R_2}{R_2^2 + (sX_2)^2}$$

The above equation relates the torque with slip and hence, it is useful to obtain the Torque-Slip characteristics.

At lower values of slip 's'(ie., at higher values of rotor speeds), the value sX_2 is very small when compared to R_2. Hence, sX_2 can be neglected. Then, torque equation becomes

$$T = \dfrac{KsV^2 R_2}{R_2^2} = \dfrac{KsV^2}{R_2}$$

Or $T \propto s$ (since, K, V and R_2 are constants)

The slip 's' is linearly related to torque. The torque increases as the slip increases at lower values of slip and it is shown in figure 4.9.

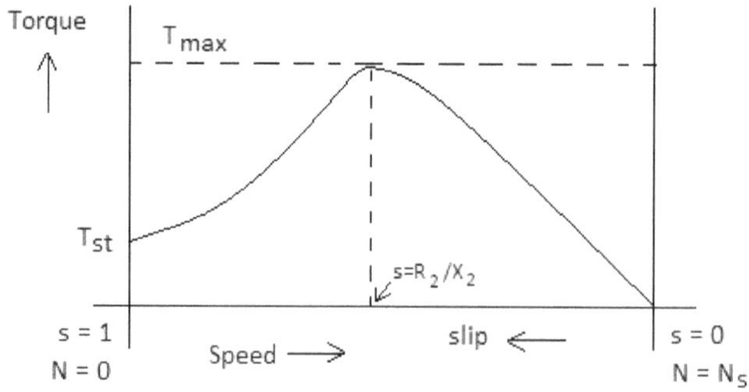

Figure 4.9 Torque-Slip Characteristics

Similarly, at higher values of slips (ie., at lower values of rotor speeds), the value sX_2 is very large when compared to R_2. Hence, R_2 can be neglected. Then, torque equation becomes

$$T = \frac{KsV^2R_2}{(sX_2)^2} = \frac{KV^2R_2}{sX_2^2}$$

Or $\quad T \propto \dfrac{1}{s}\quad$ (since, K, V and R_2 are constants)

The torque is inversely proportional to the slip at higher values of slip. The torque decreases as the slip increases which is depicted in the above figure4.9.

The torque becomes maximum at particular slip whose value is given by $s = \dfrac{R_2}{X_2}$ and the corresponding torque is called *Maximum Torque T_{max}*. Similarly, the torque exerted at starting when slip $s = 1$ is called *Starting Torque T_{st}*. The starting and Maximum Torques are also identified on Torque-Slip Characteristics shown in figure 4.9.

4.2 Synchronous Generators

The Synchronous generators generate Alternating Currents (AC). Hence, these are also called as 'A.C Generators' or 'Alternators'. The Synchronous generators convert the mechanical power into AC electrical power.

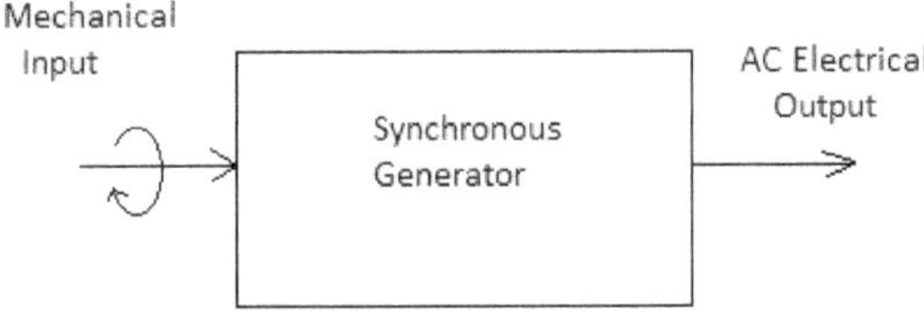

Figure 4.10 Synchronous Generator block diagram

These generators work on the same principle as that of d.c generators ie., Faraday's Laws of electromagnetic induction. The basic nature of emf induced in DC generator is also alternating and the commutator is used to convert the AC into DC. But, in AC generators, the deliberation is to generate AC output itself, *no commutator is used*. This is the major constructional difference between an Alternator and a DC generator. Another difference between these two machines is that *field winding is stationary and armature is rotating in D.C machines whereas in Synchronous machines, field winding is rotating and armature winding is stationary.*

4.2.1 Principle

Synchronous Generators work on the principle that when a conductor is rotated in presence of magnetic field or a magnetic field is moved around a conductor or whenever there is a relative motion between the conductor and magnetic field, an emf is induced in the conductor. The magnitude of this emf is proportional to rate of change of flux linkages and the direction of induced current is determined by using Fleming's Right Hand Rule.

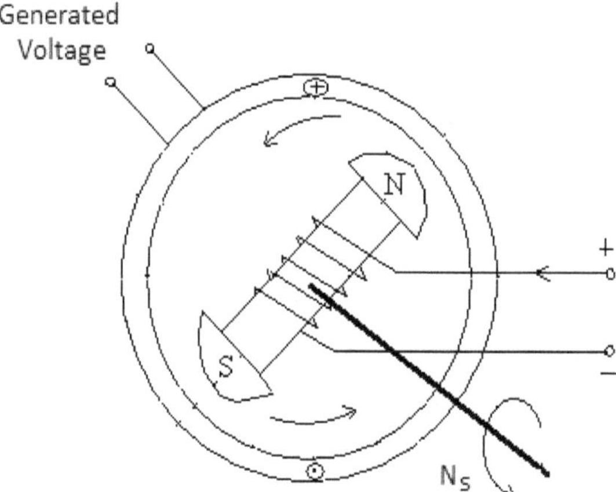

Figure 4.11 Cross sectional view of synchronous Machine

Consider a Synchronous generator whose field winding is placed on rotor and excited by DC to form the poles on rotor as shown in figure 4.11. The rotor is rotated at a speed called '*Synchronous Speed N_S*' using a prime-mover. The stator accommodates a distributed armature winding. As the rotor rotates, the flux produced by the rotor poles sweep over or link with armature conductors and induce an emf. Since, the poles which cross over the armature conductors are alternatively North and South poles, an alternating emf is generated in the conductors and it is shown in the figure 4.12.

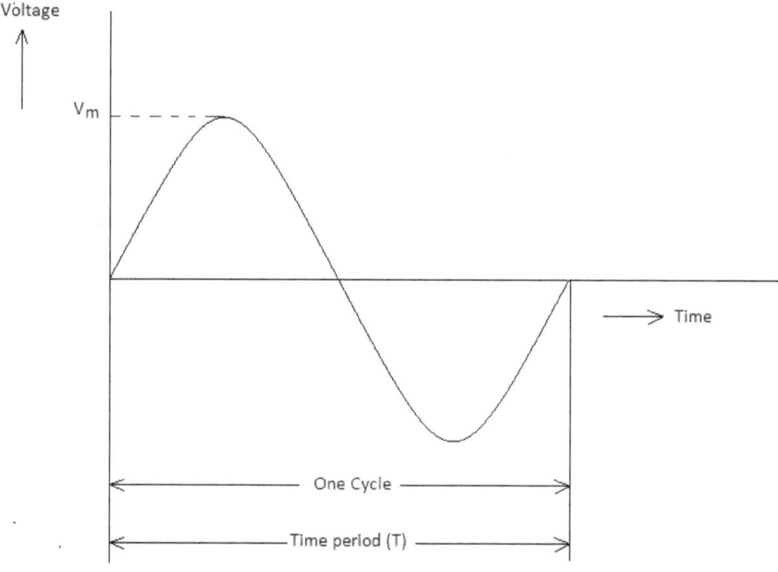

Figure 4.12 Sinusoidal Voltage

Since the air gap flux is sinusoidally distributed, the emf generated is also sinusoidal as shown in figure 4.12. The frequency of generated emf is found as follows.

In this case, the number of cycles produced in one revolution is equal to 1 as two poles have been placed. In general, the number of cycles formed in one revolution is equal to $\dfrac{P}{2}$.

Number of Cycles per Revolution $= \dfrac{P}{2}$

Frequency (f) = Number of cycles per seconds $= \dfrac{\text{No.of cycles}}{\text{revolution}} \times \dfrac{\text{No.of Revolutions}}{\text{sec}}$

$$f = \dfrac{P}{2} \times \dfrac{N_S}{60}$$

$$f = \dfrac{PN_S}{120}$$

Where 'P' is total number of poles and 'N' is the speed of the generator in rpm (revolutions per minute)

The magnitude of emf induced per phase $E = 4.44 \times \phi \times f \times N$

Where 'N' is number of armature winding turns per phase

4.3 Steeper Motor

The Stepper Motor is one of the important special machines that find many applications. The special machines are intended for special applications and the windings of these machines are generally switched on to the supply through electronic circuits. The other machines which fall under the category of special machines are BLDC (Brushless DC Motor), Servo Motor, Reluctance Motor, PMSM (Permanent Magnet Synchronous Motor) etc.

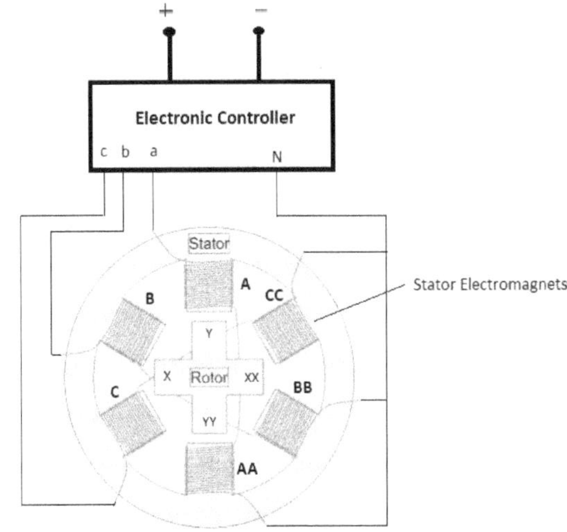

Figure 4.13 Cross sectional view of Stepper Motor

A Stepper Motor, also known as Stepping Motor rotates in equal steps instead of rotating continuously. That means, the full rotation is divided into a number of equal steps. The rotor position can be controlled and can hold at one of these steps according to the requirement. For understanding the operation of a stepper motor, a Variable Reluctance Stepper Motor is considered (The other types of stepper motors are Permanent Magnet Stepper Motor and Hybrid Synchronous Stepper Motor).

There are 6 electromagnets placed on stator and these are formed into three groups A,B and C. Each group is called a *phase*. The electromagnets A and AA form one group, B and BB form the second group and the C and CC constitute the third group. These electromagnets are energized in a sequence by the electronic controller (mostly micro controller based). When the stator electromagnets A and AA are energized through electronic controller, the flux in the machine acts along A and AA and rotor teeth Y and YY align along this axis to make the reluctance minimum. Now, the electromagnets A and AA are de-energized and B and BB are energized. Then, the flux axis shifts to B and BB and the rotor tooth X align with B and XX with BB and hence the rotor rotates in clockwise direction through an angle of 30^0. This angle is called *Step Angle*. Now, the electromagnets B and BB are de-energized and supply is given to C and CC resulting that the rotor tooth YY coincides with C and Y with CC. the rotor further rotates through 30^0 in same direction. Likewise, the rotor rotates in the steps of 30^0 as the excitation changes from A-AA to B-BB and B-BB to C-CC and so on.

Applications:

Stepper motors are extensively used in commercial applications like Flatbed Scanners, Computer Printers, 3D printers, Plotters, Slot Machines, image scanners, CD drives, intelligent lighting systems, camera lenses, CNC machines etc.

Stepper motors are also used in the field of lasers and optics as linear actuators and mirror mounts. These motors find applications in Fluid Control Systems too.

4.4 Electrical Installations and Batteries

Wiring System: Wiring system comprising of the connecting wire/cables that connects various electrical accessories like supplier meter board, controlling switches and protective devices like fuses, MCBs and electrical energy consuming devices called loads like fan, light, TV, refrigerator etc to the supply. This system is intended to distribute the electrical energy to the loads.

4.4.1 Basic lay-out of wiring in Domestic Installations

Generally, a single phase 230V or a three phase 415V AC supply is used for domestic installations. Single Phase circuit is connected between a phase and neutral so that 230V is applied to all loads.

The lay-out of wiring in a domestic installation is shown below in the figure 4.14. The supply is taken between a phase (A or B or C) and neutral from the pole using a service wire. A fuse is placed in phase on the pole itself to protect the service wire against overloading. The supply directly comes to energy meter which is provided to measure the electrical energy consumed by the consumer. The energy meter is the supplier (State Electricity Board) property and is sealed so that no tampering is done. The supply from energy meter is taken through a main MCB (Miniature Circuit Breaker) as shown. This main MCB controls the power given to entire domestic installation. The supply from main MCB is taken and given to Distribution Board (D.B).

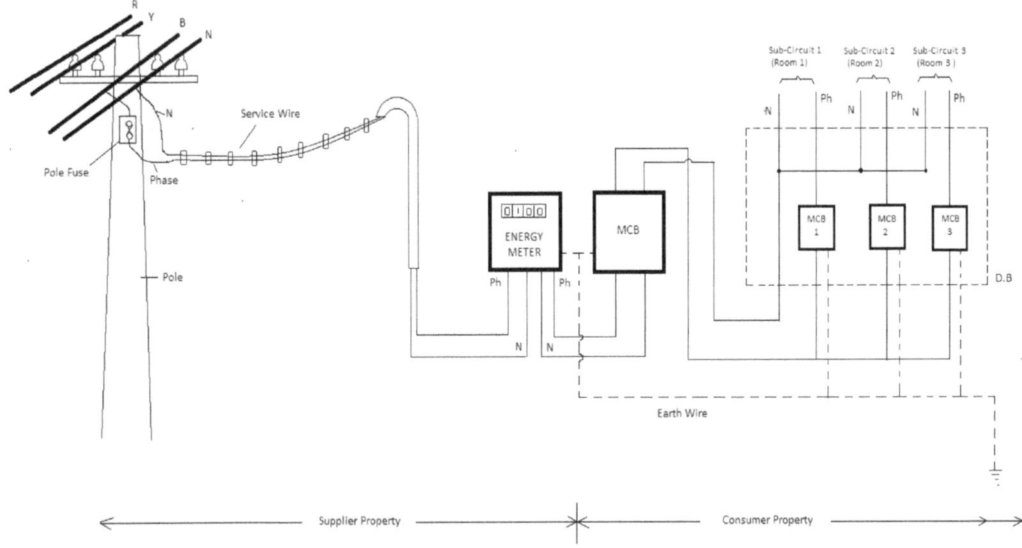

Figure 4.14 Lay-out of Wiring in Domestic Installations

In Distribution Board (D.B), the circuit is divided into many sections or sub-circuits. Each sub-circuit gets an exclusive phase and neutral. A separate MCB is placed in each sub-circuit for protection. If any problem like short circuit or over loading occurs in a section, the corresponding MCB is tripped and that particular sub-circuit is separated being other sub-circuits function normally. According to the Indian Standards, the maximum number of load points in a sub-circuit is 10 and maximum power rating is 800 W. The Fans, Lights, 5A sockets are called load points.

The metal covers of energy meter, MCBs and Distribution Board are connected to earth wire to avoid the electric shock to the consumers.

4.4.2 Types of Wiring Systems

The various wiring systems in usage are

1. Cleat Wiring
2. Wooden Casing and Capping Wiring

3. CTS or TRS Wiring
4. Lead Sheathed or Metal Sheathed Wiring
5. Conduit Wiring
   a) Surface Conduit type
   b) Concealed type

1. ***Cleat Wiring:*** PVC wires are used in this type of wiring. Porcelain Cleats are used to hold the wires or cables about 6 mm above the wall. The cleats are made into two parts. The Base and Cover. The base part is provided with grooves to house the conductors. The number of grooves depends up on the number of conductors. The distance between the cleats should not be more than 60 cms.

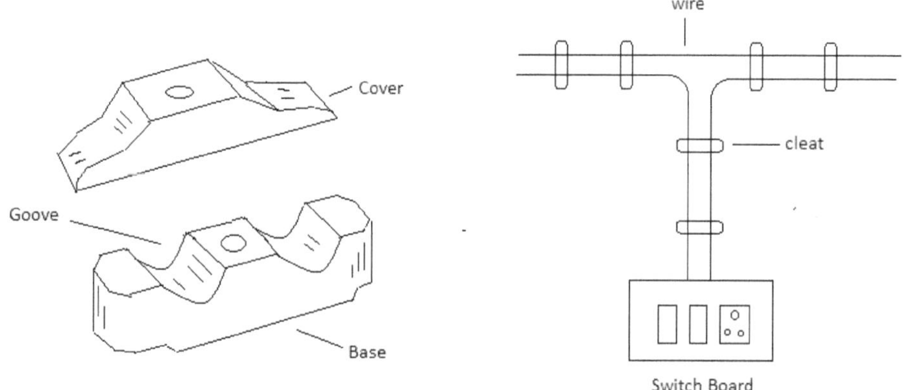

Figure 4.15 Cleat Wiring

It is cheap and used for temporary wiring. As the wires are exposed, these are more prone to mechanical injury.

2. ***Casing and Capping Wiring:*** This wiring has been used 60 years ago to provide the protection to the wire/cables. In this wiring system, a rectangular casings which have got the grooves are used. These casings are fixed to the walls and wires are placed in the grooves and a casing is fixed over it.

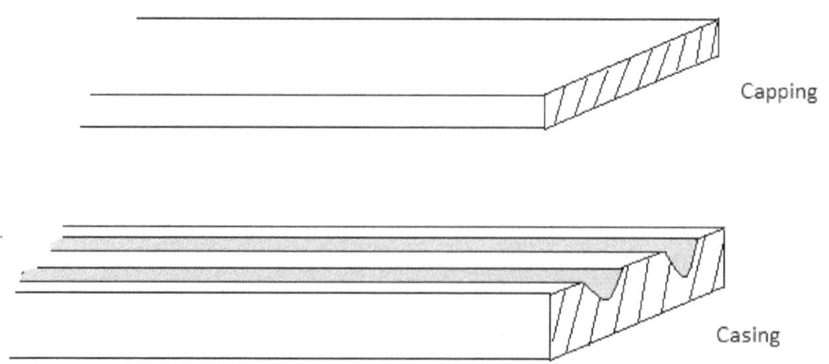

Figure 4.16 Capping and Casing Wiring

This wiring looks beautiful and has more life. No safety from fire and it requires regular maintenance.

3. ***Tough Rubber Sheath (TRS) or Cab Tyre Sheathed (CTS) Batten Wiring:*** This wiring uses TRS or CTS wires. These wires are more flexible and free from moisture absorption. These wires resist chemical, water and steam. In this wiring system, wires are fixed to wooden battens using tinned brass link clips.

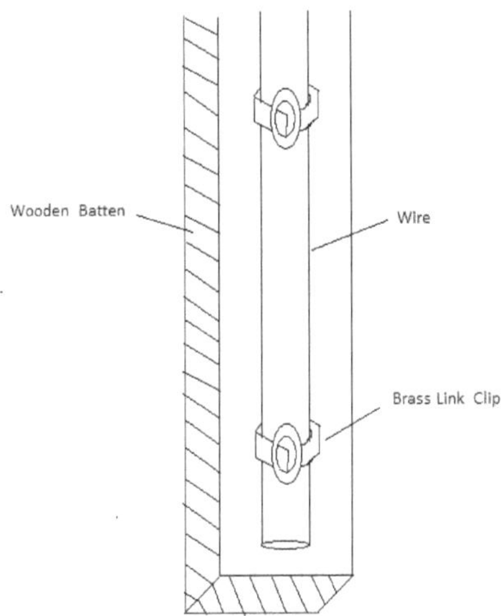

Figure 4.17 TRS/CTS Wiring

It is cheap and appear good. Its life is long and withstand chemical effects. It is not suitable in open areas.

4. ***Lead Sheathed or Metal Sheathed Wiring:*** This type of wiring uses PVC wires with an outer covering of Lead-Aluminum alloy (95% Lead and 5% Aluminum). This Lead sheath provides safety against mechanical injury and atmosphere. The wires are fixed on wooden batten using link clips.

5. ***Conduit Wiring:*** Conduits (pipes) made up of PVC are used in this type of wiring systems. The wires are run through these conduits.

In surface conduit system, the conduits are fixed on the walls using saddles and screws. This system is generally employed in workshops and factories.

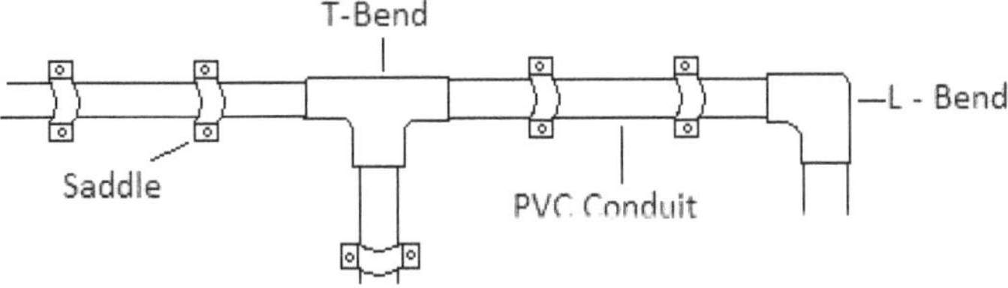

Figure 4.18 Conduit Wiring

In concealed conduit wiring, the conduits are buried in the walls and plastered. As the wiring is done in the walls, the wiring system must be done water tight. It is mostly used in offices and domestic installations. If any problem comes, the replacement of conduits becomes a difficult task.

4.4.3　Fuse

Fuse is an elementary protective device which provides the protection to an electrical circuit against the over currents. These over currents may be due to the over loading or short circuits. The essential part of the fuse is a metal strip or wire that melts when over current than a prescribed value flows through the circuit and thereby stopping the flow of current. Hence, fuse is sacrificial device and it is connected in series with the circuit. Once, the fuse is operated, it needs to be replaced by new one for a fresh operation.

The different symbols used for representing a fuse are

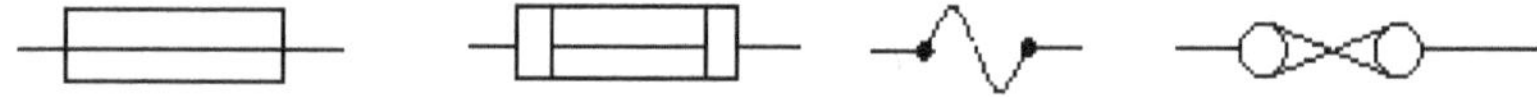

Figure 4.19 Different symbols used for Fuse

The following figure 4.20 shows the condition of a circuit during normal and abnormal conditions.

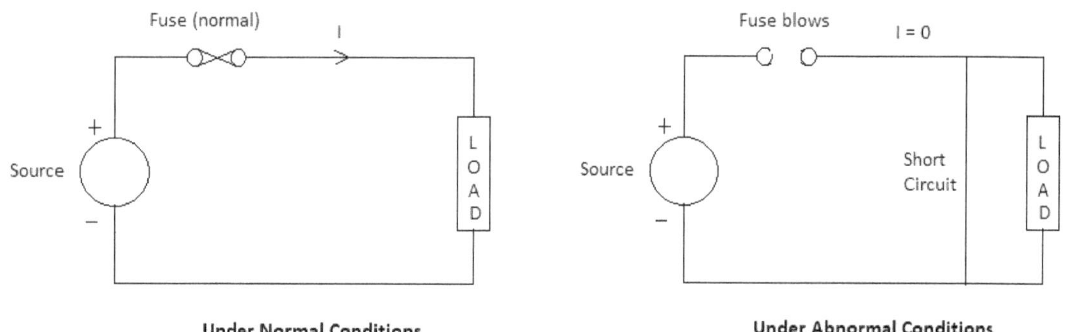

Figure 4.20 Effect of Fuse

The Short Circuits, overloading, mismatched loads, or device failure are the prime or some of the reasons for fuse operation. A fuse is an automatic means of removing power from a faulty system; often abbreviated to ADS (Automatic Disconnection of Supply). A fuse consists of a metal strip or wire fuse element, of small cross-section compared to the circuit conductors, mounted between a pair of electrical terminals, and (usually) enclosed by a non-combustible housing. The resistance of the element generates heat due to the high current flow and melts.

The fuse element is made of Zinc, Copper, Silver, Aluminium or alloys among these. The fuse ideally would carry its rated current indefinitely, and melt quickly on a small excess. The fuse element may be surrounded by air, or by materials intended to speed the quenching of the arc. The maximum current that the fuse can continuously conduct without interrupting the circuit is called the *Fuse Rated Current*.

The speed at which a fuse blows depends on how much current flows through it and the material of which the fuse is made. The operating time is not a fixed interval but decreases as the current increases. The Fuse selection depends on the load and the sensitive Semiconductor devices may require a fast or *ultrafast* fuses.

The different types of Fuses are:

1. *DC Fuses:* In a DC system, when the metallic wire melts because of the heat generated by the over current, then Arc is produced and it is very difficult to extinguish this arc because of DC constant value. So in order to minimize the fuse arcing, DC fuse are little bigger in size.

2. *AC Fuses:* On the other hand, i.e. in the AC system, the current alters at frequency of 50 Hz, the current value becomes zero 100 times every second. Hence, arc can be quenched easily than in DC circuits. Therefore, AC fuses are a little bit small in sizes as compared to DC fuses.

3. *Cartridge Fuses:* A cartridge fuse is a cylinder shaped fuse having metal caps on both the ends. The entire length of the metal meant to melt appear through a transparent glass. The entire fuse unit needs to be replaced in case of fuse blowing. The following figures 4.21 shows a few cartridge fuses in present use.

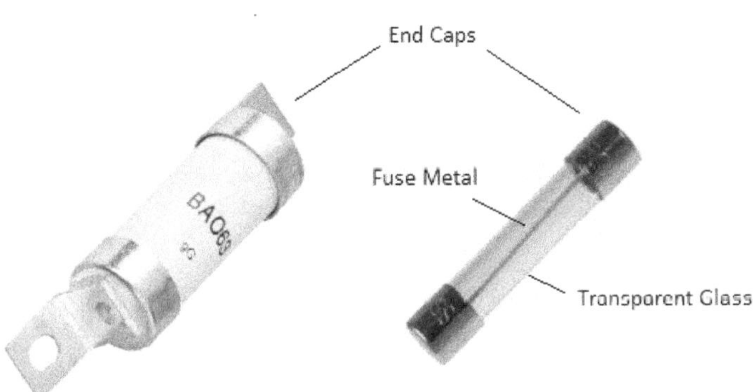

Figure 4.21 Cartridge Fuses

4. *High Rupturing Capacity (HRC) Fuses:* These fuses are extensively used for high currents. These fuses have porcelain base and covers and provide high rupturing capability for arc quenching. The commonly used HRC fuse is shown in figure 4.22.

   Merits:
   1. Simple
   2. Cheap
   3. Easy to erect and replace

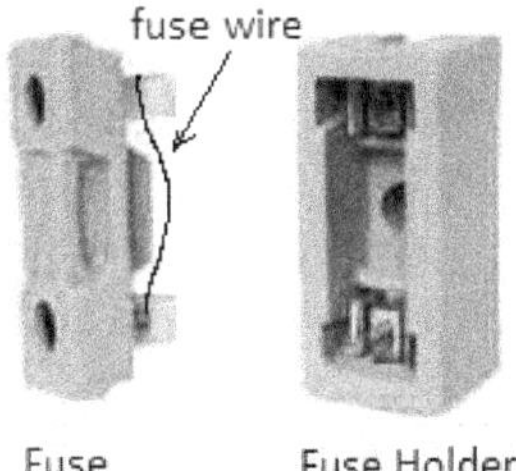

Figure 4.22 HRC Fuses

Demerits:

1. Needs to be replaced with new one after every operation
2. Not reliable when compared to Circuit Breakers
3. The power supply should be switched-off while replacing the fuse

4.4.4 Switch Fuse Unit (SFU)

Switch Fuse Unit is compact combination of switch and fuse generally enclosed in a metal cover. It is very widely used for low and medium voltages. The following figure 4.23 shows the external and internal views of SFU. The SFU is having an iron core in ring form excited by two identical coils. A Search Coil also has been placed on the same core to detect the difference in currents before and after the load point. This Search coil is connected across a trip coil.

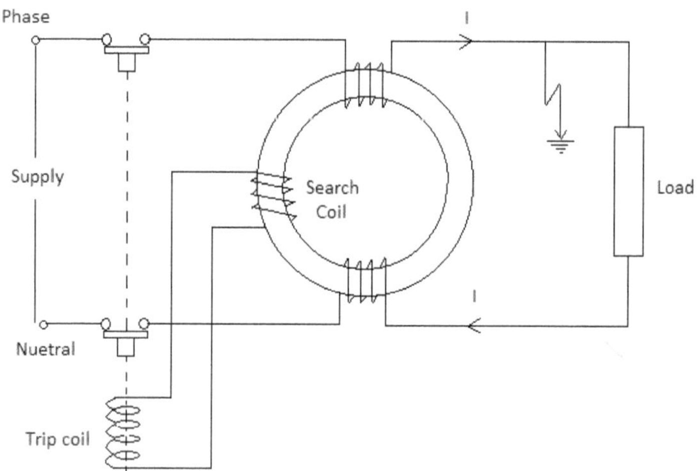

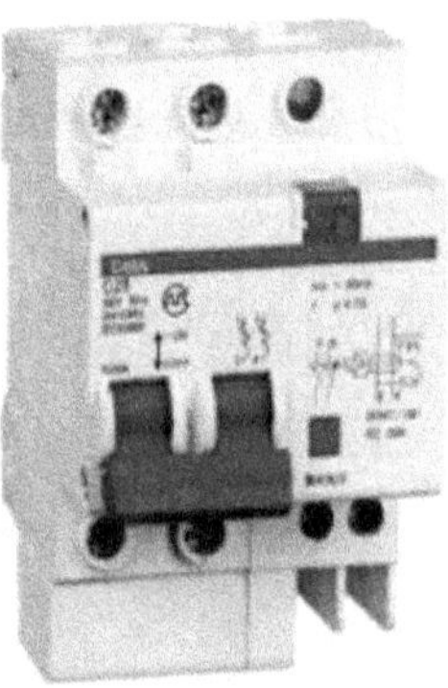

Figure 4.23 (a) SFU internal wiring diagram (b) Switch Fuse Unit

Under normal conditions, the current before and after the load is same, the fluxes produced by two identical coils cancel each other and net flux in the iron core becomes zero. No emf is induced in the Search Coil and hence trip coil does not operate. Under faulty conditions, like, a short circuit occurs at a point, there is difference in the current before and after the load, the net flux in the core is not zero, hence, finite emf is induced in the search coil. This emf operates the trip coil. The contactor or switch will be opened and protects the circuit.

4.4.5 MCB

MCB stands for Miniature Circuit Breaker. It automatically switches OFF the electrical circuit if any abnormal condition like

Figure 4.24 Miniature Circuit Breaker (MCB) (courtesy: Philips inc.)

overload, short circuit or equipment malfunction arise in the circuit. However, fuse may sense these conditions but it has to be replaced whereas the MCB can be reset. The figure 4.24 shows a typical MCB which is most commonly used.

The MCB is an electromechanical device which guards the electric wires &electrical load from over current so as to avoid any kind of fire or electrical hazards. Handling MCB is quite safer and it quickly restores the supply and it does not require any maintenance. When it comes to house applications, MCB is the most preferred choice for overload and short circuit protection. MCB works on bi-metal principle which is clearly illustrated in the figure 4.25.

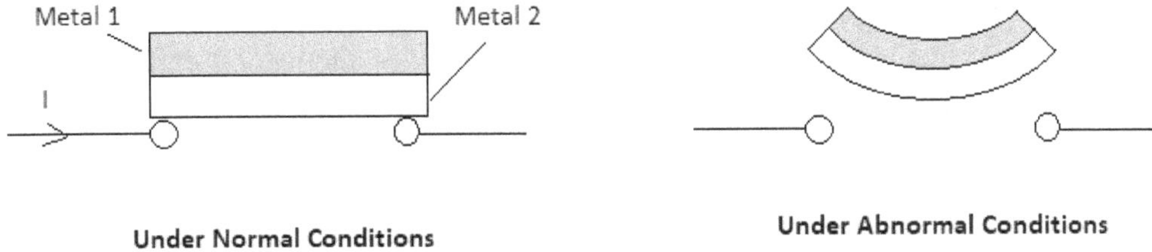

Figure 4.25 Bimetal under a) normal conditions b) abnormal conditions

The bimetal is made up two metals having different thermal expansion coefficients. Hence, they take the shape shown in figure 4.25(b) when it is subjected to heat produced by abnormal currents or over currents and opens the circuit.

4.4.6 MCCB

The MCCB stands for Moulded Case Circuit Breaker. It is another type of electrical protection device which is used when load current exceeds the limit of a miniature circuit breaker. The **MCCB** provides protection against overload, short circuit faults and is also used for switching the circuits. It can be used for higher current rating and fault level even in domestic applications. The wide current ratings and high breaking capacity in MCCB find their use in industrial applications. The **MCCB** can be used for protection of capacitor bank, generator protection and main electric feeder distribution. It offers adequate protection whenever an application requires discrimination, adjustable overload setting or earth fault protection.

Table 4.2 Difference between MCB and MCCB

S.No	MCB	MCCB
1	It stands for Miniature Circuit Breaker.	It stands for Moulded Case Circuit Breaker.
2	Rated current not more than 125 Ampere.	Rated Current up to 1600A
3	Its interrupting current rating is under 10KA	Their interrupting current ranges from around 10KA -85KA
4	Used for Domestic Installations.	Used for industrial installations.
5	Tripping current cannot be adjusted.	Tripping current can be adjusted.

4.4.7 Earthing

Earthing is the process of connecting all the metal bodies of different electrical appliances like refrigerators, mixers, grinders, electric motors etc. to ground or earth. Earthing is also called as 'Grounding'. The purpose of earthing is to ensure the safety to the consumer or operator who is using the electric equipment against the electric shock. A good conductor known as earth wire is used to connect the equipment metal bodies to the earth through earth electrodes. The need or purpose of earthing is

i)   To avoid electric shock

ii) To avoid fire accidents due to earth leakage currents

iii) To sustain the voltage generated by an alternator (neutral of alternator is earthed)

The safety provided to the consumer by earthing the appliances is well illustrated in the figure 4.26.

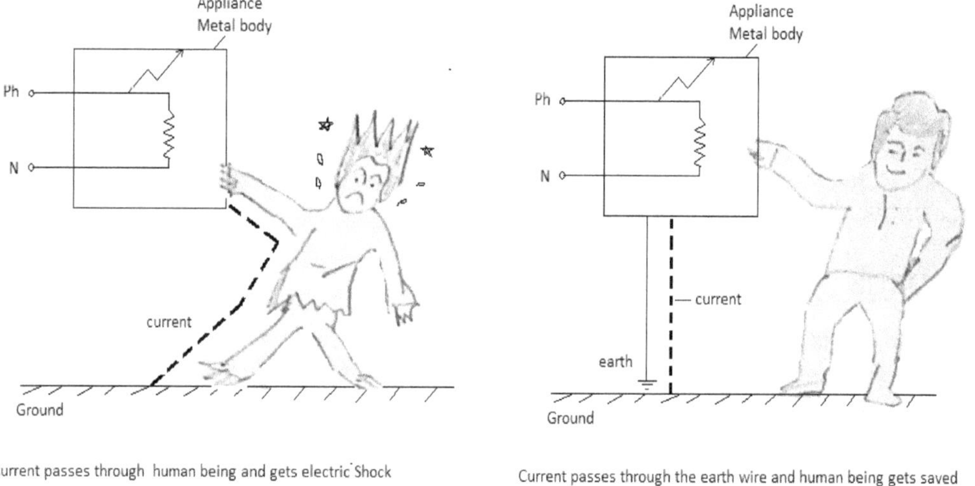

Figure 4.26 Effect of a) Not Earthing b) Earthing

When the live wire touches the metal body, the current from metal body gets diverted through earth wire instead of passing through the operator as the current always selects the low resistive path. Thus, the operator gets saved from electrocution.

Types of Earthing:

The various methods of earthing are:

1. Rod Earthing
2. Strip or Wire Earthing
3. Pipe Earthing
4. Plate Earthing

1. **Rod Earthing:** This method uses a 12.5 mm dia Copper rod or 16 mm GI rod. The minimum length required is 2.5 m.

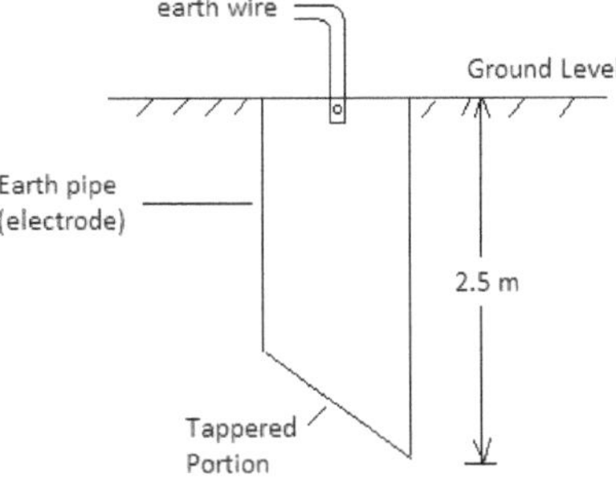

Figure 4.27 Rod Earthing

This method is very cheap and suitable for loose soil or sandy soil.

2. **Strip or Wire Earthing:** In this system, a wire or strip made up of Copper or GI is buried horizontally at the depth of minimum 0.5 m and length of 15 m.

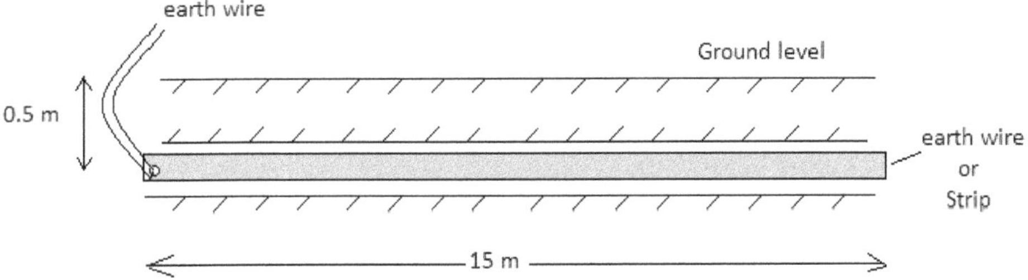

Figure 4.28 Strip or Wire Earthing

This system is used in rocky areas where is excavation work is difficult.

3. **Pipe Earthing:** It is most commonly used system in which a G.I pipe of 38 mm dia and 2.5 m long is placed vertically in an earth pit as shown in figure 4.29. The water is poured in the earth pit at regular time intervals to maintain the earth resistance low.

4. **Plate Earthing:** In this type, a G.I pipe of dimensions 60 cm x 60 cm x 6 mm is used as earth electrode.

The space around the earth plate is filled with charcoal and salt in alternate layers to reduce the earth resistance. The water is poured at regular intervals to maintain the moisture all the times. This helps to reduce earth resistance.

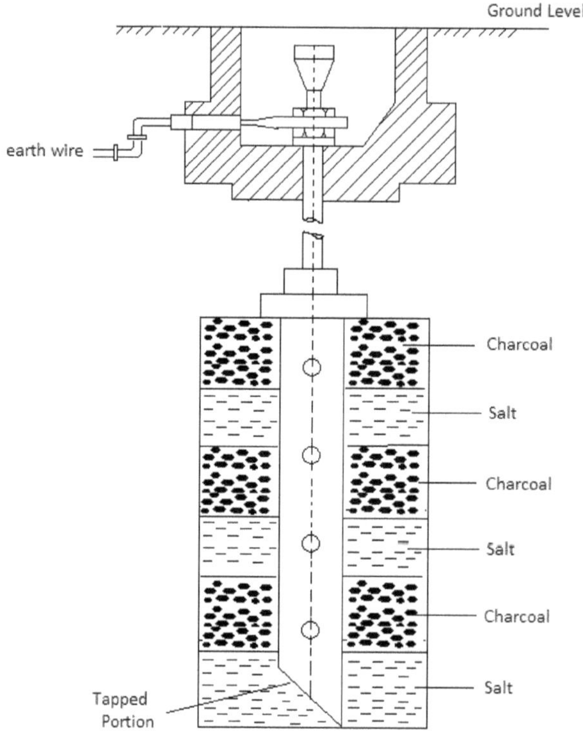

Figure 4.29 Pipe Earthing

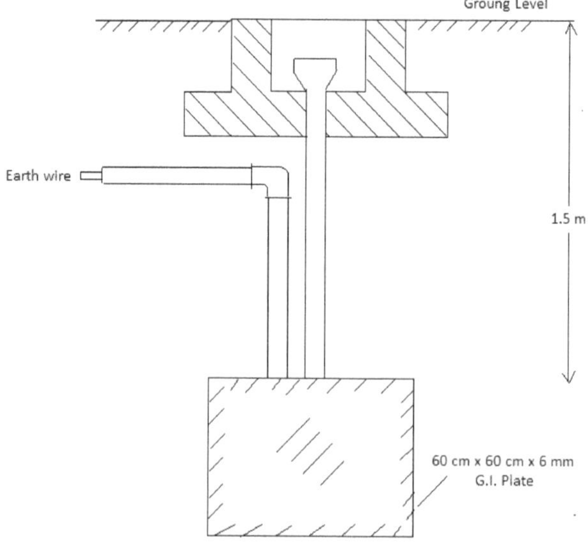

Figure 4.30 Plate Earthing

4.5 Classification of Batteries

A Battery converts the chemical energy into electrical energy. It consists of electrochemical cells connected in series. Batteries find wide range of applications and extensively used as main power supply in many cases. Batteries are also as backup source in case of outage of main power supply. Some of the applications of batteries are domestic, medical instruments, Electric Vehicles, Forklifts, flashlights, military operations etc.

Batteries are broadly classified into two types. Primary batteries and Secondary Batteries.

1. *Primary Batteries:* Primary batteries cannot be recharged and these are single use batteries. A commonly used primary battery is a Dry Cell. Zinc-Carbon Battery is an example of it. The voltage gradually decreases as the batteries are used. As the Zinc container oxidizes, the contents leak out eventually. Hence, these types of batteries should not be left in the electrical devices for longer periods.

2. *Secondary Batteries:* The Secondary batteries are rechargeable and extensively used. Lead-Acid, Nickel-Cadmium, Lithium Ion batteries are few examples of secondary batteries.

4.5.1 Lead-Acid battery

Lead Acid Battery is most commonly used battery. Though these batteries are hard and heavy, they can store high amount of energy and used in automobiles and inverters. Lead Acid batteries are cheaper and easy-to-handle. In this battery, there are two plates. The positive plate is made up of Lead Dioxide (PbO_2) and negative one with Sponge Lead (Pb). These two plates are separated using a **separator** which is an insulating material. This total construction is kept in a **hard plastic case** which has got the electrolyte (Sulphuric Acid (H_2SO_4) diluted with water in 1:3 proportion).

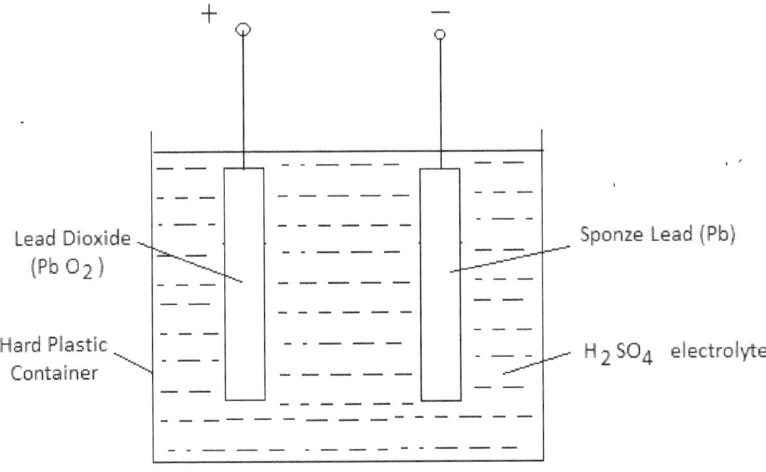

Figure 4.31 Lead- Acid Battery

A single cell typically gives 2.1V and hence, a 12V Lead Acid battery is having 6 series connected cells. These batteries are rated in **Ampere Hour (AH). The positive (PbO_2) plate and negative plate (Pb) are immersed in an electrolyte (H_2SO_4).**

The chemical reactions that take place during discharging are given below.

We have considered two moles of Sulphuric Acid $(2H_2SO_4)$.

During Discharging, One mole of H_2SO_4 gets split into $H_2^{++} + SO_4^-$

At Anode: $Pb + SO_4 \rightarrow PbSO_4$ ------Lead Sulphate is formed

At Cathode: $PbO_2 + H_2 \rightarrow PbO + H_2O$

This $PbO_2 + H_2$ reacts with the remaining mole of H_2SO_4

$$H_2SO_4 \rightarrow H_2^{++} + SO_4^-$$

$$PbO + H_2O + H_2^{++} + SO_4^{--} \rightarrow PbSO_4 + H_2O + H_2O$$

Lead Sulphate and two moles of water are formed.

The overall reaction in a Lead-Acid battery during its discharging is given by

$$Pb + PbO_2 + 2H_2SO_4 \rightarrow 2PbSO_4 + 2H_2O$$

As the battery is totally discharged, both anode and cathode get converted into Lead Sulphate with a by-product of water.

The reverse chemical process happens during charging and the corresponding chemical reaction is shown below.

$$2PbSO_4 + 2H_2O \rightarrow Pb + PbO_2 + 2H_2SO_4$$

Electrical Characteristics:

The three important features that are considered for assessing the performance of a Lead-Acid battery (in fact any battery) are a) Voltage b) Capacity and c) Efficiency

Voltage: The open circuit voltage of a fully charged cell is 2.2 Volts. This value mainly depends up on how long back it has been fully charged, specific gravity of the electrolyte and temperature. The voltage increases with increase of specific gravity and temperature. The variation of terminal voltage during a charging and discharging cycle is shown in figure 4.32 below.

The fall in terminal voltage depend up on the rate of discharge. The change in voltage is high during initial periods of discharging and maintains steady in middle periods. The voltage should not be allowed to fall below 1.8 V. This may lead to the increase of battery internal resistance due to the formation of insoluble Lead Sulphate.

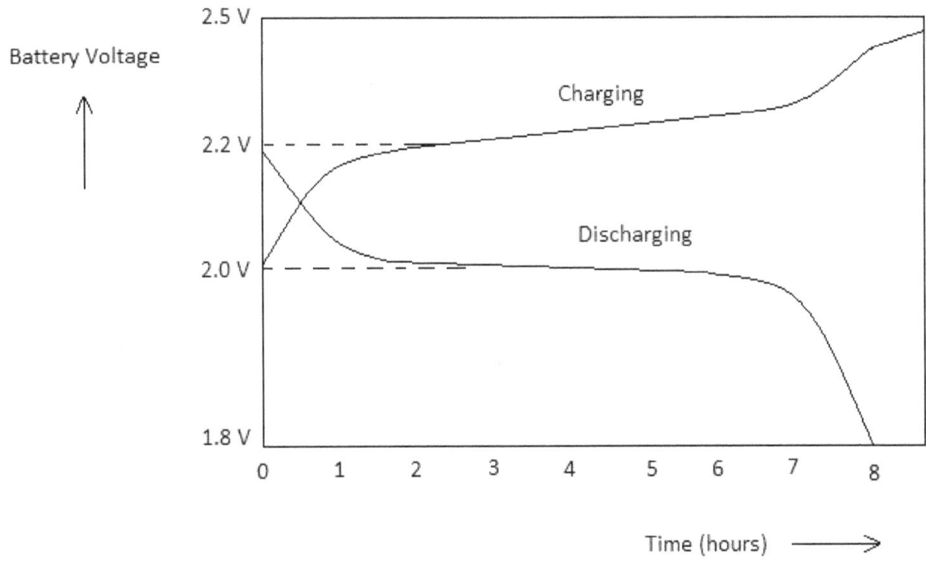

Figure 4.32 Variation in voltage during charging and discharging

(a) *Capacity:* The battery capacity is specified in Ampere-Hours (AH). For example, a 100 AH battery feeds a current of 10A for 10 hours or 20A for 5 hours or 40A for 2.5 hours. The capacity and life of the battery reduces as the discharge rate is increasing.

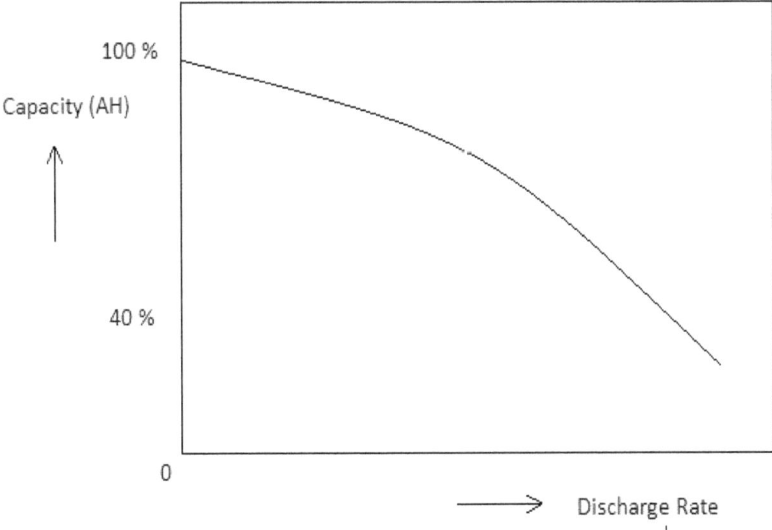

Figure 4.33 Capacity vs Discharge characteristics

(b) *Efficiency:* There are two efficiencies specified for the batteries. They are i) AH efficiency and ii) Energy efficiency or WH efficiency. The Energy efficiency is always less than the AH efficiency.

$$AH\ efficiency = \frac{AH\ delivered\ during\ discharging}{AH\ absorbed\ during\ charging} \times 100$$

This efficiency ranges between 90-95%. The reasons for the efficiency maintaining below 100% are self-discharge in the battery and leakage currents in the battery due to the usage of weak insulation.

4.6   Elementary Calculations of Energy Consumption

Generally, the electrical appliances are rated in *watts* or *kilo watts* of power consumption. When such appliances are connected to the supply of rated voltage, they take the rated current and consume certain amount of power. As these appliances remain connected to the supply line for some time (usage time), electrical energy is consumed by them. This electrical energy consumed is given by

$$Energy = Power\ X\ time$$

The electrical energy consumption is measured in Units.

One Unit $= 1\,kWH\ (kilo\ watt\ Hour)$

$$= 1 \times 1000 \times 60 \times 60 = 36 \times 10^5\ \text{watt-seconds or joules}$$

For example, to find the energy consumed monthly by 2 bulbs which are rated at 60 watts and used for 4 hours daily, the following procedure is used.

Monthly energy consumption

$$= No.of\ bulbs \times \frac{Bulb\ Power\ Rating\ in\ watts}{1000} \times Daily\ usage\ in\ hours \times 30\ \text{kWH or Units}$$

$$= 2 \times \frac{60}{1000} \times 4 \times 30 = 14.4\ \text{Units}$$

When the appliances like motors whose output is rated in H.P (Horse Power), the efficiency of the appliances needs to be considered for calculating the energy consumption.

Output Power $P_O = P_0$ in $H.P \times 735.5$ watts

$$\text{Input power } P_{in} = \frac{Output\ Power}{Efficiency} = \frac{P_o}{\eta}$$

$$\text{Energy consumption} = \frac{P_i}{1000} \times Daily\ usage\ in\ hours \times 30$$

Exercise 4.6.1: A domestic installation is having the following loads along with their usage details.

S.No	Name of the appliance	Rating	quantity	Daily usage by each item
1	Fluorescent lamps	45 W	5	4 hours
2	Ceiling fans	60 W	4	10 hours
3	Grinder	200 W	1	30 minutes
4	Washing Machine	250 W	1	2 hours
5	Heaters	1000 W	3	15 minutes

Find the total monthly electrical energy consumption in units and also the monthly energy charges in Rupees if the unit rate is Rs 6.

Solution:

1. Energy consumed by Fluorescent lamps:

$$E_1 = No.of\ lamps \times Rating\ of\ each\ lamp\ in\ kW \times Dialy\ usage\ in\ hours \times 30$$

$$= 5 \times \frac{45}{1000} \times 4 \times 30 = 27\ kWH\ or\ Units$$

2. Energy consumed by Ceiling Fans:

$$E_2 = No.of\ fans \times Rating\ of\ each\ fan\ in\ kW \times Dialy\ usage\ in\ hours \times 30$$

$$= 4 \times \frac{60}{1000} \times 10 \times 30 = 72\ kWH\ or\ Units$$

3. Energy consumed by Grinder:

$$E_3 = No.of\ Grinders \times Rating\ of\ each\ Grinder\ in\ kW \times Dialy\ usage\ in\ hours \times 30$$

$$= 1 \times \frac{200}{1000} \times 0.5 \times 30 = 3\ kWH\ or\ Units$$

4. Energy consumed by Washing Machine:

$$E_4 = No.of\ washing\ Machines \times Rating\ of\ each\ Washing\ Machine\ in\ kW$$

$$\times Dialy\ usage\ in\ hours \times 30$$

$$= 1 \times \frac{250}{1000} \times 2 \times 30 = 15\ kWH\ or\ Units$$

5. Energy consumed by Heaters:

$$E_5 = No.of\ Heaters \times Rating\ of\ each\ Heater\ in\ kW \times Dialy\ usage\ in\ hours \times 30$$

$$= 3 \times \frac{1000}{1000} \times 0.25 \times 30 = 22.5\ kWH\ or\ Units$$

Total Energy Consumption $E = E_1 + E_2 + E_3 + E_4 + E_5$

$$= 27 + 72 + 3 + 15 + 22.5 - 139.5\ kWH\ or\ Units$$

The energy charges in Rs $= No.of\ units \times Unit\ rate\ in\ Rs$

$$= 139.5 \times 6 = Rs\ 837$$

Exercise 4.6.2: A 3 phase, 5 H.P, 415 V Induction motor is operated at its full load rating daily 2 hours. The motor offers an efficiency of 82 % at full load. Calculate the monthly energy consumption by the motor in KWHs. Also calculate the energy charges if the unit rate is Rs 7.

Solution:

Full Load Output $P_O = 5H.P = 5 \times 735.5 = 3677.5$ Watts or 3.6775 kW

Full load efficiency $\eta = 82\%$

Full load input $P_{in} = \dfrac{P_O}{\eta} = \dfrac{3.6775}{0.82} = 4.485 \, \text{kW}$

Monthly Energy Consumption $= P_i \times Daily \; usage \; in \; hours \times 30$

$$= 4.485 \times 2 \times 30 = 269 \, \text{kWH or Units}$$

Monthly Energy Charges $= Unit \; Rate \times No.of \; units \; consumed \; monthly$

$$= 7 \times 269 = Rs \; 1884$$

4.7   Electrical Safety Precautions

It is very important and compulsory to follow certain safety measures or precautions while working with electrical energy. The safety should not be compromised and some ground rules need to be always followed. The vital precautions to be observed while working with electricity are:

1. Avoid the usage of water at all times when working with electricity. Never touch any electrical equipment with wet hands
2. Always use insulated tools
3. Use high quality rubber gloves and mats while working on electricity
4. Ensure that all the electrical appliances are properly earthed
5. Wet the earth-pits at regular intervals
6. Never try to repair live equipment
7. If you find or suspect a fault, stop using the equipment, disconnect from the electrical supply and label *'DONOTUSE'*
8. Avoid overloading of sockets
9. Switch off and unplug equipment before you clean it or make adjustments
10. Restrict the entry of unauthorized people into electrical plant or switch gear
11. Display the *DANGER* boards at electrical premises, especially at high voltage premises
12. Avoid working on electrical circuit during heavy lightning storm

What should one do when someone else has been electrocuted?

- Don't touch the victim if he/she is still in contact with the source of electricity.
- Don't move the victim unless he/she is in danger of further shock.

- Turn off the flow of electricity if possible. Separate the victim from live wires using a non-conducting object. Wood and rubber are both good options.
- Stay at least 20 feet away if they've been shocked by high-voltage power lines that are still on
- Keep the person warm
- Check the person's breathing and pulse. If necessary, start the CPR (Cardiopulmonary Resuscitation) until emergency help arrives.

Review Questions

Short Answer (Two marks) questions

1. Name different parts of an Induction Motor
2. Why an Induction motor is called Asynchronous Motor?
3. Define Slip
4. Define Synchronous Speed
5. What are the values of slips at a) starting and b) Synchronous Speed?
6. Show the Torque-Slip Characteristics of an Induction Motor
7. A 3 Phase, 4 –pole, 50 Hz induction motor runs at a speed of 1450 rpm. Calculate a) Synchronous Speed b) slip c) rotor frequency (*Ans: 1500 rpm, 3.33%, 1.66 Hz*)
8. Highlight the differences between a DC Generator and a Synchronous Generator
9. List the applications of a Stepper Motor
10. Name different types of wirings used in domestic installations
11. What is the purpose of a Fusc? Show its symbol
12. Differentiate between Fuse and MCB
13. Why earthing of electric appliances is done?
14. Name different types of Earthing
15. Mention the purpose of providing charcoal and salt in earth-pits
16. Differentiate between Primary and Secondary batteries
17. What are the units for electrical energy consumption?

Essay (Six marks) Questions

1. Explain the constructional aspects of Squirrel Cage and Slip Ring Induction Motors
2. Explain the working of a three phase induction motor
3. Show and explain the Torque-Slip Characteristics of a three induction motor. Also identify the maximum and starting torque on it.

4. A 3 Phase, 6 –pole, 50 Hz induction motor has the rotor leakage impedance of 2.5+j12 ohms per phase and rotor emf of 120 V per phase at stand-still. Calculate a) slip b) rotor frequency c) rotor power factor d) rotor current at i) stand-still ii) rotor speed of 960 rpm.*(Ans: i) 1, 50Hz, 0.203 lagging, 9.8A & 0.04, 2Hz, 0.9823 lagging, 3.3 A)*

5. Explain the working of a Synchronous Generator

6. Discuss on the working of a Stepper Motor.

7. Show the basic layout of wiring system in domestic installation and locate the position of Energy meter, MCB and DB

8. Discuss the merits and demerits of Cleat and Conduit Wiring systems

9. List out the differences between MCB and MCCB

10. Show and explain the Pipe Earthing Scheme

11. Explain the working of a Lead-Acid Battery

12. A 7.5 H.P, three phase, 400 V Induction motor drive is used daily 4 hours at its ¾ th full load rating. The motor offers an efficiency of 80 % at ¾ full load. Calculate the monthly energy consumption by the motor in KWHs. Also calculate the energy charges if the unit rate is Rs 6.5.(Ans: 620 units or KWH, Rs 4034/-)

CHAPTER 5

Semi-Conductor Devices

5.1   Introduction

In view of conduction of electric current, the materials are classified into Conductors, Insulators and Semi-conductors. The conductors have abundant number of free electrons and they conduct the current whereas Insulators are in the lack of free electrons and do not conduct. The conducting ability of Semi-conductors is between that of conductors and insulators. The semi-conductor materials like Silicon and Germanium are most commonly used for constructing many electronic devices and Integrated Circuits (ICs). The semi-conductor materials have the negative temperature coefficient of resistance and hence, these materials decrease the resistance as the temperature increases. These semi-conductor materials consist of four electrons in outermost orbit and these electrons are called *Valance Electrons*. These valance electrons are far from the nucleus, associated with higher energies and loosely connected with the nucleus. At absolute zero, the semiconductors behave as insulators. But, at room temperature, the valance electrons get excited, break the covalent bond and come out of influence of nucleus and become free.Hence, these electrons are called *Free Electrons* and these free electrons move randomly in conduction band. As the electron moves to conduction band, there is a deficiency of electron in the atom which is represented by a hole. At room temperature, semiconductors are able to conduct.

The semiconductors in their purest form have equal number of free electrons and holes. These pure semiconductors are called *'Intrinsic Semiconductors'*. When impurities are added to a pure semiconductor material by a process called *'Doping'*, the intrinsic semiconductors get converted into *Extrinsic Semiconductors*. An Extrinsic Semiconductor can be a 'P' type or 'N' type.

Consider an Intrinsic semiconductor material having three pairs of free electrons and holes shown in figure 5.1.

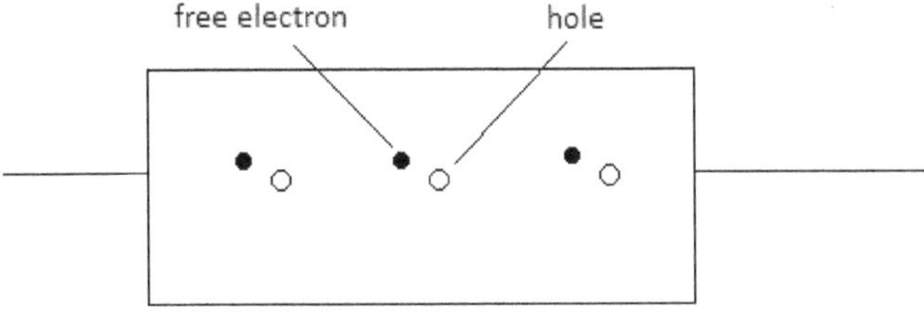

Figure 5.1 Intrinsic Semiconductor

When three free electrons along with their immobile positive ions are added, it gets converted into an Extrinsic semiconductor. As the number of electrons in this is more than holes or the electrons are majority carriers, this material is called 'N type' semiconductor. It is observed in the figure 5.2 that the total number of free electrons is 6 and the number of holes is 3. (But, the net charge of a N type material is zero or neutral as there are three more immobile positive ions also).

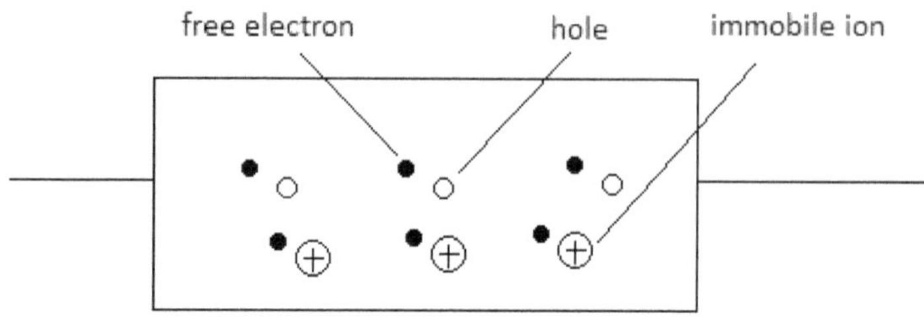

Figure 5.2 N type Semiconductor

Similarly, when three holes along with their immobile negative ions are added, it gets converted into an Extrinsic Semiconductor. As the number of holes is more than the electrons, or the holes are majority carriers, this material is called *'P type' semiconductor*. It is observed in the figure 5.3 that the total number of holes is 6 and the number of free electrons is 3. (But, the net charge of a P type material is zero or neutral as there are three more immobile negative ions also).

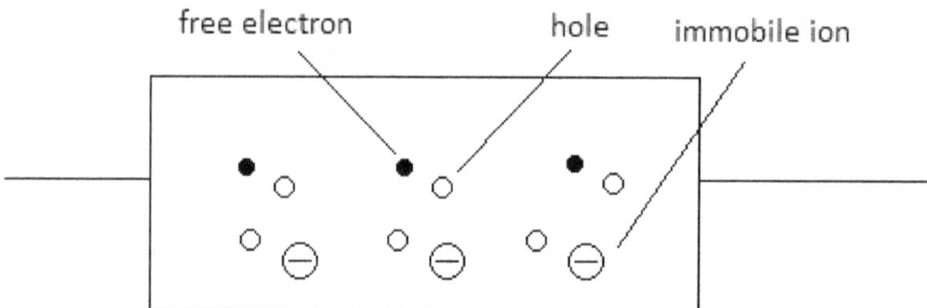

Figure 5.3 P type Semiconductor

Finally, in 'P' type semiconductor, the holes are majority carriers and free electrons are minority carriers. But in 'N' type, the free electrons are majority carriers and holes are minority carriers.

5.1.1   P-N Junction Diode

If we join a P type material with a N type material such that crystalline structure remains continuous at the boundary, a P-N junction is formed. Such a P-N Junction is called *'P-N*

Junction Diode' or *'Semiconductor Diode'*. The figure 5.4 shows the P-N junction immediately after formed.

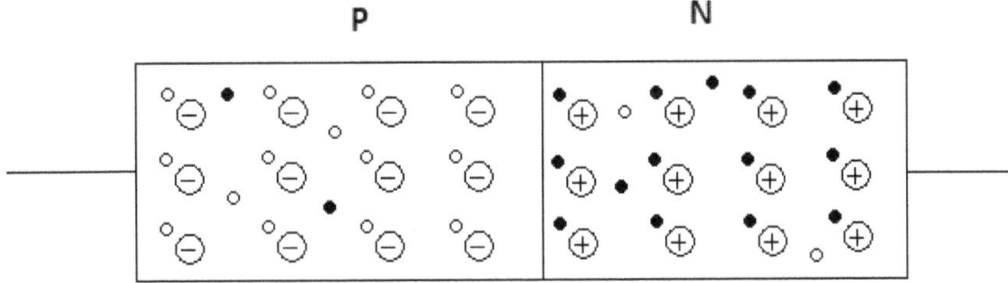

Figure 5.4 P N Junction

The electrons and holes which are nearby junction diffuse across the junction and combine. For every recombination, one pair of electron-hole disappears. This happens for short time and some region around the junction is left with only immobile ions.

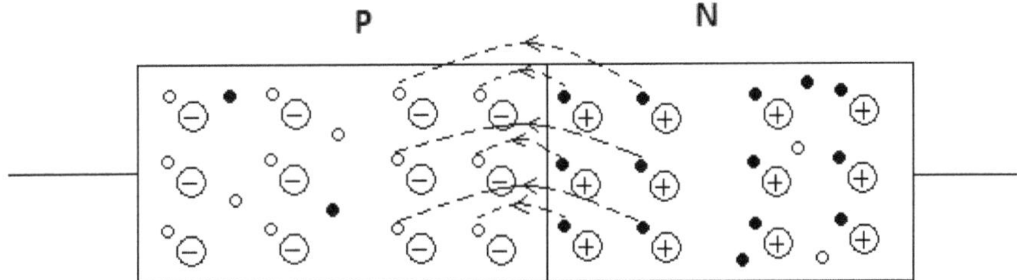

Figure 5.5 Drifting of holes and electrons across the junction

There will be no charge carriers (holes and electrons) in this region. This region which is depleted of charge carriers is called 'Depletion Region' and it shown in the figure 5.6.

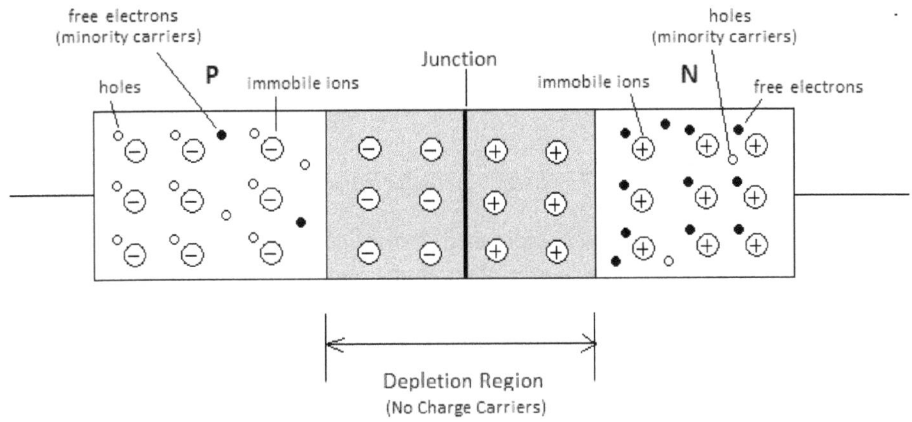

Figure 5.6 Formation of Depletion Region

A restoring force is developed by this depletion region and it does not allow the further recombination of holes and electrons. This force is called *Barrier* and it is 0.7V for Silicon and 0.3 V Germanium. The PN Junction thus formed exhibits a peculiar property of conducting current in one direction (when forward biased) and not conducting in opposite direction (when reverse biased). This property is exploited by the PN junction for rectification of AC into DC. Then, it is called as *PN Junction Diode*.

PN Junction with Forward Bias:

When the P type material is connected to the positive terminal of the battery and N type material to the negative terminal of the battery as shown in figure 5.7, the PN Junction Diode is said to be *Forward Biased*. Under this condition, the holes in the P type material repelled by the positive terminal of the battery and compelled to move towards the junction. Similarly, the free electrons in N type material are pushed towards the junction by repulsive force offered by the battery negative terminal. Hence, the width of the depletion region becomes narrow as shown. With this reduced depletion region under forward biased conditions, diode is ready for conduction.

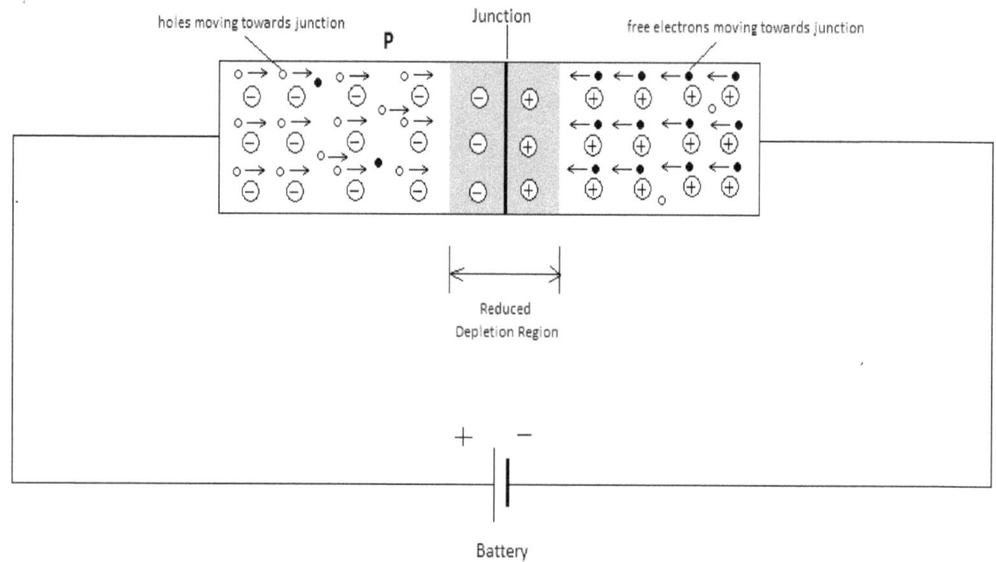

Figure 5.7 PN junction under Forward Bias

As the voltage across PN junction is increased further, the majority carriers (holes in P type material and free electrons in N type material) acquire enough energy and cross the depletion region easily and thus participates in heavy conduction. The current through PN junction increases as the voltage increases and this current continues as long as the battery exists in the circuit.

PN Junction with Reverse Bias: When the P type material is connected to the negative terminal of the battery and N type material to the positive terminal of the battery as shown in figure 5.8, the PN Junction Diode is said to be *Reverse Biased*. Under this condition, the holes in the P type material attracted by the negative terminal of the battery and compelled to move away from the junction. Similarly, the free electrons in N type material are moved away from the junction by attractive force exerted by the positive terminal of the battery. Hence, the width of the depletion region increases as shown.

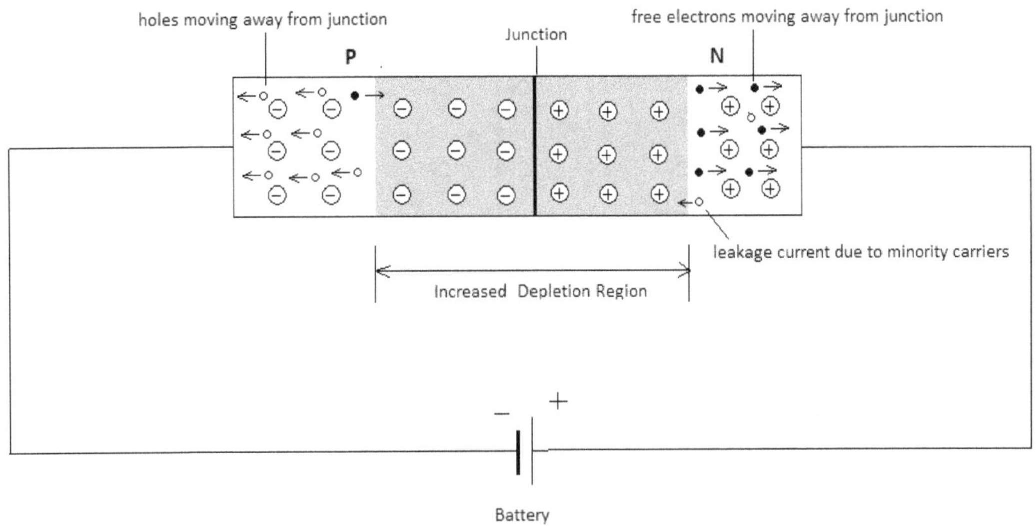

Figure 5.8 PN Junction under Reverse Bias

The widened depletion region does not allow any majority carriers to cross the junction and hence, a reversed biased PN junction does not conduct any current. However, a small leakage current due to minority carriers (holes in N type and free electrons in P type materials) flows across the junction. This current is called *Reverse Leakage Current* and it increases with the temperature. The symbol of PN junction diode is shown in figure 5.9.

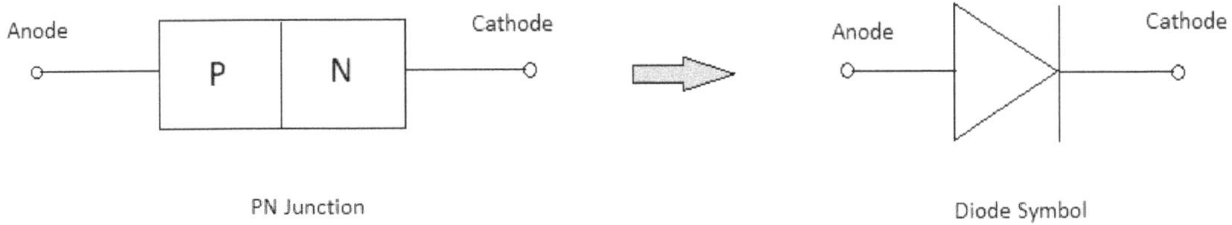

Figure 5.9 a) PN Junction   **b)** Diode symbol

5.1.2   V-I Characteristics of PN Junction Diode

The operational behavior of a PN Junction Diode in a circuit under different conditions is assessed through its V-I characteristics. The V-I characteristic is drawn between the voltage applied across the diode and current passing through it. There are two operating regions of Diode. They are a) Forward Biased Region b) Reverse Biased Region.

(a) *Forward Biased Region:* In this bias condition, the anode of the diode is connected to the positive terminal of the battery and cathode is connected to negative terminal as shown in figure 5.10 (a).

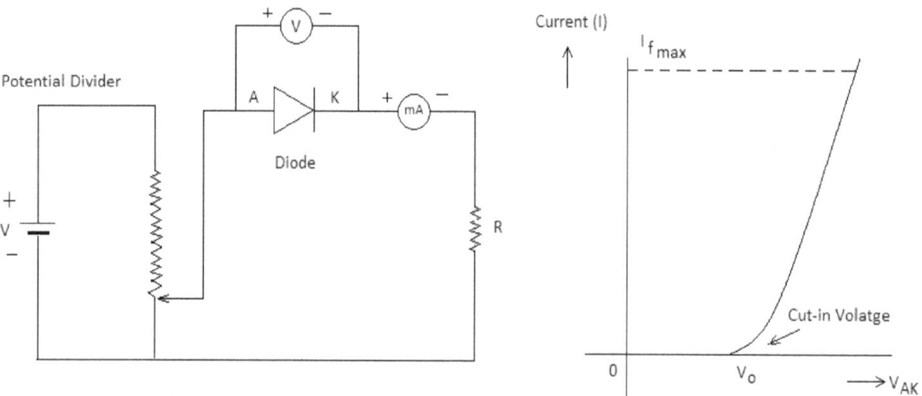

Figure 5.10 (a) circuit setup for Diode forward bias   **(b)** Diode Forward characteristics

Potential divider is used to supply variable voltage to the diode. As the voltage is increased gradually from zero, the diode remains forward biased, but no current flows initially. The depletion region is narrowed down. As the voltage across diode V_{AK} overcomes the barrier of depletion region, the diode starts conducting the current. This voltage at which the diode starts conducting in forward bias condition is called *Forward Cut-in Voltage or Knee voltage (V_0)*. If the voltage across diode is increased beyond this point, the current steeply increases as shown in figure. An additional resistance R is connected in the circuit to limit the excess current. The typical values of V_o for Silicon and Germanium diodes are 0.7 V and 0.3 V respectively.

(b) *Reverse Bias Region:* The circuit shown in figure 5.11 (a) is used to obtain the Reverse Bias characteristics. The P type material is connected to the negative terminal of the battery and N type material is connected to the positive terminal of the battery. The milli-ammeter is replaced with a micro-ammeter to measure the leakage current.

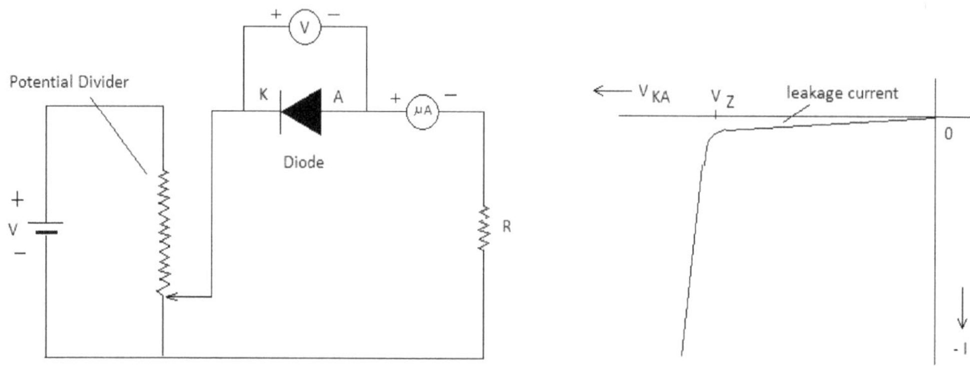

Figure 5.11 (a) circuit setup for Diode Reverse bias   **(b)** Diode Reverse characteristics

The current passing through the diode during reverse bias is very small and this is due to the flow of minority carriers. As the voltage exceeds a particular value, the current abruptly increases. This voltage is called *'Reverse Break down Voltage V_Z.'*

5.1.3 Diode Applications

PN Junction Diodes are widely used in rectifiers to convert the Alternating Current (AC) supplies to Direct Current (DC) supplies. The DC supply is required for the operation of different electronic devices. The unidirectional conduction property of diode is exploited for rectifier operation.

A. Half-Wave Rectifier:

The following figure 5.12 shows the circuit of a Half-Wave Rectifier. A transformer is used in input to bring the ac voltage to a desired level before rectification. It is also useful for getting isolation between power circuit and rectifier circuit.

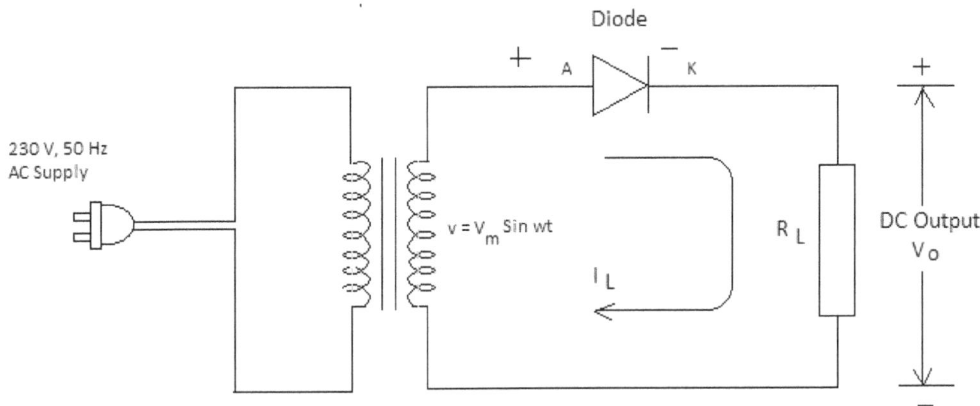

Figure 5.12 Half-Wave Rectifier Circuit

The voltage across the secondary winding of the transformer is given by

$$v = V_m \, Sin\omega t$$

And it is represented against time in the figure 5.13 (a). V_m is the peak value of this alternating voltage.

During positive half cycle of the supply voltage, the diode gets forward biased and conducts the current. The input voltage is conveyed to the output as it is. The voltage across the load resistance R_L at any instant is equal to the supply voltage.

During negative half cycle, the diode is reverse biased as the potential at cathode is greater than the potential at anode. The diode doesn't conduct and hence the output voltage is zero as shown in figure 5.13 (b).

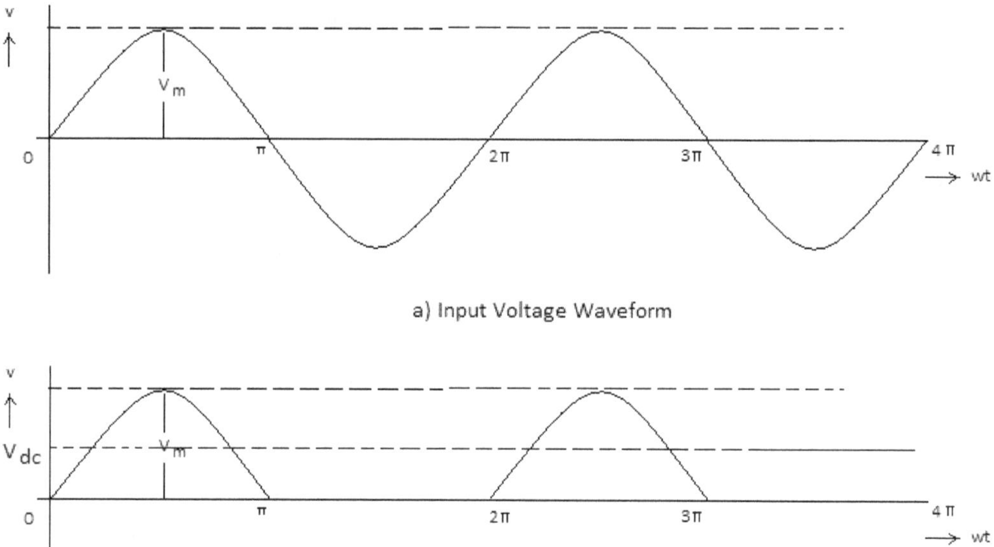

Figure 5.13 (a) Input and (b) Output voltages of Half-Wave Rectifier

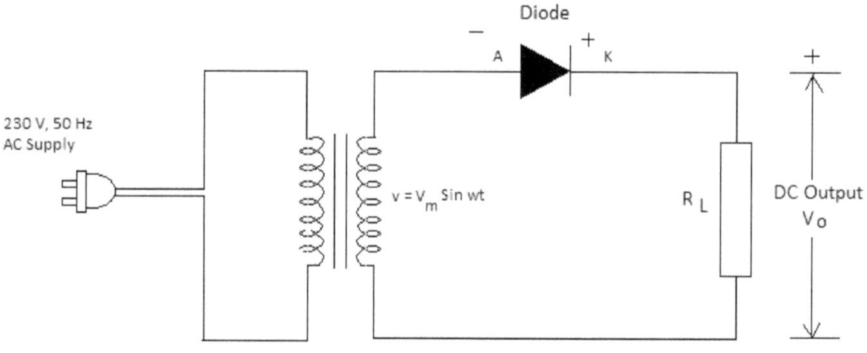

Figure 5.14 Half wave Rectifier under Reverse Bias condition

The DC output voltage or Average Voltage is given by

$$V_0 = \frac{1}{2\pi} \int_0^{\pi} V_m Sin\omega t.dt = \frac{V_m}{2\pi} \left[-Cos\omega t\right]_0^{\pi}$$

$$= \frac{V_m}{2\pi} \left[-\left(Cos\pi - Cos.0\right)\right] = \frac{V_m}{2\pi} \left[-\left(-1-1\right)\right]$$

$$= \frac{V_m}{\pi}$$

B. Full Wave Rectifier:

Half wave Rectifier utilizes only the half cycles (say only positive half cycles) of the input voltage whereas the Full wave Rectifier uses both the positive and negative half cycles and supplies more dc output voltage. There are two types of Full wave Rectifiers. A) Centre Tap Rectifier and B) Bridge Rectifier

(i) *Centre Tap Rectifier:* It uses a center tapped transformer and two diodes D_1 and D_2 as shown in figure 5.15. During positive half cycle, the point 'a' becomes positive with respect to point 'x', the diode D_1 gets forward biased and the diode D_2 is reverse biased as the point 'b' is negative with respect to 'x'. the current flows through top half of the Centre tap transformer, D_1 and load resistance R_L.

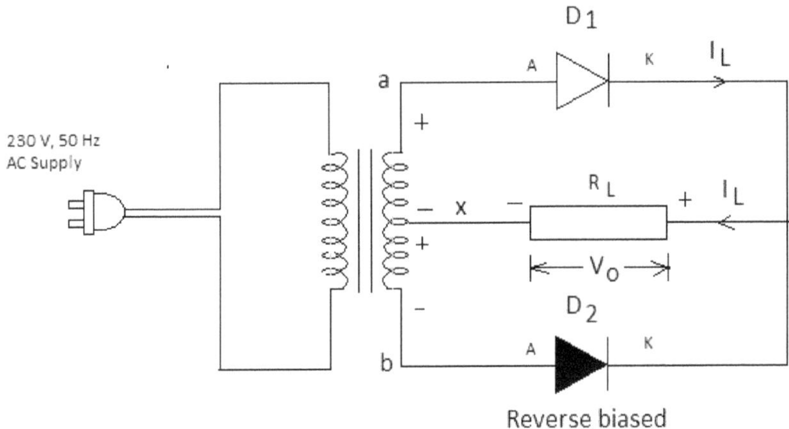

Figure 5.15 Center tapped Rectifier – positive half cycle

During negative half cycle, the point b becomes positive with respect to point x and the diode D_2 gets forward biased. The diode D_1 is reverse biased as the point a is negative with respect to x and it is shown in figure 5.16.

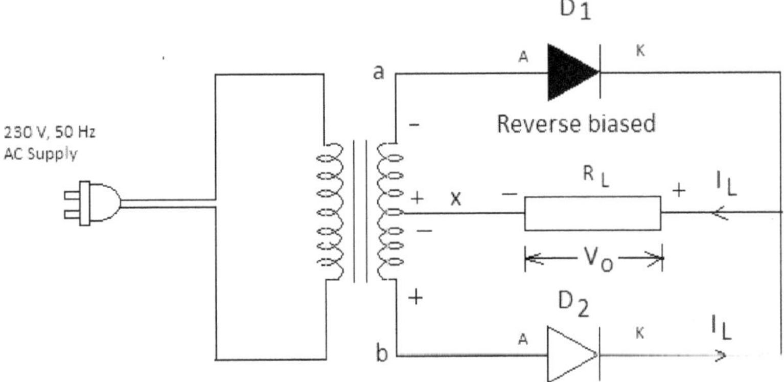

Figure 5.16 Center tapped Rectifier – Negative half cycle

Now, the current passes through lower half of the Centre tap transformer, diode D_2 and the load resistance R_L. The current direction through the load resistance is always found to be unidirectional and the corresponding voltage waveform is shown below in figure 5.17.

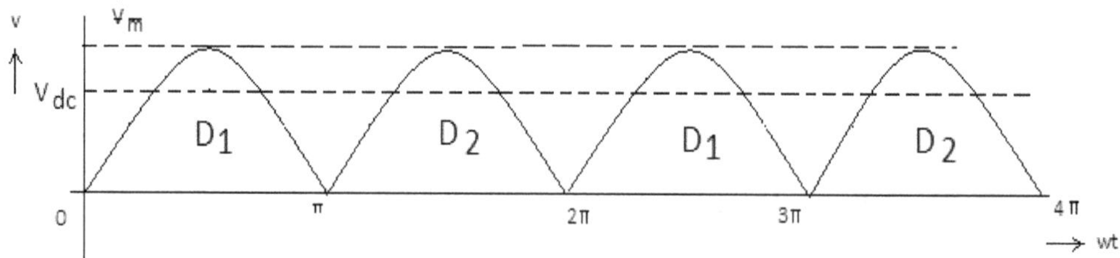

Figure 5.17 Output voltage

The instantaneous value of voltage in any half of center tapped transformer secondary winding is given by $v = V_m \, Sin\omega t$

The DC output voltage or Average Voltage is given by

$$V_0 = \frac{1}{\pi}\int_0^{\pi} V_m \, Sin\omega t.dt = \frac{V_m}{\pi}\left[-Cos\omega t\right]_0^{\pi}$$

$$= \frac{V_m}{\pi}\left[-\left(Cos\pi - Cos.0\right)\right] = \frac{V_m}{\pi}\left[-\left(-1-1\right)\right]$$

$$= \frac{2V_m}{\pi}$$

(ii) **_Bridge Rectifier:_** It is most widely used Full Wave Rectifier. This rectifier requires four diodes arranged in a bridge pattern as shown in figure 5.18. It eliminates the need of bulky and costly center tapped transformer.

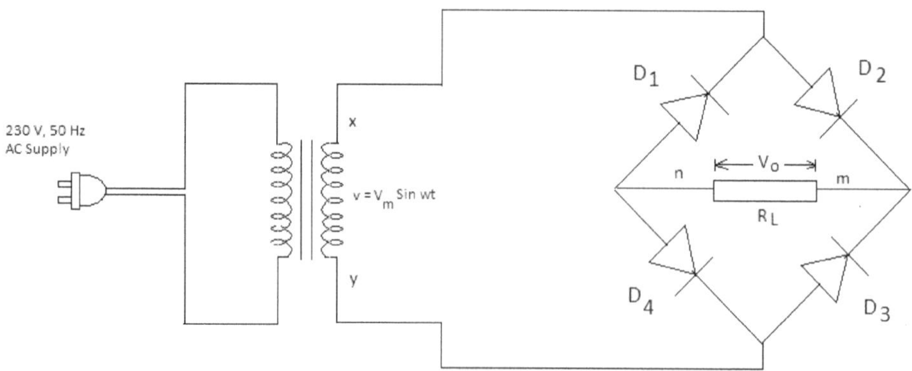

Figure 5.18 Diode Bridge Rectifier

During positive half cycle, the point 'x' becomes positive, the diodes D_2 and D_4 are forward biased & the diodes D_1 and D_3 are reverse biased. The current flows through secondary winding of transformer, D_2, load (from point 'm' to 'n') and D_4.

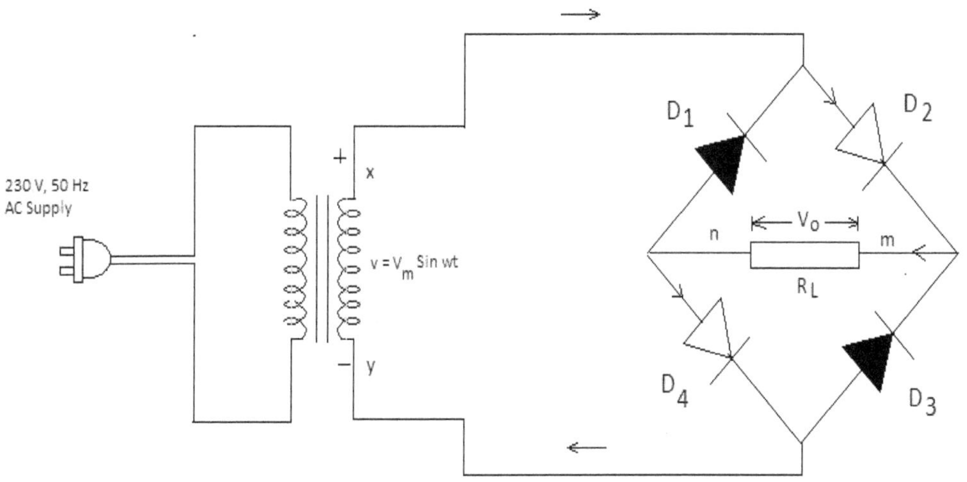

Figure 5.19 Bridge Rectifier during positive half-cycle

Similarly, during negative half cycle, the point 'y' becomes positive, the diodes D_3 and D_1 are forward biased & D_2 and D_4 are reverse biased. The current passes through the secondary winding, D_3, the load (from the point 'm' to 'n') and the D_1.

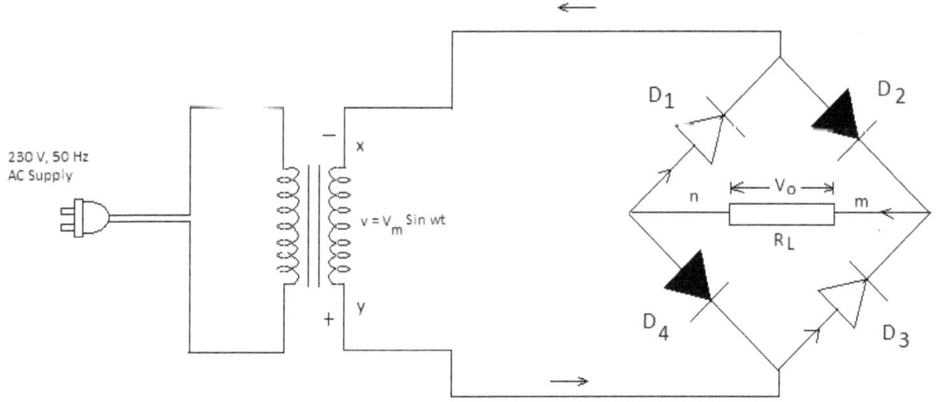

Figure 5.20 Bridge Rectifier during negative half-cycle

The current passing through the load resistance is always unidirectional ie., from the point 'm' to 'n' and hence, the voltage across it is also unidirectional. The output voltage waveform is shown in the figure 5.21.

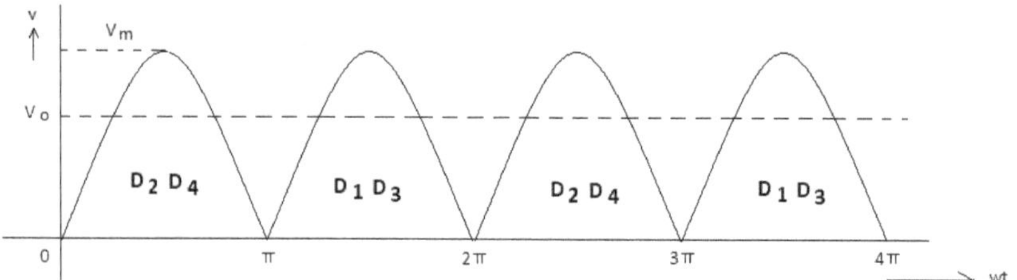

Figure 5.21 Bridge Rectifier output voltage

The instantaneous value of voltage across the secondary winding of transformer is

$$v = V_m\ Sin\omega t$$

The DC output voltage or Average Voltage is given by

$$V_0 = \frac{1}{\pi}\int_0^\pi V_m\ Sin\omega t.dt = \frac{V_m}{\pi}\left[-Cos\omega t\right]_0^\pi$$

$$= \frac{V_m}{\pi}\left[-\left(Cos\pi - Cos.0\right)\right] = \frac{V_m}{\pi}\left[-\left(-1-1\right)\right]$$

$$= \frac{2V_m}{\pi}$$

5.2 Bipolar Junction Transistor

The *'Bipolar Junction Transistor (BJT)'* which is popularly known as *'Transistor'* was invented in 1948 and this invention has completely revolutionized the electronics industry.

5.2.1 Construction

A transistor is basically made-up of Silicon or Germanium semiconductor materials arranged in three layers. These transistors are NPN type or PNP type. The middle region is called *Base* and the outer regions are known as *Emitter* and *Collector*. The figures 5.22and 5.23 show the basic structure of a NPN and PNP type transistors respectively.

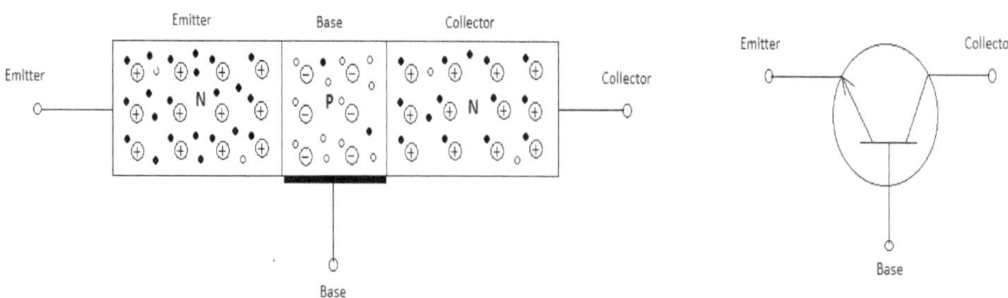

Figure 5.22 (a) Structure of NPN transistor (b) NPN transistor symbol

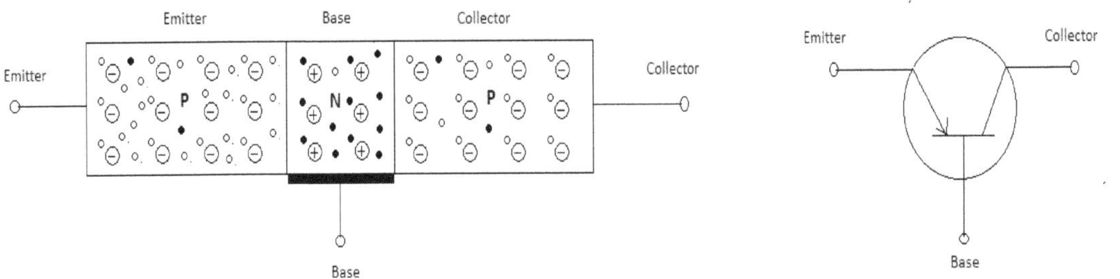

Figure 5.23 (a) Structure of PNP transistor   (b) PNP transistor symbol

The BJT is a three layer and two junction device. These three layers are called *Emitter, Base and Collector*. In NPN transistor, the Emitter (N type material) emits the free electrons into the Base and these electrons are collected by the Collector (N type material). Similarly, in PNP transistor, the Emitter (P type material) emits the holes into the Base and these holes are collected by the Collector (P type material). The Emitter is heavy doped whereas the Collector is lightly doped. The doping level of Base is in between that of Emitter and Collector.

The junction between Emitter and Base is called *Emitter-Base Junction* or simply *Emitter Junction*. The junction between Base and Collector is called *Collector-Base Junction* or simply *Collector Junction*.

5.2.2 Biasing of Transistors

Biasing a transistor means preparing it ready to operate in a required region by applying appropriate voltages across the junctions. The transistor has two junctions, Emitter – Base Junction and Collector-Base Junction and hence there are four possibilities of biasing a transistor. These biasing types are:

1. *Forward-Reverse Bias (FR Bias):* In this type of biasing, the Emitter-Base Junction is forward biased and Collector-Base Junction is reverse biased. The circuit for this bias of a NPN transistor is shown below in the figure 5.24.

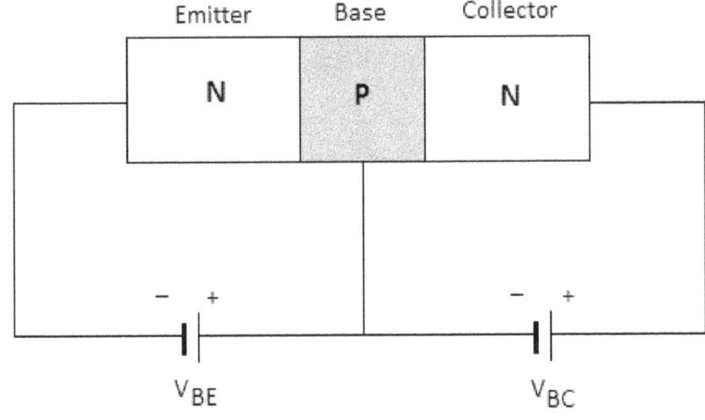

Figure 5.24 Forward-Reverse Biasing

For forward biasing the Emitter-Base Junction, the Emitter (N type) is connected to the negative terminal of the battery and Base (P type) connected to the positive terminal of the battery. Similarly, for reverse biasing the Collector-Base junction, the Collector (N type) is connected to positive terminal of the battery and Base (P type) connected to the negative terminal. The transistor operates in *Active Region* and can fulfill the purpose of an *Amplifier*.

2. ***Forward-Forward Bias (FF Bias):*** In this type of biasing, both the Emitter-Base Junction and Collector-Base Junction are forward biased. The circuit for this bias of a NPN transistor is shown below in the figure 5.25.

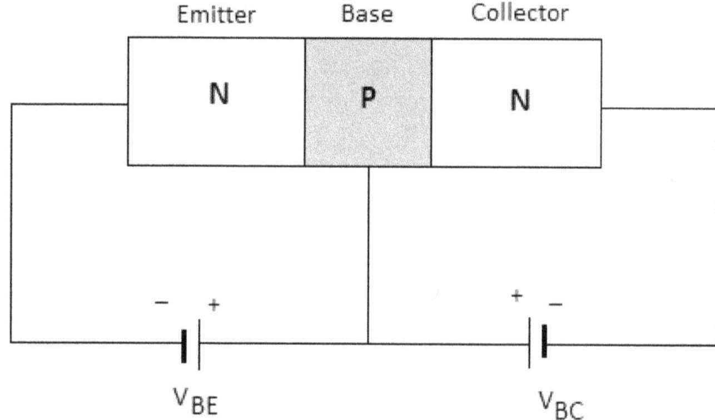

Figure 5.25 Forward-Forward Biasing

For forward biasing the Emitter-Base Junction, the Emitter (N type) is connected to the negative terminal of the battery and Base (P type) connected to the positive terminal of the battery. Similarly, for forward biasing the Collector-Base Junction, the Collector (N type) is connected to negative terminal of the battery and the Base (P type) connected to the positive terminal. The current at Collector-Base junction becomes independent of emitter current and the transistor operates in *Saturation Region*. The transistor acts as a *Closed Switch* in this region.

3. ***Reverse-Reverse Bias (RR Bias):*** In this type, both the Emitter-Base Junction and Collector-Base Junction are reverse biased. The circuit for this bias of a NPN transistor is shown below in the figure 5.26.

For reverse biasing the Emitter-Base Junction, the Emitter (N type) is connected to the positive terminal of the battery and the Base (P type) connected to the negative terminal of the battery. Similarly, for reverse biasing the Collector-Base junction, the Collector (N type) is connected to positive terminal of the battery and the Base (P type) connected to the negative terminal. The depletion regions at both junctions get widened and block the currents. Hence, the transistor behaves as an *open circuited switch* in this biasing and operates in the *Cut-off Region*.

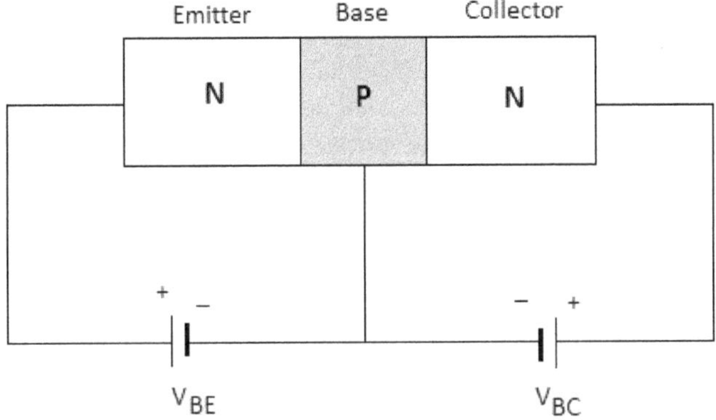

Figure 5.26 Reverse-Reverse Biasing

4. ***Reverse-Forward Bias (RR Bias):*** Here, the Emitter-Base junction is reverse biased and the Collector-Base junction is forward biased. The circuit for this bias of a NPN transistor is shown below in the figure5.27.

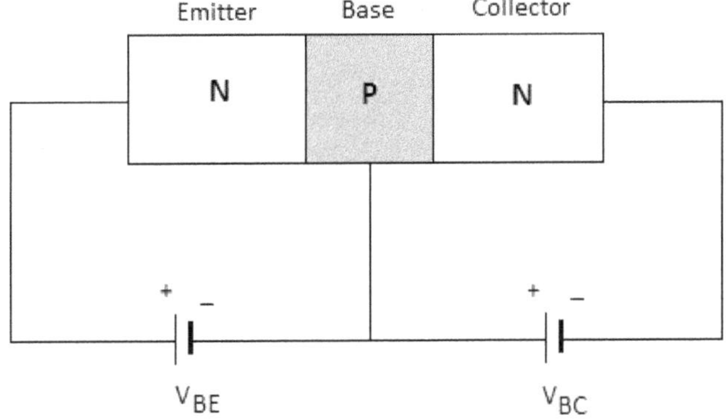

Figure 5.27 Reverse-Forward Biasing

The transistor operates in *Inverted Region* under this biasing. This operation is quite different from normal active operation and rarely used.

The following table 5.1 summarizes the operating regions for different biasing technics.

Table 5.1 Different Biasing methods of a BJT

S.No	Biasing type	Emitter-Base Junction	Collector-Base Junction	Operating Region
I	FR	Forward Biased	Reverse Biased	Active Region (Amplifier)
II	FF	Forward Biased	Forward Biased	Saturation (Closed Switch)
III	RR	Reverse Biased	Reverse Biased	Cut-off (Opened Switch)
IV	RF	Reverse Biased	Forward Biased	Inverted Region

5.2.3 Types of Transistor Configurations

As we know that a transistor has three terminals or electrodes, namely., Emitter, Base and Collector. Out of these 3 terminals, two are used as input and output ends. The third terminal is made common for both input and output terminals. Based on the terminal which is made common, we have three transistor configurations.

1. Common Base (CB) Configuration

2. Common Emitter (CE) Configuration

3. Common Collector (CC) Configuration

In all these configurations, the Emitter –Base Junction is forward biased and Base-Collector Junction is reverse biased to operate the transistor in active region. The configurations are shown below only in input ac signal point of view. The DC biasing is not shown.

1. ***Common Base (CB) Configuration:*** In this configuration, as shown in figure 5.28, the terminal 'Base' has been made common for both input and outputs.

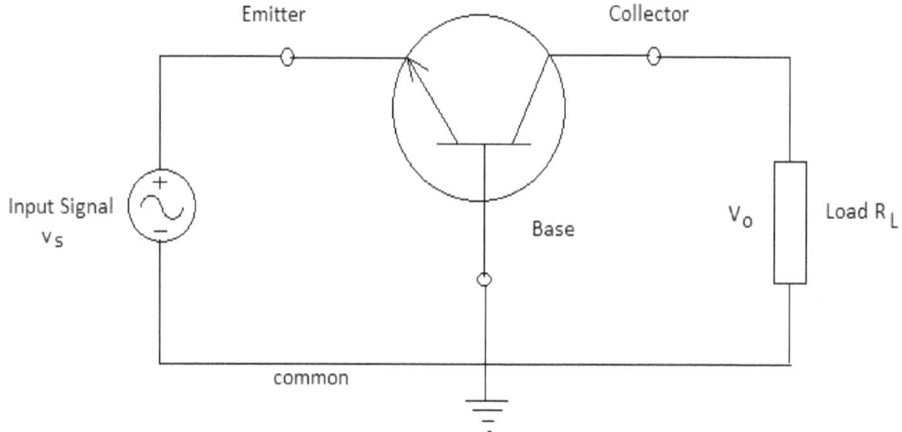

Figure 5.28 Common Base Configuration

The input signal is fed between emitter and base and the output is taken between collector and base terminals. Since, the emitter junction is forward biased and collector junction is reverse biased, the input resistance is very low and output resistance is very high in this configuration. Also, the collector current I_C is slightly less than the emitter current I_E, the current gain which is the ratio of output current (I_C) to the input current (I_E) is less than unity.

2. ***Common Emitter (CE) Configuration:*** In this configuration, as shown in figure 5.29, the terminal 'Emitter' has been made common for both input and outputs.

The input signal is fed between Base and Emitter and the output is taken between Collector and Emitter terminals. This configuration gives low input resistance and high output resistance. As the output Collector Current I_C is much larger than the input Base Current I_b, the current gain which the ratio of I_b to I_C is much larger than unity.

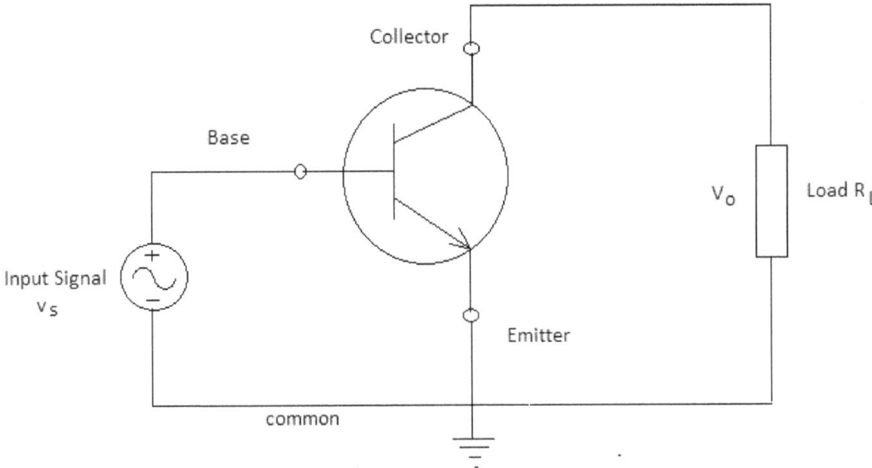

Figure 5.29 Common Emitter Configuration

3. ***Common Collector (CC) Configuration:*** In this configuration, as shown in figure 5.30, the terminal 'Collector' has been made common for both input and outputs.

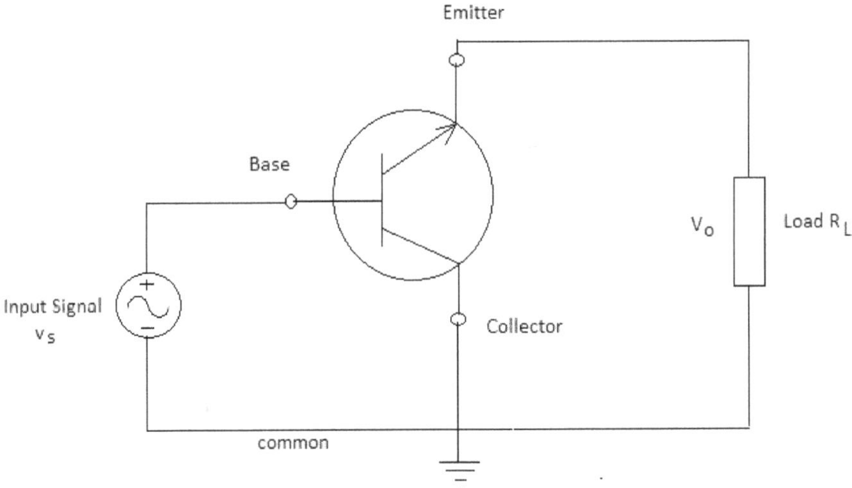

Figure 5.30 Common Collector Configuration

The input signal is fed between Base and Collector and the output is taken between Emitter and Collector terminals. This configuration is popularly known as *'Emitter Follower'*. The voltage gain in this configuration cannot become more than one. But, it has very high input impedance and low output impedance and it finds wide applications.

5.2.4 Working of a Transistor

Consider a NPN transistor biased for its active region operation. The emitter junction is forward biased by the voltage source V_{EB} and the collector junction is reverse biased by the battery V_{CB}. Since the emitter junction is forward biased, the depletion region at emitter junction becomes almost narrow. The electrons in emitter (N type) repelled by the negative terminal of battery V_{EB} and diffused into base. And also, few holes move from base (P type) recombine with the electrons in emitter and contribute for the emitter current. As the base is lightly doped, this current due to holes movement is very small when compared to the electroncurrent. The ratio of electron current to the total emitter current is called *'Emitter Injection Ratio'* or *'Emitter Efficiency'* and it is denoted by ϒ. The typical value of ϒ is 0.995.

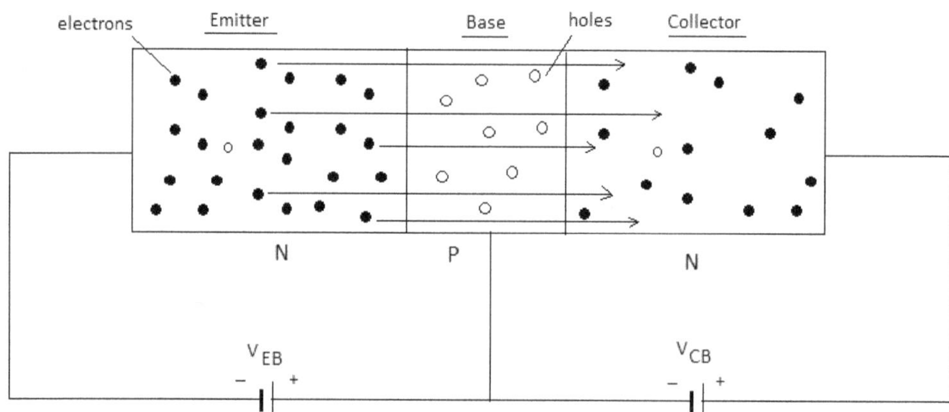

Figure 5.31 Forward Reverse Biasing of NPN Transistor

Most of the electrons emitted by the emitter into the base reach collector and contribute for collector current. Very few electrons recombine with the holes that present in base while travelling though base towards collector. It is possible as the base is made very thin. The ratio of electrons arriving at collector to the number of electrons emitted by the emitter is known as *'Base Transportation Ratio'* and it is denoted by β'. The typical value of β' is 0.995. The difference of emitter current and the collector current is the base current I_B.

$$I_E = I_C + I_B$$

The collector current is slightly less than the emitter current. The ratio of collector current I_C to the emitter current I_E is called *'DC Alpha'* of the transistor $D.C\ Alpha\ \alpha_{dc} = \gamma \times \beta'$. The typical value of DC alpha is 0.99.

5.2.5 Transistor as an Amplifier

Though transistor is used for many purposes, its main application is for Amplifiers to amplify the weak signals. The Common Emitter Configuration is best suited for amplifier applications as it offers high input resistance, low output resistance and higher values of current, voltage and power gains. The figure 5.32 shows CE Amplifier circuit which uses a NPN transistor. The batteries V_{BB} and V_{CC} are used to forward bias the emitter junction and reverse bias the collector junction respectively. The resistance R_B is used to limit the base current as emitter base junction is forward biased and the R_C is the load resistance from which an amplified output signal is tapped.

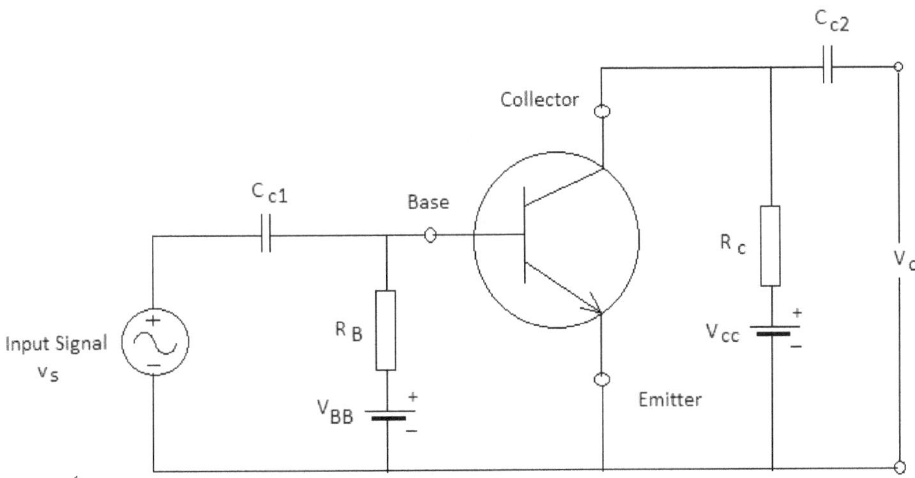

Figure 5.32 Transistor as an Amplifier in CE Configuration

The signal to be amplified V_s is given in the base circuit through a coupling capacitor C_{C1}. This capacitor does not allow DC from biasing voltage V_{BB} to input signal source. Similarly, another coupling capacitor C_{C2} is used to pass only amplified ac to output terminals. To understand the operation of amplifier circuit, the concept of DC load line is needed.

The condition of the circuit having no input signal is called *Quiescent Condition*. Under these conditions, the voltage equation in collector circuit is given by

$$V_{CC} = I_C R_C + V_{CE}$$

Or the collector current is given by $I_C = \left[-\dfrac{1}{R_C}\right]V_{CE} + \dfrac{V_{CC}}{R_C}$ which can be compared with an

equation for straight line $y = mx + c$.

$$y = I_C, \quad m = -\dfrac{1}{R_C}, \quad x = V_{CE} \quad \text{and} \quad C = \dfrac{V_{CC}}{R_C}$$

This line is called *DC Load Line* as the slope $\left(m = -\dfrac{1}{R_C}\right)$ of this line is decided by the load

resistance R_C. To draw this line two points are required. One point on I_C is obtained by making

I_C zero in above equation. $I_C = \left[-\dfrac{1}{R_C}\right]V_{CE} + \dfrac{V_{CC}}{R_C}$ or $0 = \left[-\dfrac{1}{R_C}\right]V_{CE} + \dfrac{V_{CC}}{R_C}$ or $V_{CE} = V_{CC}$.

Similarly, the second point is obtained by making V_{CE} zero in above equation.

$$I_C = \left[-\frac{1}{R_C}\right]V_{CE} + \frac{V_{CC}}{R_C} \quad \text{or} \quad _C = \left[-\frac{1}{R_C}\right] \times 0 + \frac{V_{CC}}{R_C} \text{ or } I_C = \frac{V_{CC}}{R_C}$$

This DC Load line is plotted on output characteristics of transistor in CE configuration. The output current I_C is almost equal to emitter current I_E and is independent of the voltage V_{CE}. This current increases as the base current increases and it is shown in figure for different base currents.

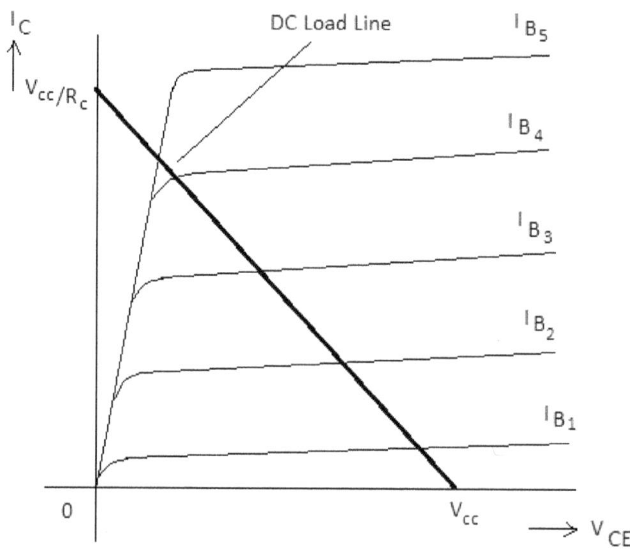

Figure 5.33 Output Characteristics of a NPN Transistor in CE Configuration

The operating point will be shifting on the DC load line according to the base current. The operating point on DC load line under quiescent condition is called *Quiescent Operating Point* or simply *Q point*. The Q point is selected to be in the middle of the active region so that the output amplified signal does not get distorted.

When the input signal V_s is applied to input (base) of the circuit, the base current varies as per V_s. As the base current varies, the instantaneous operating point of transistor moves on the dc load line. Thus, the instantaneous values of collector current I_C and the voltage V_{CE} vary according to the input signal (base current). As the input base current increases, the operating point on dc load line moves up from Q, the current I_C increases and V_{CE} decreases. Similarly, the current I_C decreases and V_{CE} increases as the operating point moves downwards from Q point with the decrease in base current.

The changes in the current I_C and the voltage V_{CE} are shown below in the figure 5.34.

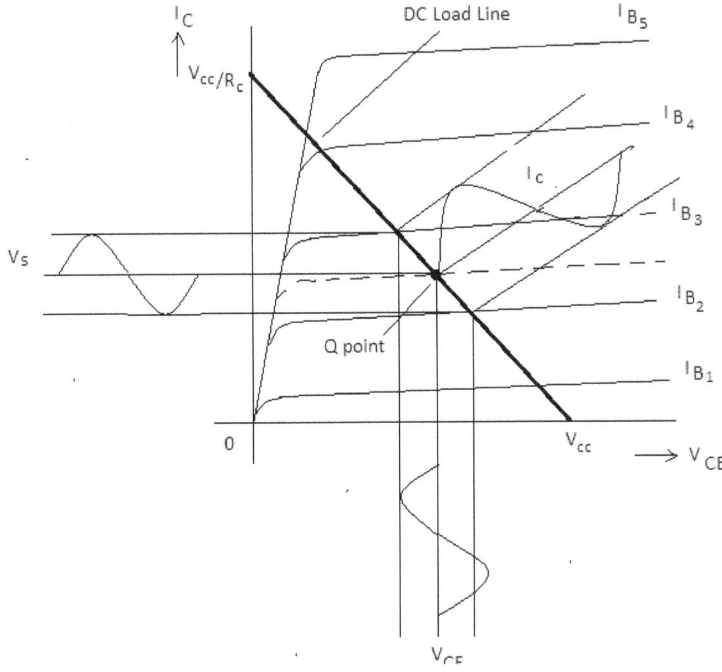

Figure 5.34 Transistor as an Amplifier

The variation in Collector voltage is many times larger than the variations in input signal. The collector voltage reaches the output end through the coupling capacitor C_{C2}.

5.3 Power Converters

Power Electronics deals with the use of low power electronics for controlling and conversion of high electrical powers. The power electronic converters mainly use a power semiconductor device called *'Silicon Controlled Rectifier (SCR)'*. This is also called as *'Thyristor'*.

The SCR is a 4-layer, 3 terminal controllable device. The SCRs are available up to the power rating of 10 MW, 2 KA and 1.8 KV. The following figure 5.35 shows the basic lay-out and symbol of SCR.

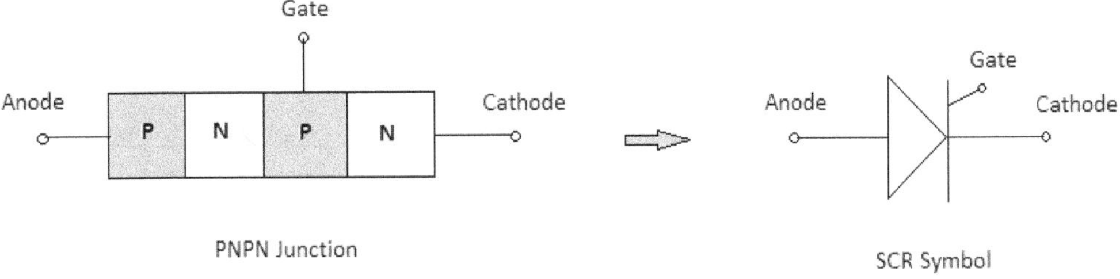

Figure 5.35 (a) PNPN junction (b) Symbol of SCR

The three terminals are identified as Anode, Cathode and Gate. When the Anode is connected to the positive terminal and Cathode to negative terminal of the battery, the SCR is said to be forward biased. This forward biased SCR conducts only when its gate terminal is given a signal or pulse which is called *Firing or Triggering* of the gate. Hence, the conduction of SCR can be controlled from Gate end and thus, it can be used as a *Controlled Switch*. The SCR does not conduct under reverse biased conditions (ie., its anode is connected to the negative terminal and cathode to positive terminal of the battery) even its gate is triggered.

The SCRs are mainly used in Power Converters as switching devices. The usage of Power Coveters in industry increases the productivity, efficiency and control over the process. The Power Converters are classified into the following broad categories.

1. AC to DC Converters

2. DC to DC Converters

3. DC to AC Converters

5.3.1 AC to DC Converters

These Converters convert input AC into DC output. These are basically Rectifiers whose DC output voltage can be controlled. Hence, these controllable rectifiers are called *Phase Controlled Rectifiers*.

Figure 5.36 AC to DC Converter

These converters convert fixed AC voltage into variable dc output voltage. The diode rectifiers' supply fixed dc output voltage whereas the Phase controlled Rectifiers which use SCRs supply variable output dc voltage. The outputs of a Diode Bridge Rectifier and a Phase Controlled Rectifier are shown in the figures 5.37 and 5.38.

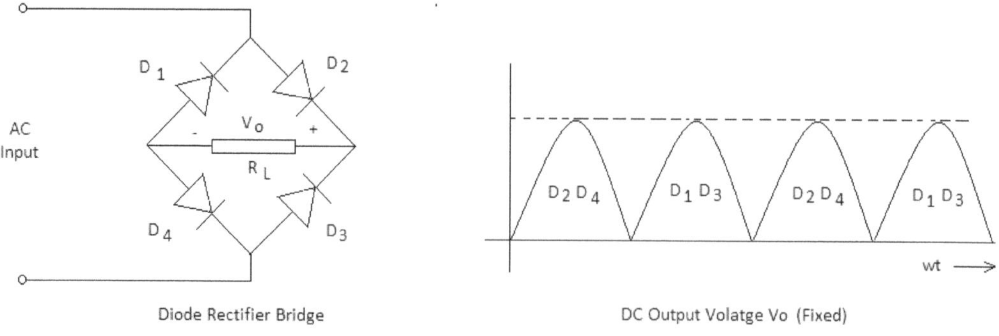

Figure 5.37 (a) Diode Bridge Rectifier (b) its DC Output

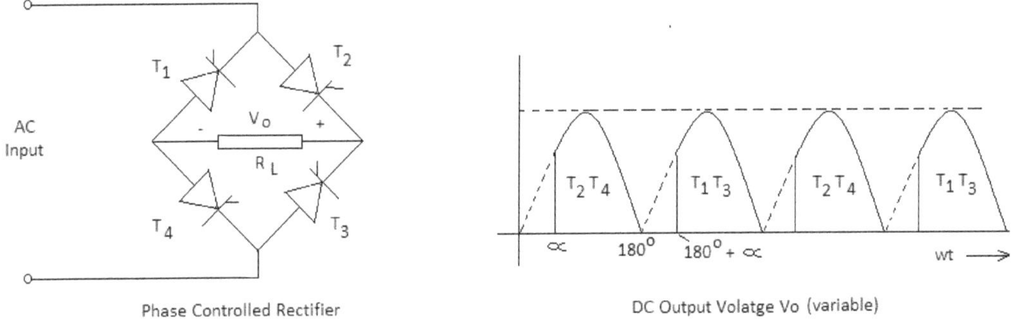

Figure 5.38 (a) Phase Controlled Rectifier (b) its DC Output

Where α is the firing angle where the gate terminals are triggered by giving pulses. By controlling the angle α, the dc output can be controlled.

Applications:

a. Battery Charging

b. D.C motor drives

c. Regulated Power Supplies (RPS)

d. HV DC transmission

e. Wind Generator Converters

5.3.2   DC to DC Converters

These converters convert a fixed dc input into a variable dc output. These are called *Choppers*.

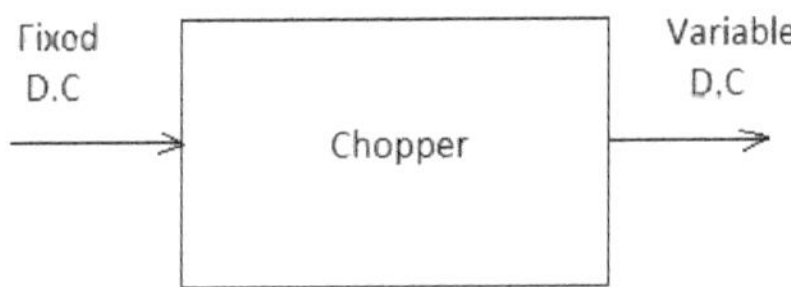

Figure 5.39 DC-DC Converter

It is like a DC Transformer and its circuit in basic form is shown in the figure 5.40.

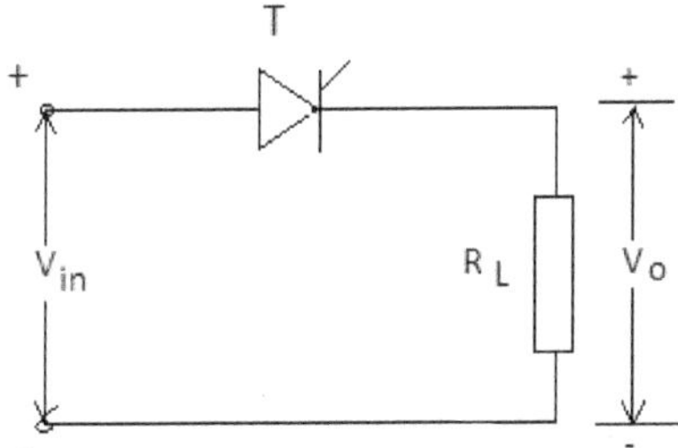

Figure 5.40 Basic Chopper Circuit

By controlling the time of 'ON' of SCR through its gate terminal, its average value output can be controlled.

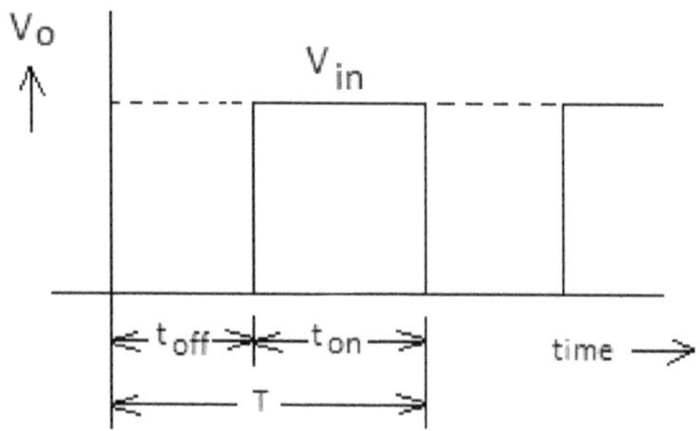

Figure 5.41 Output Voltage waveform from Chopper

$$\text{Duty Ratio } \alpha = \frac{t_{on}}{t_{on} + t_{off}} = \frac{t_{on}}{T}$$

The output voltage is given by $V_o = \alpha \times V_{in}$

Applications:

a. DC motor drives

b. Subway cars

c. Battery Driven Vehicles

d. Electric Traction

e. Switch Mode Power Supplies (SMPS)

5.3.3   DC to AC Converters

These converters convert fixed DC into variable AC voltage (both magnitude and frequency of ac signal can be varied). These converters are called *Inverters*.

Figure 5.42 DC-AC Converter

An inverter is having a DC voltage input voltage from battery or Solar PV cell. The output is alternating voltage which is depicted in the figure 5.43.

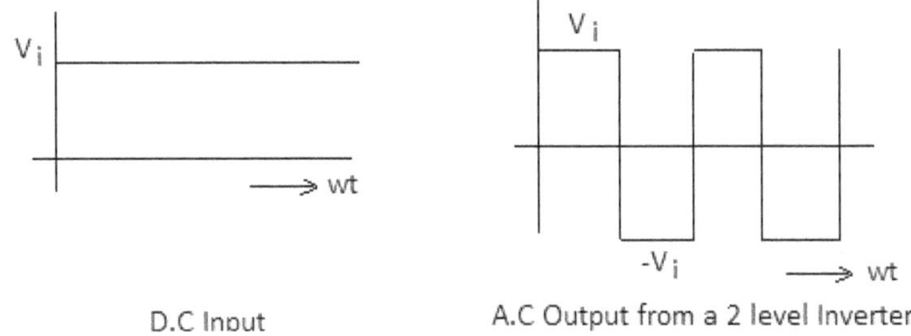

Figure 5.43 (a) DC input of inverter (b) Output of Inverter

Applications:

a. Flash Lights in Photography

b. UPS (Uninterruptible Power Supply)

c. Air Craft and Space Power Supplies

d. Induction Motor drives

e. Induction Heating

5.4 Uninterruptible Power Supply (UPS)

Uninterruptible Power Supply (UPS) ensures the continuity of power supply under all conditions. The UPS is used in critical applications like medical intensive care systems, chemical plant process control, Safety monitors etc. where a temporary loss of power would result for severe consequences.

The figure 5.44 shows the block diagram of a typical UPS system.

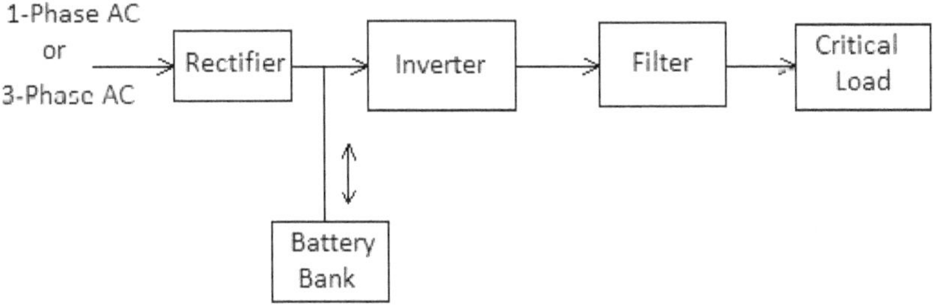

Figure 5.44 UPS Block Diagram

The Rectifier converts AC into DC and it is supplied to the Inverter. The dc output of rectifier is also used to charge the Battery under normal conditions. Whenever, the AC supply goes off, the inverter converts the dc power supplied by battery into AC and delivers to the load after necessary filtering.

Thus, UPS provides quality power to AC loads. It also regulates the output load voltage against the input voltage fluctuations.

Review Questions

Short Answer (Two marks) Questions

1. What do you mean by Negative Temperature Coefficient of Resistance?
2. Differentiate between Valance Band and Conduction Band
3. Differentiate between Intrinsic and Extrinsic Semiconductors
4. Differentiate between P type and N type materials
5. What is the effect of forward bias and reverse bias on the depletion region of a PN Junction Diode?
6. Show the circuit symbol of a PN Junction Diode
7. Define Cut-in Voltage and Reverse Breakdown Voltage of a diode
8. Show the circuit symbol of a Bipolar Junction Transistor and name the terminals.
9. Compare the doping levels of Emitter, Base and Collector of a BJT
10. Specify the applications of Forward- Reverse biasing and Forward-Forward biasing of BJT
11. Show the circuit for Common Base Configuration of a NPN Transistor
12. What do you mean by Quiescent Condition of a transistor?
13. Why DC load line has got that name?
14. Show the circuit symbol of a Silicon Controlled Rectifier (SCR) and name the terminals
15. Name different Power Converters generally used
16. List the applications of a DC-DC Converter

Essay (Six marks) Questions

1. Explain the process of forming PN junction and Depletion Region
2. With the help of a neat sketch, explain the V-I Characteristics of a PN Junction Diode
3. Explain the working of Half Wave rectifier. Show the relevant waveforms.
4. How a PN junction diodes are useful for Full Wave Rectifier applications? Explain with the help of relevant waveforms.
5. Discuss on different biasing methods used for a Bipolar Junction Transistor
6. What are the configurations used for a BJT? Briefly explain.
7. Explain the working of a transistor as an Amplifier in Common Emitter Configuration
8. What do you mean by DC-AC Converter? Explain its necessity with the help of a block diagram
9. Show the block diagram of a typical Uninterruptible Power Supply (UPS) and briefly explain its operation.

CHAPTER 6

Op-Amps, Transducers and Data Acquisition

6.1   Operational Amplifier

The Op-Amp is a versatile device that is used to amplify both dc and ac input signals. The Operational Amplifier was originally meant for performing mathematical operations like addition, subtraction, multiplication etc. and later for amplification. Hence, it has got the name 'Operational Amplifier' which is abbreviated as 'Op-Amp'. The other uses of Op-Amps are as Oscillators, Comparators, Regulators, Active Filters etc.

An Operational Amplifier, in short, Op-Amp is a direct coupled high gain amplifier. The input signal which is to be amplified is directly fed to the amplifier. It has been organized as one or more differential amplifiers and usually followed by a level translator and an output stage. The output is a Push-Pull Amplifier and the Op-Amps are available in a single IC (Integrated Circuit) chip.

The following figure 6.1 shows the block diagram of a typical Op-Amp.

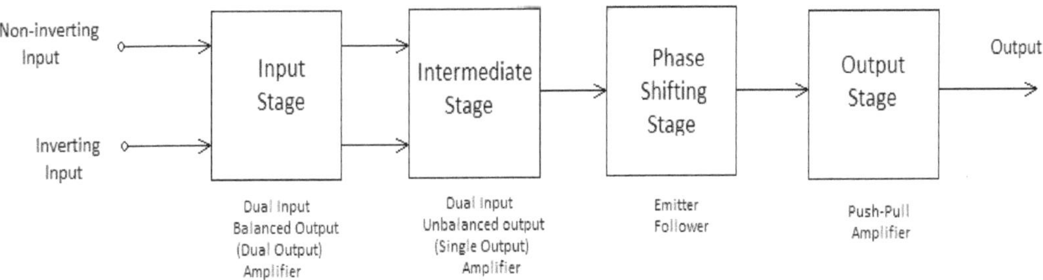

Figure 6.1 Block diagram of typical Op-Amp

The Op-Amp has two input terminals called Inverting and Non-inverting terminals. A Dual input, Balanced Output (Dual Output) differential amplifier is used as input stage. This stage constitutes for most of amplifier gain and provides high input impedance which is highly desired in Cascade Systems. The output of the first stage is fed to the intermediate stage which is another Differential Amplifier. This is a Dual Input-Single Output (Unbalanced) Amplifier. As the direct coupling is used in this amplifier, the dc voltage at output of intermediate stage is well above the ground level. To bring this potential down to zero, so that the original signal shape is not distorted, a Level Translator(shifting circuit) is used. The output stage is a Push-Pull Complementary Amplifier and provides the low output impedance.

Schematic Symbol of Op-Amp:

The figure 6.2 shows the symbol of an Op-Amp.

The output signal will have the same phase of input signal (of course, the magnitude increases) fed at non-inverting input terminal. And, for the input signal fed at inverting input terminal, the output will be of 180^0 out of phase (the polarity of output signal gets reversed).

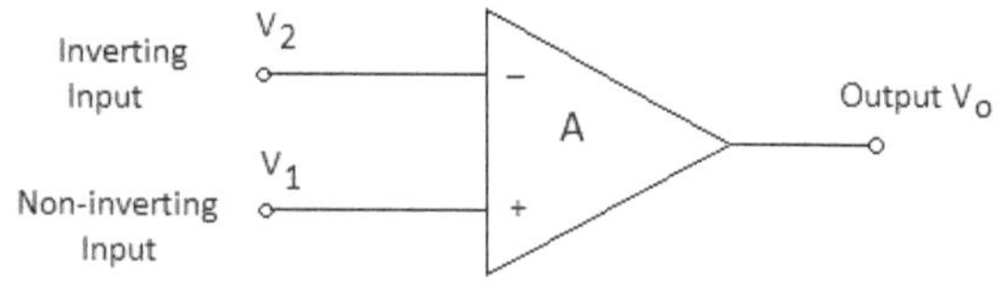

Figure 6.2 Schematic of an Op-Amp

6.1.2   Ideal Operational Amplifier

An Ideal Operational Amplifier is having the following characteristics.

1. Infinite Voltage Gain $\dfrac{V_o}{V_{in}}$

2. Infinite Input Impedance Z_{in}

3. Zero Output Impedance Z_{out}

4. Zero Noise

5. Infinite Bandwidth and Infinite Slew Rate (It is the rate at which an Op-Amp can detect voltage changes

6. Zero Input Offset Voltage

7. Infinite Common Mode Rejection Ratio

6.1.3   Commercial IC 741 Op-Amp

The IC 741 is a practical Op-Amp which is a monolithic Integrated Circuit (IC). The high input impedance and very small output impedance makes IC 741 almost an ideal voltage amplifier. This IC has 7 functional pins, out of that 4 can be used for inputs and 1 for output. It can provide very high voltage gain for wide range of voltages. It is the best option for Integrator, Summing Amplifier, Short Circuit Protection and General Feedback applications. The figure 6.3 shows the pin out diagram of IC 741.

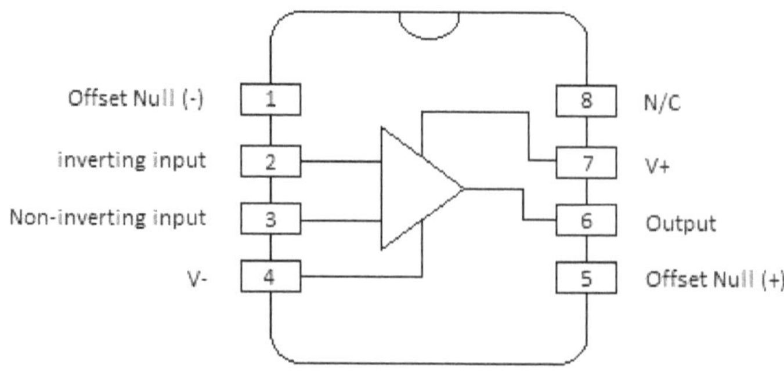

Figure 6.3 Pin-out diagram of IC 741

The functions of different pins are:

Pins 4 and 7 (Power Supply): The power supply is needed for the operation of Op-Amp. The power supply is given between the pins 4 and 7. The pins 4 and 7 are connected to the negative and positive terminals of the battery respectively. The supply voltage can range between 5 and 18 V.

Pins 2 and 3 (Inputs): The pins 2 and 3 are intended to give inputs to Op-Amp. The inverting input is given at pin 2 and the non-inverting input is given at pin 3. If the voltage at Pin2 is greater than the voltage at Pin3, the output signal stays low. Similarly, if the voltage at Pin3 is greater than the voltage at Pin2, the output goes high.

*Pin 6 (Output):*This pin provides the output. The output at this pin depends up on the inputs given at pins 2 and 3. The output is said to be high if the voltage at this pin is equal to the positive voltage of the power supply and low if the output is equal to the negative supply voltage.

*Pin 1 and 5 (Offset Null):*The manufacture defects and external disturbances may cause for very slight difference in voltages fed at input inverting and non-inverting terminals. This slight difference in voltages gets amplified by high voltage gain provided by this IC and may affect the output. To nullify or offset this effect, a finite voltage is applied between the pins 1 and 5.

*Pin 8 (N/C):*This pin is not connected to any circuit inside 741 IC. It's just a dummy lead used to fill the void space in standard 8 pin packages

The gain of IC 741 is not maintained constant at all the frequencies of input voltage signal. It varies as shown below in figure 6.4.

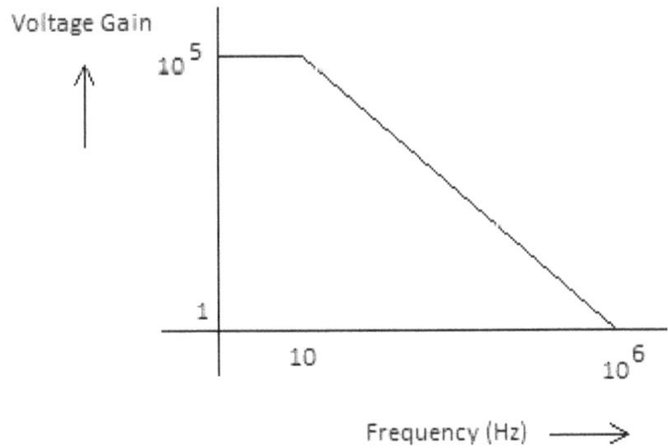

Figure 6.4 variation of Voltage Gain with frequency

The gain is maintained constant up to the input frequencies of 10 Hz and it reduces after 10 Hz. It becomes unity when the input signal frequency increases to 100,000 Hz.

Applications:

1. Amplifiers

2. For many mathematical operations like integration, differentiation, summing etc

3. Rectifiers

4. Oscillators

5. Comparators etc.

6.2 Remote Control and Monitoring

Remote Monitoring and Control (M&C) systems are used to control the larger and complex amenities like industries, power stations, network operation centers, airports, space crafts etc., to some degree of automation. These Remote Monitoring and Control Systems (M&C systems) receive the data from sensors/transducers and these systems are mostly closed loop systems. Both Sensors and Transducers are used to sense the changes in a physical quantity. But, a sensor will give an output in the same units and a transducer will convert the measured physical quantity into an electrical signal. In this section, the working of transducers used to sense strain, temperature, acceleration and light have been presented.

6.2.1 Transducer used to Sense Strain

When an external force (either tensile or compressive force) is applied to an object, the object gets subjected to some deformation. This deformation is called the Strain. The 'Strain Gauge' is used to measure the strain.

Figure 6.5 a) Strain with tensile force b) strain with Compressive force

As the dimensions change with strain, the electrical resistance of the object also changes. This phenomenon is used in Strain Gauges for measuring the strain.

A **Strain gauge** is basically a resistor and is sensitive to small changes occur in the geometry of an object. By measuring the change in resistance of the object, the amount of strain and induced stress are estimated. As the change in resistance is normally very small, a long thin metallic strip made into a zig-zag pattern is used to enhance the change in resistance for a given strain.

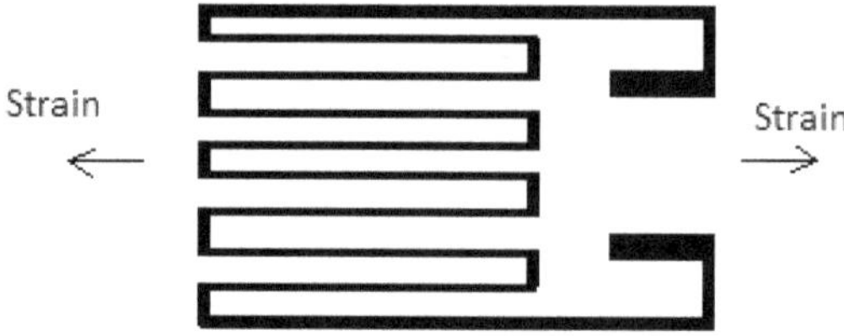

Figure 6.6 Strain Gauge

The strain gets enhanced by the parallel lines and the sensitivity of strain gauge increases. The Strain Gauge is connected in a bridge network as shown in figure 6.7.

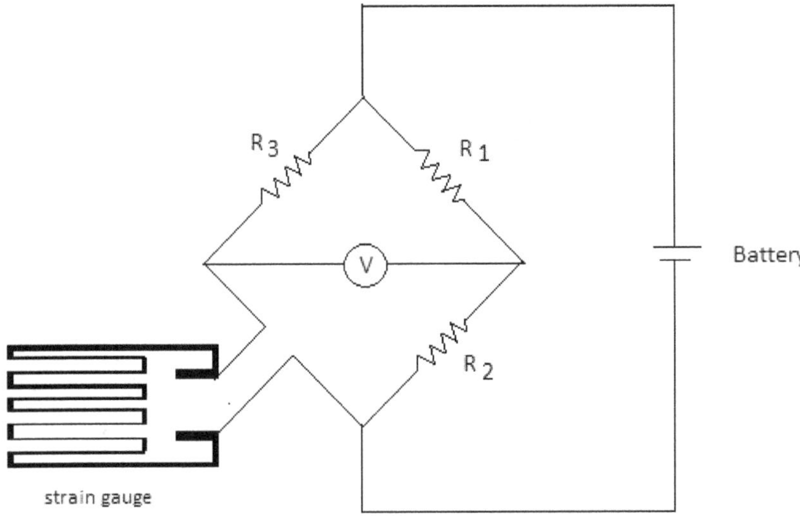

Figure 6.7 Strain Gauge in Bridge Network for strain measurement

The Resistances R_1 and R_2 are designed to be equal. The resistance R_3 is a variable resistance and it is selected to be equal to the resistance of strain gauge under unstrained conditions. The bridge is balanced under these conditions and it is ensured by the zero reading of the voltmeter. When the strain occurs, the resistance of the strain gauge changes and the bridge gets unbalanced and produces a deflection in the voltmeter. Since, this voltage is very small, it is fed to an amplifier. The amplified voltage signal is proportional to strain and hence, it is measures the strain.

The Strain Gauges are used for measuring stress in railway lines, aircraft wings, structural components of bridges and buildings. These are also used for Aircraft component testing and Automotive testing. The Strain Gauges also find applications for measuring the rotational strain in turbines, wheels, fans, propellers and motors.

6.2.2   Transducer used to sense the Temperature

Thermistor is most widely used transducer to sense the temperature. The Thermistor is derived from Thermal Resistor. Thermistors are made with semiconductor materials, mostly compressed metal oxides. These devices change their resistance as the temperature changes and hence useful for measuring the temperatures. These devices have very high sensitivity than thermocouples which are also used for temperature sensing and thus thermistors are generally called as Ideal Temperature Transducers.

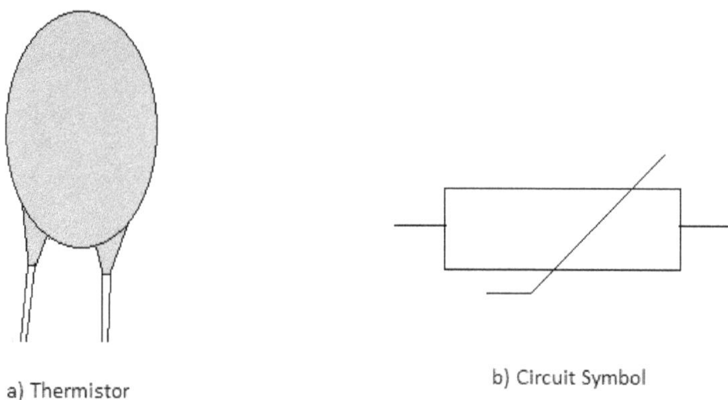

Figure 6.8 a) Thermistor b) Circuit symbol

Thermistors have the quality of Negative Temperature Coefficient ie., its resistance decreases as the temperature increases. These devices are used to measure the temperatures ranging from -60^0 to 150^0 C.

Thermistors are extensively used in fire alarms, ovens and refrigerators. They are also used in digital thermometers and in many automotive applications to measure temperature. Some more commercial uses for thermistors include Industrial and Medical Electronics, Food Handling and Processing, Aerospace, Communication and Instrumentation.

6.2.3   Transducer to Sense the Acceleration

We know that Acceleration is the time derivative of velocity. Hence, a Velocity transducer followed by a differentiator circuit is enough to measure the acceleration. A velocity transducer/sensor consists of a moving coil suspended in the magnetic field of a permanent magnet. The velocity is given as the input, which causes the movement of the coil in the magnetic field. This causes an emf to be generated in the coil. This induced emf will be proportional to the input velocity and thus, is a measure of the velocity.The output of the velocity transducer is given as input to the differentiator circuit. The figure 6.9 shows an accelerator transducer.

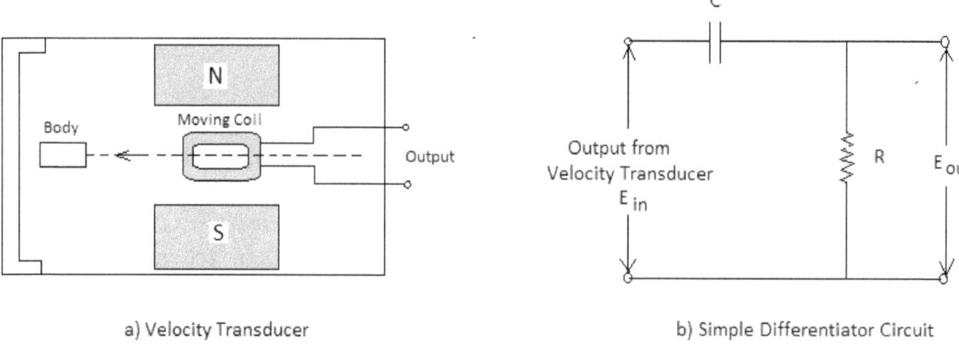

Figure 6.9 a) Velocity Transducer b) Differentiator Circuit

The output voltage of a differentiator can be written as

$$E_{out} = R \times i$$

$$= R \times \frac{E_{in}}{R + jX_C} \text{ (where } X_C = \frac{1}{2\pi fC} \text{)}$$

At lower values of frequency, the term R can be neglected in comparison with X_C. Then, the equation can be written as

$$E_{out} = R \times \frac{E_{in}}{(1/2\pi fC)}$$

Or $\qquad E_{out} = j2\pi fRC \times E_{in}$

The operator j represents that the output voltage leads input voltage by 90^0. Also, the frequency $f = \frac{1}{T}$ and $2\pi RC$ is a constant, the E_{out} is the derivative of E_{input}. Hence, the E_{out} is proportional to the acceleration.

6.2.4 Transducer to Sense the Light

A light sensor is a photoelectric device that converts light energy (photons) detected to electrical energy (electrons). The different light sensors available are Photo-resistor, Photodiodes, and Phototransistors. The operation of Photo-resistor light sensor is explained in this section.

Photo-resistors (LDR) Sensor:

This is the most commonly used light sensor and it is also known as Light-Dependent Resistor (LDR). The resistance of this LDR varies as light or illumination on it varies. The symbol and the variation of its resistance with respect to the changes in illumination are shown in figure 6.10.

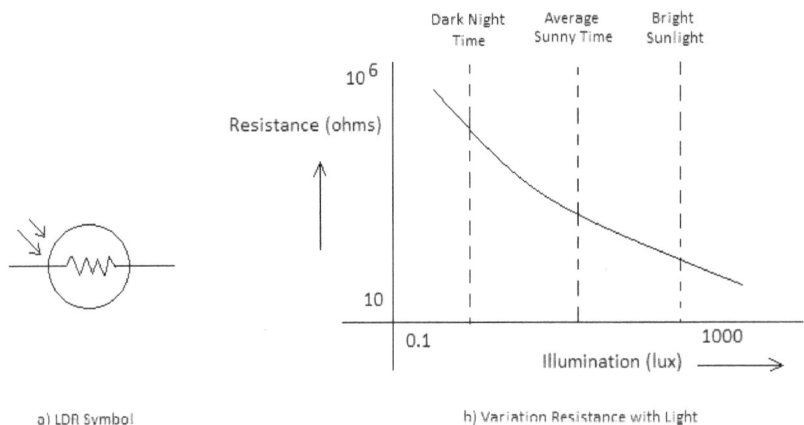

Figure 6.10 a) LDR Symbol b) Variation of Resistance with light

A Photo-resistor or Light-Dependent Resistor is composed of photo-conductor material. When the light hits the LDR, the valance electrons get excited and move from valance band to conduction band. The more free electrons in the conduction band of the resistor, the less is the LDR resistance. Photo-resistors are made with high resistive Semiconductor material called Cadmium Sulphide cells. These cells are highly sensitive to visible and near-infrared light. The working of LDR is explained with the help of figure 6.11.

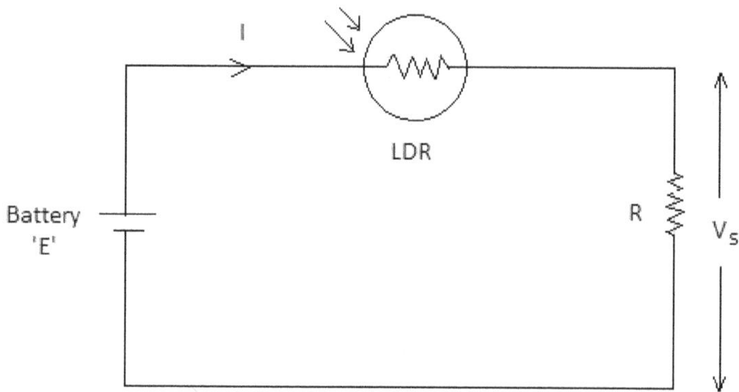

Figure 6.11 Circuit to sense the light

When there is no much lighting or illumination, the LDR offers very high resistance and it drops the entire battery voltage E across it. The voltage across the output resistance R is very small. As the illumination on LDR increases, its resistance reduces, the voltage across LDR decreases and the voltage drop across the resistance R increases for a given battery voltage E. Hence, the voltage across the output resistance R is proportional to the illumination.

Applications:

- Used in Street Lamp Control
- Automatic screen brightness of mobiles
- In automobiles to support the drivers' field of vision
- To detect the sunlight levels in agriculture field
- For Security applications

6.3   A/D and D/A Converters

Need for Analog to Digital (A/D) and Digital to Analog (D/A)Converters:

The Digital Computers play a vital role in automation, remote control and monitoring of industrial processes and other systems. The digital computers receive and process the data in digital or binary form. A digital quantity will have a value that is specified as one of two possibilities such as 0 or 1, LOW or HIGH, true or false, and so on. Most of the industrial processes involve physical quantities like temperature, pressure, light intensity, audio signals, position, rotational speed, and flow rate etc. and these are converted into electrical signals using transducers. Since, the Digital computers perform all of their internal operations with digital data using digital circuitry, the analog output from transducers is converted into digital form and is inputted to a digital computer system. For this an Analog to Digital Converter or A/D Converter is needed. Similarly, the output of digital computer is in digital form and it is converted into analog form using Digital to Analog converter or D/A Converter. The role A/D and D/A converters is well explained for the control of water temperature in the figure 6.12.

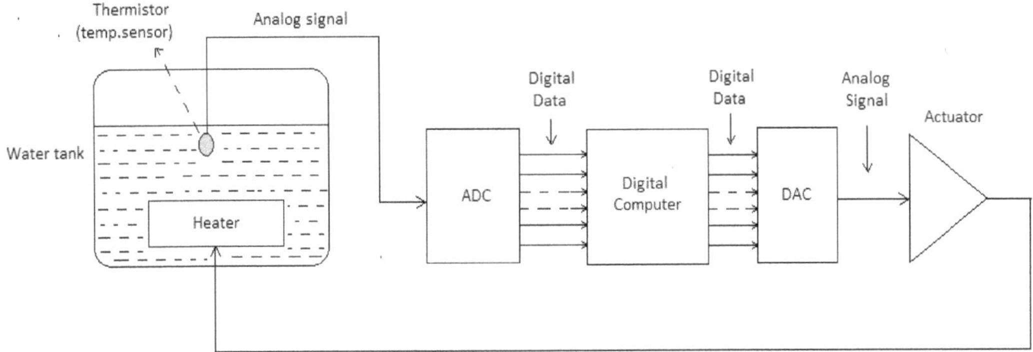

Figure 6.12 Role of A/D and D/A Converters in water temperature control

The electrical analog output of the transducer (a thermistor to sense the temperature of water) serves as the input to the ADC. The ADC converts this analog input to a digital output. This digital output consists of a number of bits that represent the data in the form of 0 and 1. This digital data is fed to the digital computer which stores and processes the data according to a set of instructions called programming. The digital output from the computer is converted into an analog electrical signal using D/A Converter. This analog signal from the DAC is often connected to some power electronic circuit that serves as an actuator to control the heater which in turn regulates the water temperature.

Thus, the ADCs and DACs function as interfaces between a completely digital system, like a computer, and the analog world.

6.3.1 Digital to Analog Converter (D/A Converter)

The Digital-to-Analog Converter converts the digital data expressed in terms of 0s and 1s into an analog voltage or current signal which is proportional to the digital input. The following figure 6.13 shows a R/2R Ladder type D/A Converter which is mostly used DAC. It only uses two different values of resistances, R and 2R.

The binary word applied to the switches produces a proportional output voltage. The binary inputs B_3, B_2, B_1, B_0 control the states of the switches. The position of these switches decides the output current I_{out}. This current is allowed to flow through an op-amp current-to-voltage converter to deliver an output voltage V_{out}.

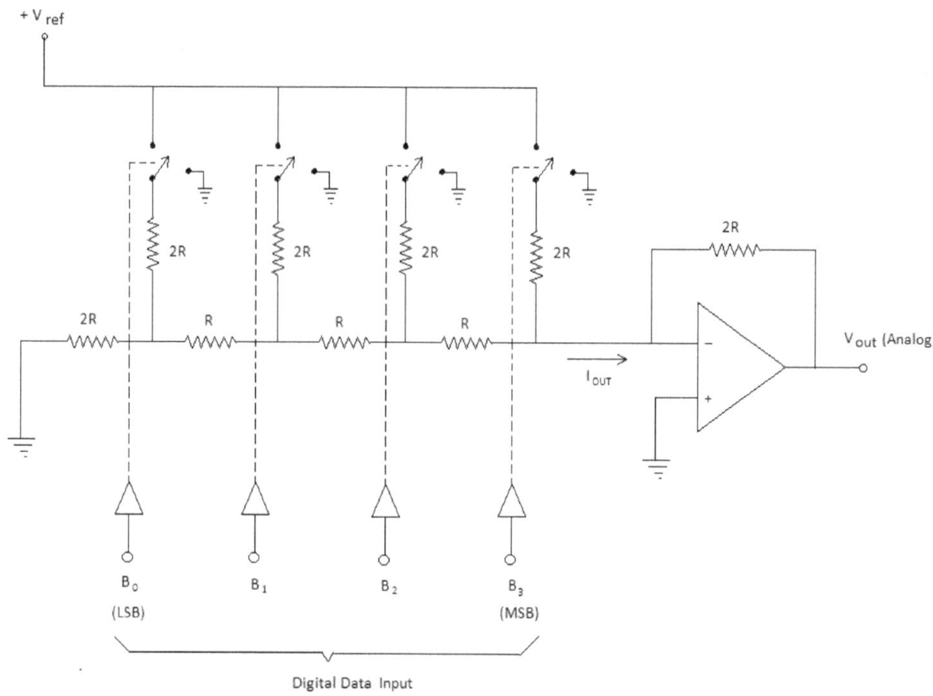

Figure 6.13 R/2R Ladder DAC

The value of V_{out} is given by the expression,

$$V_{out} = \frac{-V_{ref}}{8} \times B$$

where B is the value of the binary input, which can range from 0000 (0) to1111 (15). The negative sign is present because the summing amplifier is a polarity inverting amplifier, but it will not concern us here. For example, if the digital data input is 1110, the equivalent decimal value (B) is given by, (starting from LSB (Least Significant Bit) to MSB (Most Significant Bit),

$$\left[0 \times 2^0\right] + \left[1 \times 2^1\right] + \left[1 \times 2^2\right] + \left[1 \times 2^3\right] = 0 + 2 + 4 + 8 = 14$$

Then, V_{out} for a V_{ref} of 5V is given by

$$V_{out} = \frac{-V_{ref}}{8} \times B = \frac{-5}{8} \times 14 = -8.75 \text{ V}$$

6.3.2   Analog to Digital Converter

An Analog to Digital Converter (ADC) converts the analog input voltage into a digital output code which is proportional to the analog input. The A/D conversion process is generally more complex and time-consuming than the D/A conversion process. Several important types of

ADCs utilize a DAC as a part of their circuitry. The figure 6.14 shows a general block diagram of ADC.

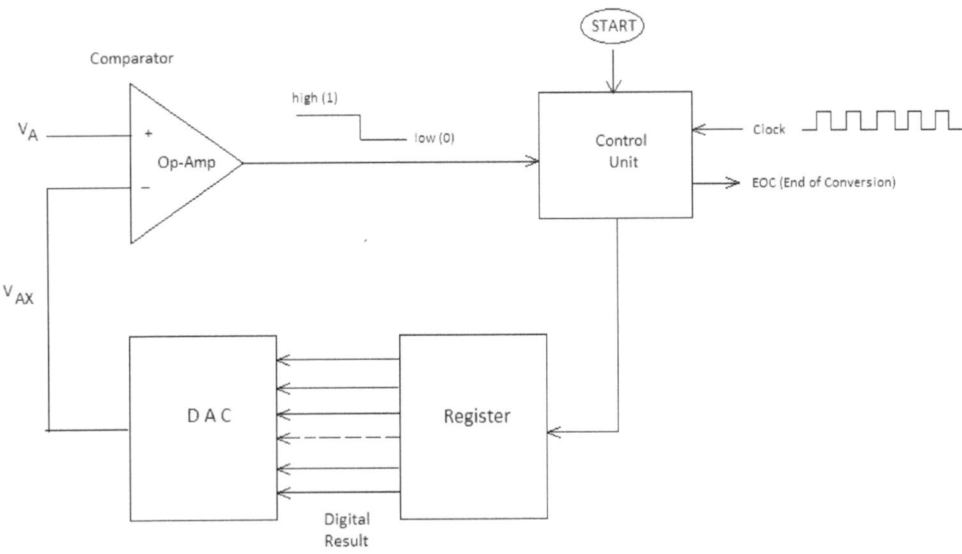

Figure 6.14 Block Diagram of ADC

The analog signal which needs to be converted into digital form is fed at non-inverting end of the Op-Amp. The Op-Amp is operated as a Comparator. The START command at Control Unit initiates the conversion process. The control unit having the logic circuitry continually increments the binary number at a rate determined by the clock and it is stored in the Register. The digital number in the register is continuously converted into an analog voltage signal V_{AX} using a Digital to Analog Converter (DAC). This analog voltage signal V_{AX} is fed at inverting input of Op-Amp. The Op-Amp compares the V_{AX} with V_A and stays HIGH as long as V_A is greater than V_{AX}. When V_{AX} exceeds V_A by a small threshold voltage V_T, the comparator output goes LOW and stops the process of incrementing the digital number by control unit. At this point, the digital number stored in the Register is the digital equivalent of analog signal V_A. The Control Unit activates the 'End of Conversion' signal (EOC) when the conversion is complete.

6.4 Data Acquisition and Control

The automatic collection of data from sensors, instruments and devices in a factory, laboratory or in the field is called *Data Acquisition*. Data Acquisition and Control technologies are very common in industrial settings, such as factories and warehouses. Data acquisition is the process of sampling a real world signal and converting it into a digital data that can be recorded in a computer system. Once the real world signal is captured and/or recorded, it can be used for Control and analysis of it. The figure 6.15 shows the typical layout of a PC based Data Acquisition System.

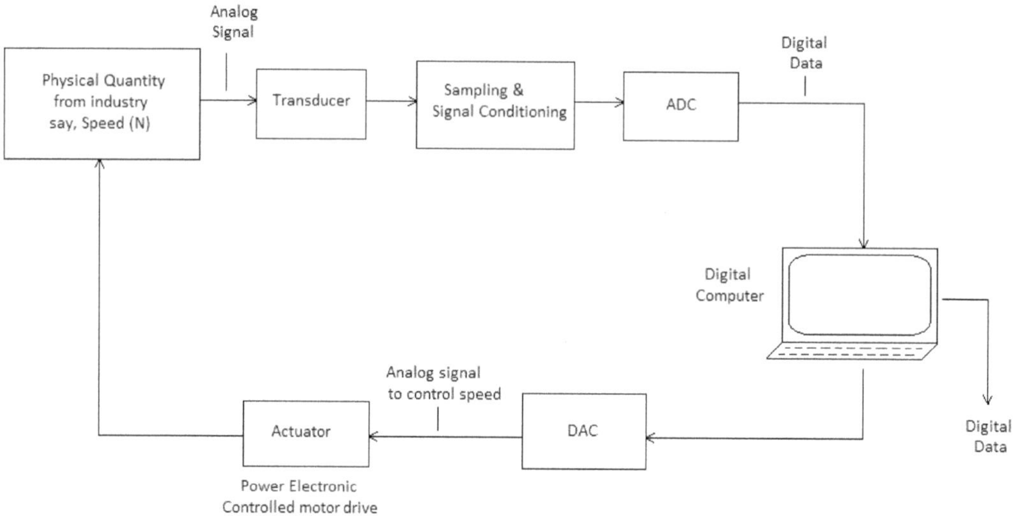

Figure 6.15 Data Acquisition System

Since the Computer stores and processes the data in digital form, the sampled real world signal must first be converted to a compatible digital format. Hence, the data acquisition system mainly includes the basic elements of Sensing, Sampling, Conversion, Filtering, Amplification, Processing and Recording. These elements are illustrated in the figure 6.15.

A simple example of a real world signal is speed of a process in an industry. We can capture speed with the help of a speed transducer. The transducer provides an analog voltage signal proportional to the speed. The captured voltage signal is sampled and conditioned and is fed to an Analog to Digital Converter (ADC) for necessary digital output. The digital data is processed and stored in the digital computer. The Computers uses this data to control the actuators through D/A Converters. Actuators are devices that correct the physical quantity to desired value. Here, a power electronic circuitry is employed to adjust the speed to its required value.

Computers can allow us to predict or understand the conditions of real world signals by their ability of rapid analysis. A revolutionary new technology that depends on large data set analysis is Artificial Intelligence and Machine Learning (AIML). The Computers have become the decision makers by these advanced technologies. From washing machines to self-driving cars, Mars Rover explorers and smart machines depend on the Data Acquisition, Analysis and the Control processes.

Review Questions

Short Answer (Two marks) Questions

1. How 'Op-Amp' has got that name?

2. Show the schematic symbol of an Op-Amp

3. Mention the qualities of an Ideal Operational Amplifier

4. Show how the voltage gain of an IC 741 commercial Op-Amp changes with frequency?

5. List the applications of Op-Amps

6. Show the circuit symbols of Thermistor and Light Dependent Resistor (LDR)

7. What is the need of A/D and D/A Converters?

Essay (Six marks) Questions

1. Show and explain the block diagram of a typical Op-Amp

2. Explain the characteristics of an ideal Operational Amplifier

3. Show the Pinout diagram of IC 741 Commercial Op-Amp. Specify the function of each pin

4. Discuss on the working of Strain Gauge

5. Briefly explain the operation of transducers used for a) Light and b) temperature

6. How a velocity transducer can be converted into an Acceleration Transducer? Explain

7. What do you mean by R/2R Ladder DAC? Explain its operation

8. With the help of a block diagram, explain the operation of ADC.

9. What is Data Acquisition and Control? Explain its principles with a real time application.

Index

A

Active Element, 7, 8
A.C Fuses, 205
A.C Generator, 197
A.C through Pure Resistance, 89
A/D Converters, 252
AC through Pure Inductance, 90
AC through Series R-C Circuit, 98
AC through Series R-L Circuit, 96
Active Power, 90, 95, 107
Active Region, 232, 236
Alternating Current (A.C),
    73, 158, 186, 197, 225
Alternator, 122, 197, 208
Apparent Power, 98, 106, 119
Armature Control Method, 183
Armature Core, 164
Armature Winding, 161, 164, 170
Asynchronous Motor, 190
Average Value, 80, 242

B

Back Emf, 173
Balanced System, 113
Barrier, 222
Base, 202, 231, 234, 238
Base Transportation Ratio, 236
Batteries, 200, 211
Biasing of Transistor, 231
Bilateral Network, 48
Bimetal, 207
Bipolar Junction Transistor, 230
Bridge Rectifier, 228
Brushes, 164

C

Cab Tyre Sheathed (CTS) Batten Wiring, 203
Calculations of energy consumption, 214
Capacitance, 7, 93, 108
Capacitive Reactance, 95, 111, 127
Cartridge Fuses, 205
Casing and Capping Wiring, 202
Centre Tap Rectifier, 227
Choppers, 241
Circuit Symbol of DC Generator, 165
Cleat Wiring, 202
Collector, 230, 235
Common Base (CB) Configuration, 234
Common Collector (CC) Configuration, 235
Common Emitter (CE) Configuration, 234
Commutator, 162, 197
Compound Generators, 170
Compound Motors, 177
Long shunt, 177
Short shunt, 177
Condition for maximum efficiency, 151
Conduit Wiring, 203
Constant Losses, 150
Construction of Transformers, 132
Controlled Sources, 9
Controlled Switch, 240
Copper Losses, 149
Current Controlled Current Source, 11
Current Controlled Voltage Source, 11
Cut-off Region, 232
Cycle, 74, 199, 225

D

D/A Converter, 252, 253
Data Acquisition and Control, 255

K

L

M

N

O

P

Q

R

S